ACS SYMPOSIUM SERIES **899**

Environmental Fate and Safety Management of Agrochemicals

J. Marshall Clark, Editor
University of Massachusetts

Hideo Ohkawa, Editor
Fukuyama University

Sponsored by the
**ACS Division of Agrochemicals and
the Pesticide Science Society of Japan**

American Chemical Society, Washington, DC

Library of Congress Cataloging-in-Publication Data

Pan-Pacific Conference on Pesticide Science (3rd : 2003 : Honolulu, Hawaii)
    Environmental fate and safety management of agrochemicals / John Marshall Clark, editor ; Hideo Ohkawa, editor.
        p. cm.— (ACS symposium series ; 899)

    Papers presented at the 3rd Pan-Pacific Conference on Pesticide Science, held June 1-4, 2003 in Honolulu, Hawaii.

    "Sponsored by the ACS Division of Agrochemicals and the Pesticide Science Society of Japan."

    Includes bibliographical references.

    ISBN 0-8412-3910-X (alk. paper)

    1. Pesticides—Environmental aspects—Congresses. 2. Pesticides—Safety measures—Congresses. 3. Agricultural chemicals—Environmental aspects—Congresses. 4. Agricultural chemicals—Safety measures—Congresses.

    I. Clark, John Marshall, 1949- . II. Ohkawa, Hideo. III. Nihon Noyaku Gakkai. IV. American Chemical Society. Division of Agrochemicals. V. Title. VI. Series.

TD196.P38P36   2003
628.5'29—dc22
                                                            2004062672

The paper used in this publication meets the minimum r equirements o f A merican N ational Standard for Information Sciences—Permanence of Paper for Printed Library Materials, ANSI Z39.48-1984.

Copyright © 2005 American Chemical Society

Distributed by Oxford University Press

All Rights Reserved. Reprographic copying beyond that permitted by Sections 107 or 108 of the U.S. Copyright Act is allowed for internal use only, provided that a per-chapter fee of $30.00 plus $0.75 per page is paid to the Copyright Clearance Center, Inc., 222 Rosewood Drive, Danvers, MA 01923, USA. Republication or reproduction for sale of pages in this book is permitted only under license from ACS. Direct these and other permission requests to ACS Copyright Office, Publications Division, 1155 16th Street, N.W., Washington, DC 20036.

The citation of trade names and/or n ames o f m anufacturers i n t his p ublication is not to be construed as an endorsement or as approval by ACS of the commercial products or services referenced herein; nor should the mere reference herein to any drawing, spec-ification, chemical process, or other data be regarded as a license or as a conveyance of any right or permission to the holder, reader, or any other person or corporation, to manufacture, reproduce, use, or sell any patented invention or copyrighted work that may in any way be related thereto. Registered names, trademarks, etc., used in this publi-cation, even without specific indication thereof, are not to be considered unprotected by law.

PRINTED IN THE UNITED STATES OF AMERICA

Foreword

The ACS Symposium Series was first published in 1974 to provide a mechanism for publishing symposia quickly in book form. The purpose of the series is to publish timely, comprehensive books developed from ACS sponsored symposia based on current scientific research. Occasionally, books are developed from symposia sponsored by other organizations when the topic is of keen interest to the chemistry audience.

Before agreeing to publish a book, the proposed table of contents is reviewed for appropriate and comprehensive coverage and for interest to the audience. Some papers may be excluded to better focus the book; others may be added to provide comprehensiveness. When appropriate, overview or introductory chapters are added. Drafts of chapters are peer-reviewed prior to final acceptance or rejection, and manuscripts are prepared in camera-ready format.

As a rule, only original research papers and original review papers are included in the volumes. Verbatim reproductions of previously published papers are not accepted.

ACS Books Department

Contents

Preface ix

Overview

1. Agrochemical Industry's Contribution to Sustainable Agriculture and Environmental Health 2
   Shigehiro Oba

Residue Analysis

2. Evolution of Residue Analysis and Its Role in Improving the Safety of Agrochemicals 14
   James N. Seiber

3. Multiresidue Analysis of Pesticides in Foods Using Liquid Chromatography/Mass Spectrometry 28
   Futoshi Sato, Hitoshi Iwata, Takakazu Nomura, and Kazuhiro Komatsu

4. Multiresidue Monitoring of Pesticides in Domestic and Imported Foods Collected in Hyogo Prefecture, Japan 38
   Yumi Akiyama, Naoki Yoshioka, and Kiyoshi Teranishi

5. Long-Term Fate of the Herbicide Azimsulfuron in a Rice-Grown Lysimeter over Six Consecutive Years 50
   J. K. Lee, F. Führ, S. H. Park, E. Y. Lee, Y. J. Kim, and K. S. Kyung

6. Quick Analysis of Fipronil and Its Metabolites in Gauze and Soil Samples 62
   Sonia Campbell and Qing X. Li

Environmental Fate

7. Understanding the Tropospheric Transport and Fate of Semivolatile Pest Management Chemicals 70
   Vincent R. Hebert

8. Ecology of Pesticide-Degrading Bacteria: Degradation
   of Organophosphorus and Carbamate Insecticides 82
   Masahito Hayatsu, Kanako Tago, Mitsuru Fukui, and Emi Sekiya

9. Polymerization of the Herbicide Bentazon and Its Metabolite
   with Humic Monomers by Oxidoreductive Catalysts 92
   Jong-Soo Kim and Jang-Eok Kim

10. Predicting Pesticide Volatilization from Bare Soils 101
    Scott R. Yates, Sharon K. Papiernik, and William F. Spencer

Environmental Risk Assessment

11. Concentrations of Herbicides Used in Rice Paddy Fields in River
    Water and Impact on Algal Production 112
    S. Ishihara, T. Horio, Y. Kobara, S. Endo, K. Ohtsu, M. Ishizaka,
    Y. Ishii, and M. Ueji

12. Ecotoxicological Risk Assessment of Atrazine in Amphibians 124
    Keith R. Solomon, J. A. Carr, L. H. du Preez, J. P. Giesy,
    T. S. Gross, R. J. Kendall, E. E. Smith, G. J. Van Der Kraak

13. Dietary Risk Assessment of the Organophosphate
    Insecticide/Acaricide Methamidophos 138
    Derek W. Gammon, Wesley C. Carr, Jr., and Keith F. Pfeifer

14. Evaluation of Estrogen Receptor Binding Affinity
    of DDT-Related Compounds and Their Metabolites 159
    Masahiro Miyashita, Takahiro Shimada, Shizuka Nakagami,
    Norio Kurihara, Hisashi Miyagawa, and Miki Akamatsu

Metabolism

15. Metabolism of Pesticides by Plants and Prokaryotes 168
    Robert M. Zablotowicz, Robert E. Hoagland, and J. Christopher Hall

16. Comparative In Vitro Metabolism of [^{14}C]Methoxychlor
    in Vertebrate Species Using Precision-Cut Liver Slice Technique 185
    K. Ohyama

17. Herbicide Detoxification: Herbicide Selectivity in Crops and
    Herbicide Resistance in Weeds 195
    Christopher Preston

18. Identification of Thiolactic Acid Conjugated Metabolites of Fungicide Diethofencarb in Grape (*Vitis vinifera* L.) and the Mechanism of Their Formation in Plant and Rat 205
Takuo Fujisawa, Luis O. Ruzo, Yoshitaka Tomigahara, Toshiyuki Katagi, and Yoshiyuki Takimoto

19. The Role of Plant Glutathione *S*-Transferases in Herbicide Metabolism..216
Dean E. Riechers, Kevin C. Vaughn, and William T. Molin

Resistance and Management

20. Sodium Channel Point Mutations Associated with Pyrethroid Resistance in the Head Louse, *Pediculus humanus capitis* 234
Takashi Tomita, Noboru Yaguchi, Minoru Mihara, Noriaki Agui, and Shinji Kasai

21. Molecular Characterization of Resistance to Acetolactate Synthase Inhibitors in *Lindernia micrantha*: Origin and Expansion of Resistant Biotypes 244
Hiroyuki Shibaike, Akira Uchino, and Kazuyuki Itoh

22. Molecular Characterization of Acetolactate Synthase in Resistant Weeds and Crops 255
Tsutomu Shimizu, Koichiro Kaku, Kiyoshi Kawai, Takeshige Miyazawa, and Yoshiyuki Tanaka

23. Glyphosate-Resistant Horseweed (*Conyza canadensis* L. Cronq.) in Tennessee 273
Thomas C. Mueller, Joseph H. Massey, Robert M. Hayes, and Chris L. Main

24. Resistance Management Strategies for Fungicides............................280
Hideo Ishii

Advances in Formulation and Application Technology

25. Effects of Surfactant Concentration on Stability of Dispersion 290
Tatsuo Sato

26. Pesticide Formulation Technology Innovation 297
Curtis Elsik, Scott Tann, and Andrew Kirby

27. **Dinotefuran Release Properties from Controlled Release Formulations** 309
    Daisuke Kishi and Seiichi Shimono

28. **Advances in Pesticide Applications and Their Significance to the Agrochemical Industry Servicing Tropical Farming** 320
    William Taylor and Per Gummer Andersen

29. **Tape Formulations and Their Application Technology** 328
    Masao Inoue, Satoshi Nakamura, and Toshiro Ohtsubo

Author Index 339

Subject Index 341

Preface

Pesticide science provides human society with the crop and disease vector protectorant products necessary to sustain the food, fiber, and health that it requires in an environmentally safe, sustainable, and affordable manner. This critical task has been made even more daunting by the rising costs of product registration and increasing environmental concerns. Nevertheless, recent technologies and approaches have fundamentally altered how we approach pest management and the control of vectors of communicable diseases. In the forefront of all new agrochemical developments are the criteria of human safety, environmental stewardship, and resistance management.

With these goals in mind, approximately 200 pesticide scientists convened the Third Pan-Pacific Conference on Pesticide Science, which was jointly hosted by the Pesticide Science Society of Japan and the American Chemical Society (ACS) Division of Agrochemicals June 1–4, 2003, in Honolulu, Hawaii. Researchers from 14 countries (Australia, Canada, China, Germany, India, Iran, Japan, New Zealand, Nigeria, South Korea, Taiwan, Thailand, the United Kingdom, and the United States of America) presented 70 invited papers, 68 posters, and 4 panel discussions that dealt with two main topics: New Discoveries in Agrochemicals and Environmental Fate and Safety Management of Agrochemicals.

This ACS Symposium Series book deals with the second topic. Twenty-nine invited and peer-reviewed chapters from internationally recognized pesticide experts are divided into six sections: Residue Analysis, Environmental Fate, Environmental Risk Assessment, Metabolism, Resistance and Management, and Advances in Formulation and Application Technology.

Acknowledgments

We thank all the authors for their presentations at the conference and for their contributed chapters. Special thanks go to our keynote speakers (H. Strang and S. Oba), the conference co-chairs (N. K. Umetsu and B. Cross), topic organizers (I. H. Yamamoto, K. Solomon, K. Sato, R. Toia, C. Hall, H. Ishii, C. Staetz, T. Ohsubo, and J. Zabkiewicz). In particular, we extend our deepest appreciation to the many expert colleagues who provided helpful and necessary critical reviews. We thank Bob Hauserman and Stacy Vanderwall in acquisitions and Margaret Brown in editing and production of the ACS Books Department for all their help, suggestions, and encouragement. Lastly, we thank the Pesticide Science Society of Japan and the ACS Division of Agrochemicals and their benefactors, contributors, and donors, whose financial support made this book possible.

J. Marshall Clark
Department of Veterinary and Animal Science
University of Massachusetts
Amherst, MA 01003

Hideo Ohkawa
Research Center for Environmental Genomics
Kobe University
Kobe, Hyogo 657–8501
Japan

Overview

Chapter 1

Agrochemical Industry's Contribution to Sustainable Agriculture and Environmental Health

Shigehiro Oba

Sumitomo Chemical Company, Ltd., Chuo-ku, Tokyo 104-8260, Japan

Sustainable agriculture plays a crucial role in sustainable development and is a cornerstone of our management philosophy. Environmental health products, which increase the safety and comfort of our living environment, must also be promoted in a manner consistent with the principals of sustainable development. The agrochemical industry can make a key contribution to sustainable development over the course of this century by offering agrochemicals and environmental health chemicals. Agrochemicals companies are expected to provide safer and more environmentally friendly chemicals, and R&D efforts must be targeted toward that goal.

Introduction

Harmonizing development with environmental conservation is one of mankind's most important challenges. The World Commission on Environment and Development (WCED), in its 1987 report "Our Common Future," advocated the concept of "sustainable development" defined as development that meets the needs of the present without compromising the ability of future generations to meet their own needs (*1*). In 1992, the United Nations Conference on Environment and Development (UNCED) in Rio de Janeiro, Brazil adopted the "Rio Declaration on Environment and Development," which makes sustainable development a basic philosophy, and "Agenda 21," a comprehensive plan of action towards its realization (*2*). In 2002, ten years after this conference, "the World Summit on Sustainable Development" (WSSD) was held in Johannesburg, South Africa. "The Johannesburg Plan of Implementation," which is aimed at the implementation of Agenda 21, was adopted at this summit. Sustainable development has thus been recognized as an important common goal with serious challenges.

The world's population exceeds 6 billion and is estimated to reach about 9 billion by the year 2050 (*3*). Today, more than 800 million people are suffering from undernourishment (*4*). Reduction of hunger through an increase of food production is one important challenge for sustainable development.

Changes in climate brought about by global warming affect people's lives and all animal and plant ecosystems. The habitat of tropical infectious disease-carrying mosquitoes is projected to cover wider areas as global temperatures increase. People are concerned that malaria and dengue fever will spread from the tropical and subtropical regions to the temperate zones. Malaria transmitted by *Anopheles* mosquitoes infects at least 300 million people each year, of which over one million people, mostly children less than five years of age, die from the disease (*5*). Furthermore, West Nile Virus, the vector of which is *Culex* and other mosquito species, reached New York in 1999. In 2002, over 200 people died of West Nile fever in the United States. Eradication of such insect-borne infectious diseases is also an important challenge for sustainable development.

In the challenge to continuously improve food supplies and the living environment in accordance with the philosophy of sustainable development, both sustainable agriculture and the promotion of sustainable environmental health have crucial roles to play. Sustainable agriculture can help to eradicate starvation through increases in food production and the supply of safer foods. It is also important to protect people from various infectious diseases and to construct more comfortable living environments through the practice of sustainable environmental health.

The agrochemical industry can contribute to sustainable development in the course of this century by offering agrochemicals and environmental health chemicals. In this paper, the author first discusses the mission of the agrochemical industry for human welfare with respect to plant protection and environmental health business. He also describes his ideas on how the agricultural chemical business should be advanced by focusing on the example of Sumitomo Chemical. He then outlines the current situation and future outlook of the environmental health business.

Human Welfare Mission of the Agrochemical Industry

Agrochemicals are used as indispensable ingredients to increase food production in sustainable agriculture. Environmental health chemicals are useful for control of disease vectors such as mosquitoes and other hazardous insects, for protection of people's health and safety, as well as for the improvement of the quality of life and the environment. It should be added that agrochemicals and environmental health chemicals have to be safe and eco-friendly in order to play their important roles appropriately. This is a requirement for materials that promote sustainable development.

The mission of the agrochemical industry for human welfare in sustainable development is: the protection of human health and safety; food security; and environmental conservation. Specifically, it contributes to: alleviating the threat of starvation by increasing food production; supplying safer foods; protecting people from infectious diseases; and constructing more comfortable living environments. Under this mission, the agrochemical industry is responsible for supplying customers with agrochemical and environmental health chemical products as well as pest control systems for sustainable development. For that purpose, it needs to develop new products that are safer and more eco-friendly. It is also important to address the problem of chemical resistance.

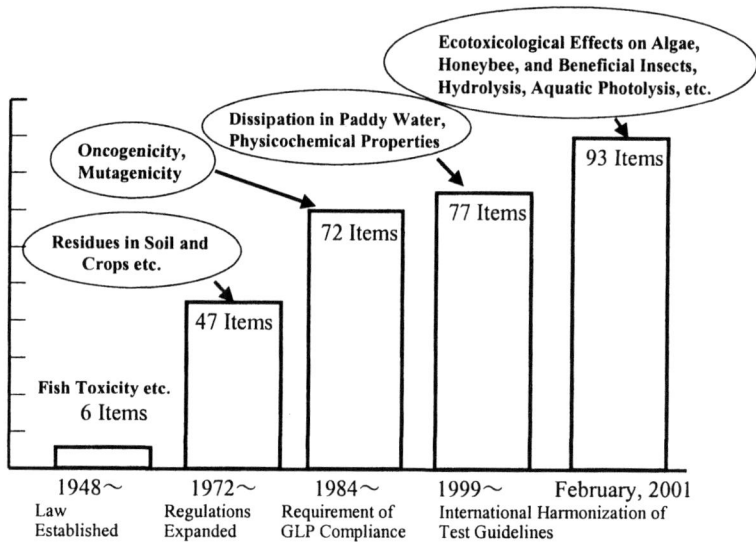

Figure 1. *Regulation of Agricultural Chemicals in Japan: Number of Mandatory Test Items for Registration of New Chemicals*

In many industrialized countries, regulations on agrochemicals and environmental health chemicals are becoming more comprehensive with respect to safety and environmental compatibility. One example of agrochemicals regulation in Japan is shown in Figure 1. A law was established in 1948, identifying 6 items, including fish toxicity, as test items. The number of test items has been increasing over time, with 93 test items required as of February 2001. The number has approximately doubled since 1972. Only active ingredients that meet the stringent standards for each of these tests can be registered as new agrochemicals.

Also by the re-evaluation program for existing active chemical substances in the European Union, it has been already decided that many chemicals are excluded from Annex I of the EU agrochemical registration Directive (91/414). Because of cost-performance considerations of re-registration, the number of agrochemicals being registered has been decreasing.

Performance of the Agricultural Chemicals Business

Agrochemical companies are expected to provide customers with new products that are safer and more eco-friendly. They must have a clear vision and firm business strategies for strong and stable corporate management in order to meet these expectations.

If the above–mentioned points are considered in terms of business strategies and R&D strategies, it can be said that the two sets of strategies are both key pillars of management and must be consistent with each other. R&D for new products must bring profits to business. If not, the future viability of the company will be threatened. The company must be able to earn sufficient profits from its R&D activities to justify its investment in R&D. In order for a company's business to be linked to its R&D activities, the intellectual property resulting from its R&D must be appropriately protected through intellectual property rights (IPR) (see Figure 2). Intellectual property rights serve as a driving force for business and must not be infringed upon.

In the global agricultural chemical industry, competition has been continuously intensifying. Giant U.S. and European agrochemical companies have become even larger through M&A and other strategic alliances. In this environment, Sumitomo Chemical, which is a medium-sized agrochemical firm, is focusing on plant protection and environmental health business. The company will pursue its business by emphasizing the importance of research and development as the foundation of corporate management.

Early launch and sales promotion of new products serve as the keys to developing a thriving business. Sumitomo Chemical has been developing safe and eco-friendly products, and supplying them to customers, in order to strengthen its business. To enrich and strengthen its pipeline of compounds, the company aims to develop new active ingredients as well as introduce compounds from other companies. It strives for an intellectual property strategy that is in accordance with its business strategy.

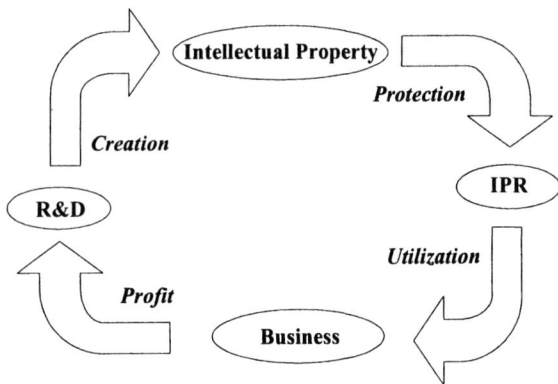

Figure 2. Business, R&D, and Intellectual Property

Integrated Pest Management in Sustainable Agriculture

The system called "Integrated Pest Management," or IPM, serves as an important crop protection mechanism when practicing sustainable agriculture. According to the definition in the FAO Code of Conduct on the Distribution and Use of Pesticide, IPM means "the careful consideration of all available pest control techniques and subsequent integration of appropriate measures that discourage the development of pest population and keep pesticides and other interventions to levels that are economically justified and reduce or minimize risk to human health and the environment" (6). IPM as defined here emphasizes "the growth of a healthy crop with the least possible disruption to agro-ecosystems" and encourages "natural pest control mechanisms."

CropLife International member companies and associations, one of which is Sumitomo Chemical, are committed to IPM as defined in the FAO Code of Conduct, as an economically viable, environmentally sound and socially acceptable approach to crop protection (7).

The key point of IPM is to utilize an appropriate combination of biological, chemical, physical, and agronomical pest control measures. A chemical insecticide used in IPM is required to have a high selectivity between the targeted pest insects and natural enemy insects.

Many IPM-compatible products have already been launched in the global market. Sumitomo Chemical markets products for biological pest control including Bt formulations such as Florbac^{TR}(*Bacillus thuringiensis* spp. *aizawai*) and natural predator insects such as *Orius strigicollis* (Oristar-A^{TR}). Pyriproxyfen, which is already on the market, and pyridalyl, which is under development, are examples of chemical insecticides (see Figure 3).

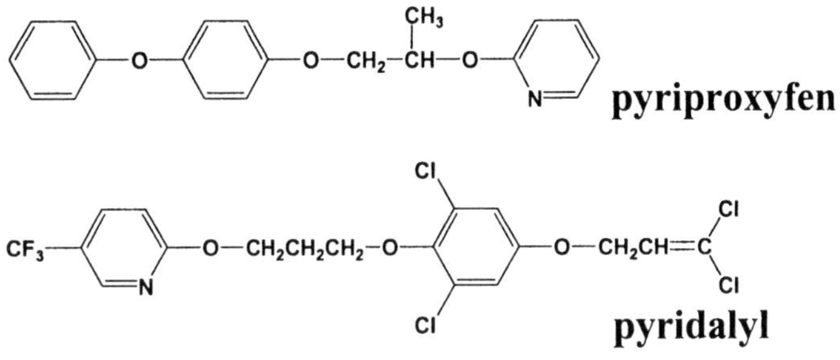

Figure 3. Chemical Structure of Pyriproxyfen and Pyridalyl

Pyridalyl controls a wide variety of lepidopterous pest strains such as the diamondback moth (*Plutella xylostella*), which is resistant to various existing insecticides (*8*). This insecticide is also effective against thysanopterous pests. Moreover, it does not exert an adverse effect on various beneficial arthropods such as natural predator insects (e.g., Oristar-A^{TR}) and honeybees. Therefore, this novel insecticide is expected to be a useful material for controlling lepidopterous and thysanopterous pests in IPM and insecticide resistant management programs.

Environmental Health

The environmental health business focuses on developing products that promote a safe and comfortable living environment (see Figure 4). One of the primary roles is offering measures to protect people from insect-borne infectious diseases, such as malaria and dengue fever. Other important roles are offering products for regional public health and products that contribute to the creation of a comfortable living space. One important issue in the environmental health business is the increase and proliferation of insect-borne tropical diseases as a result of global warming. Two other important issues are: maintaining public health standards in the face of rising world population levels, and the increasing desire for comfortable living space in accordance with higher living standards.

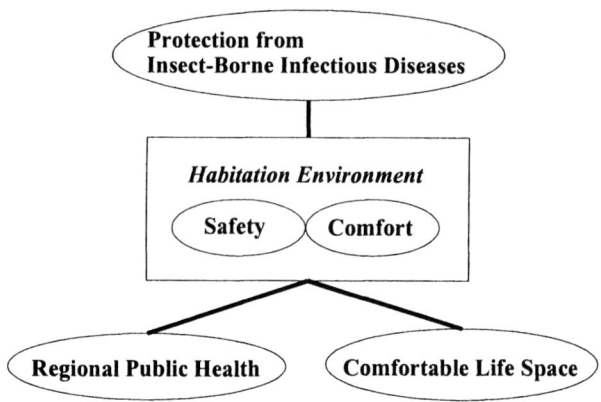

Figure 4. Concept of Environmental Health Business

The outline of the environmental health business is shown in Table I. This field includes indoor consumer products such as household insecticides, outdoor consumer products such as those for pest control, and public-health related products for mosquito abatement and professional vector control. All these products are required to have superior safety characteristics. The size of the world market for the environmental health business in terms of retail sales is also shown in Table I. The market size is estimated to be roughly 7.2 billion dollars overall, of which about 70%, or 5.0 billion dollars, represents the indoor use insecticide market. The size of the outdoor use insecticide and public health insecticide markets is 1.4 billion and 800 million dollars, respectively.

The present condition of the environmental health business can be summarized as follows. In advanced countries such as the U.S., Europe, and Japan, there is a strong demand for high value-added products that comply with very stringent safety and environmental standards. In developing countries in such regions as Africa, products for controlling vectors like malaria-carrying mosquitoes are greatly needed. Over the past several years, there has been an acceleration in the global consolidation among technical product makers.

Furthermore, in all locations generic product companies have gained market share in the technical product business.

In 2001, Sumitomo Chemical acquired the household insecticide business of Aventis Environmental Sciences. The predecessor of this company was the environmental health business of Schering, Roussel Uclaf, and Wellcome. As a result of this acquisition, Sumitomo Chemical succeeded in building a strong position as a principal manufacturer in the environmental health field. Other major makers include Bayer CropScience, Syngenta, and FMC.

Table I. Outline of Environmental Health Business

Field	Product type	Market size ($M)[a]
Indoor consumer use	Household insecticides, Textile protection, etc.	5,000
Outdoor consumer use	Nuisance insect control, Lawn & garden care, etc.	1,400
Public health	Mosquito abatement, Professional vector control, etc.	800

[a] retail sales

Consumer Insecticides

Consumer insecticide products are distributed through three tiers globally. Active ingredients supplied by technical product makers such as Sumitomo Chemical are formulated by formulators. Then, the formulated products are sold to consumers at retail.

There are many challenges facing companies in the consumer insecticide business. One challenge is to ensure a stable supply of safe and high cost-performance technical products and the end products derived from them. R&D targeted toward such products is essential. It is important to develop new products to promote a more comfortable living environment, and new materials for environmental conservation and public health. It is also important to offer new solutions to the problem of chemical-resistant pests.

Various structural evolutions have led to a constant stream of new products with increased performance features.

Sumitomo Chemical is now developing a novel pyrethroid named metofluthrin for environmental health use. The company expects to start wide-spread sale of this pyrethroid beginning in 2004. This insecticide has high knock-down activity against mosquitoes and is highly safe to mammals. In addition, it has high vapor action. For example, it was about 40 times more active than *d*-allethrin against tropical house mosquito (*Culex quinquefasciatus*) in a mosquito coil test. This pyrethroid is applicable to various existing devices like mosquito mats, and new devices such as one that can volatilize chemicals at room temperature.

Public Health Insecticides

In the public health business, promoting effective epidemic prevention in accordance with global epidemic prevention policies, which are advocated by the United Nations, is of central importance. For that purpose, it is necessary to develop epidemic prevention methods that are extremely safe and cost-effective, and to secure a stable supply of the active ingredients used in these new methods.

The World Health Organization (WHO) and its partners initiated a project named "Roll Back Malaria" in 1998 (*9*). Its objective is a 50% reduction of the malaria burden by 2010. To reach its objective, the Roll Back Malaria project wanted to promote the development of a long-lasting insecticidal net. A long-lasting insecticidal net was important because re-treatment rates after washing are very low even in urban areas, and it is difficult to access remote areas.

Sumitomo Chemical developed a long-lasting insecticidal net for the WHO project. Its product, Olyset^{TR} net remains effective for more than 4 years under normal use conditions. Furthermore, ventilation during sleep is good and no re-treatment of the insecticide after washing is necessary. This is the only product authorized by the WHO. Technologies based on know-how for premixing the resin with the insecticide, and spinning the premixed resin into fiber are utilized in the manufacturing process for Olyset^{TR} net.

Concluding Remarks

In conclusion, the author wishes to stress that the agrochemical industry aims at the contributing to sustainable development by providing agrochemicals and environmental health chemicals to society. To achieve this goal, he believes that the agrochemical industry needs to develop new chemicals and formulations which are safe and eco-friendly, and give high cost-performance.

For that purpose, it is very important to clarify first the biological, physiological, and ecological characteristics of target pests like agriculturally noxious insects and mosquitoes, identify differences between those characteristics and those of non-targets like mammals and fishes, and then try to make effective use of such findings in R&D of new chemicals. In other words, by taking advantage of such findings, (1) it is important to determine where the target pests should be aimed at, on the tissue level, cell level, or the receptor level. In this process, it is expected that genome information of insects, etc. would be useful. Then, the principal issues become: (2) what compounds to synthesize, (3) how to synthesize them, and (4) how to evaluate and analyze the biological activities of the compounds thus synthesized. Various advanced technologies for molecular design, compound synthesis, and evaluation and analysis of biological activities play an important role in carrying out such agrochemical research. More specifically, efficient asymmetric synthesis technologies, computer chemistry, etc. have already been put in practical use. Combinatorial chemistry and high throughput screening technologies have also proved to be useful.

After all, development of "agrochemical science" and "agrochemical R&D technology" and their close collaboration are essential ingredients for the agrochemical industry to contribute to sustainable development. For that reason, the author believes that close cooperation between industry and academia is very important in various stages.

Acknowledgements

This paper was prepared with the assistance of Drs. Chiyozo Takayama, Masakazu Miyakado, and Nobuaki Mito. Their contribution was invaluable. The author sincerely thanks his colleagues at Sumitomo Chemical for providing him with extensive material.

References

1. *Our Common Future*: The World Commission on Environment and Development; Oxford University Press: Oxford, UK, 1987.
2. United Nations Division for Sustainable Development: http://www.un.org/esa/sustdev/.

3. *World Population Prospects: The 2002 Revision*: United Nations Population Division: New York, 2003.
4. *The State of Food Insecurity in the World*: FAO: Rome, 2002.
5. *The World Health Report 2002*: WHO: Geneva, 2002.
6. *International Code of Conduct on the Distribution and Use of Pesticides (Revised Version)*: FAO: Rome, 2002.
7. *Integrated Pest Management- The Way forward for the Plant Science Industry*: CropLife International: Brussels, 2003.
8. Saito, S.; Isayama, S.; Sakamoto, N.; Umeda, K.; Kasamatsu, K.: *Proc. BCPC Conf.- Pests & Diseases 2002, Vol. 1*, 33-38.
9. Roll Back Malaria: http://www.rbm.who.int/newdesign2/.

Residue Analysis

Chapter 2

Evolution of Residue Analysis and Its Role in Improving the Safety of Agrochemicals

James N. Seiber

Western Regional Research Center, U.S. Department
of Agriculture, Agricultural Research Service, Albany, CA 94710
Department of Environmental Toxicology, University
of California, Davis, CA 95616

Residue analysis plays an important role in meeting challenges associated with pesticides and related toxicants. Residue analytical techniques underpin the regulatory system for pesticides, including the setting and enforcing of tolerances, action levels, buffer zones, worker exposure standards, and acceptable intakes for chemicals in foods and the environment. The field has developed by adapting analytical techniques and instrumentation to the unique problems of ultra-low level analysis in complex matrices. Chromatography with selective detectors and mass spectrometry have been mainstays for determination. Molecular biology is providing new techniques which will lead to more biologically-based analytical methods in the future.

Introduction

In many respects it is a great time to be involved in pesticide residue chemistry and residue analysis. The data from good residue studies is the foundation of exposure assessment and ultimately risk assessment. It is sought by regulatory agencies and committees of the National Academy of Sciences and World Health Organization, among others. As society becomes more interested in the details of how contaminants enter the food supply, as well as the environmental compartments of water, air and wildlife, residue chemistry provides critical answers. Total exposure, from all sources, is the rule of the day in risk assessment requiring ever more and better quantitative residue data, from many types of samples and differing matrices.

And the instruments and other technologies available to residue chemistry are far superior to the past. In addition to the now-usual Gas Chromatography (GC) and High Performance Liquid Chromatography (HPLC), often coupled with mass spectrometry, developments of Enzyme-Linked Immunosorbent Assay (ELISA) and biosensors and related techniques relying heavily on new developments in molecular biology are finding more use in residue chemistry. The evolution of residue chemistry amounts to a 'revolution' in methodologies, generating more data, richer in information content and higher in quality than ever before.

Historical Development

Residue chemistry traces its early growth to the lack of critical information with the second-generation, primarily synthetic organic pesticides introduced in large scale in agriculture in the 1940s and after (*1*). More than the first generation materials, arsenicals, copper sulfate, Paris Green and other inorganic and botanical pest control agents, the second generation agents had several characteristics which heightened consumer concern, regulatory attention, and monitoring activity:

1. They were widely used, some would argue overused, compared with the first generation products.
2. They were applied at such lower rates that their residues were not visible or detectable to the consumer, and so required new sensitive methodologies for detection and quantitation.
3. Many were acutely and/or chronically toxic to humans, domestic animals, and wildlife. The ability of some to cause tumors in experimental animals was of particular concern.
4. Their residues were mobile, environmentally in air, water, soil, and food chains, and for some, systemically within plants.
5. Because of their relatively complex organic structures, many were degraded/metabolized to multiple products which had different toxicity and dissipation characteristics, and analytical characteristics, from the parents.

Regulation developed in the 1950s and 1960s, including legal limits (tolerances) for residues on foods and in feeds and, with time, in water and air. Analytical methods of increasing sophistication and sensitivity were needed, and residue chemists responded magnificently to the challenge.

Early efforts at regulating residues confounded the situation, but in some respects stimulated developments in residue analysis. A 'zero tolerance' concept was advanced for pesticides that produced cancer in experimental animals, for certain agricultural and food products such as milk and butter, and for some chemicals that could be concentrated during food processing (*2*). Key congressional mandates were introduced in Congress in 1953, by Representatives James Delaney (D, N.Y.) and A.L. Miller (R, Nebr) (*3*). But zero tolerance meant not one molecule could be allowed to remain in food products for consumption, which was basically indefensible from a scientific viewpoint and from practicality considerations also. As methods became more sensitive, what was formerly zero from an analytical viewpoint, ie, non-detectable using the Best Available Technology, became detectable as new, better methods evolved. In effect, 'zero tolerance' served as a magnet for attracting and stimulating better, more sensitive, residue methods.

Such was the case for aminotriazole residues on cranberries, detected by a new method just a week before Thanksgiving in 1959. Both fresh cranberries and canned products from prior years were confiscated and 1959 was mostly a cranberry-less Thanksgiving in the U.S. (*2*). A second case came about in 1960 when chlorinated insecticides, such as heptachlor and DDT, were found in butter shipped from the mainland to Hawaii, using new paper-chromatography–based (*4*, *5*) and colorimetric assays (*6*). It was clear that small residues could be transferred to animal products such as milk, butter, and meat if residues were present, intentionally or inadvertently, on feed.

The publication of Silent Spring in 1962 revealed the potential of residues to contaminate food and wildlife to a previously largely unaware public, both in foods, as well as in the wildlife food chain (*7*). The rest is history. The U.S. Environmental Protection Agency (EPA) was formed in 1970, DDT was banned in 1972 (followed by other organochlorine insecticides later), and zero tolerance was replaced by finite limits that seem always to go down, never up, with new toxicology findings and the evolution of probabilistic risk assessment.

The benefits of insecticides like DDT and heptachlor were deemed unimportant in light of the threats to public health and the environment posed by persistent residues, and even the manufacturing and formulation facilities were often shunned as 'toxic waste sites', attracting further attention by the residue analytical community.

With this backdrop, pesticide residue analysis grew and matured from, roughly, the 1950s to the present. Advances were published in such primary outlets as the *Journal of the Association of Official Analytical Chemists, Journal of Agricultural and Food Chemistry, Bulletin of Environmental Contamination and Toxicology, Pesticide Monitoring Journal,* and *The Analyst*. Secondary references and compendia were introduced, by Francis Gunther and Roger Blinn (*8*), Gunter Zweig (*9*), Gunther's Residue Reviews (*10*) and several others. The Public Health Service and U.S. Food and Drug Administration (FDA) became

17

very active and remain so to this day, in advancing new methods of residue analysis for monitoring the safety of U.S. foods and also in publishing the Pesticide Analytical Manual, which is a key reference for updated methods for analysis of foods for residue content (*11*).

An early article by Gunther and Blinn (*12*) outlined the basic principles of pesticide residue analysis, including the possibilities for systematization and standardization of analytical methods. In many respects this thinking led to the evolution of today's methods, both single residue (SRM) and particularly integrated multi-residue methods (MRMs) that can include several hundred individual analytes, both parent compounds and significant breakdown products (*13*).

A cadre of agricultural analytical chemists emerged, within federal agencies such as the FDA, EPA, USDA's Food Safety Inspection Service, Agricultural Marketing Service (or its contractors), and Agricultural Research Service, and the Fish and Wildlife Service. Several states, particularly large agricultural states like California and Florida followed suit, as did major land-grant universities, many of which are now linked together through the IR-4 Minor Use Program network. Although not as well known, because much of the work is not published in peer review outlets, were residue groups in the food processing industry, such as Del Monte, Campbells, Kraft, and General Mills. Chemical Companies (Dow, DuPont, Monsanto, Stauffer, FMC, etc.) required strong residue chemistry groups in order to provide data for registration, including workable methods for analysis of residues. Similar developments occurred internationally, particularly in Europe, Japan, Australia, and Canada.

Recent Improvements in Residue Analysis

Gas chromatography proved to be a particularly useful tool for residue analysis, and was adopted by residue chemists by about 1960 (*14*). The Aerograph Hy-Fi was a reasonably good tritium-based electron-capture GC instrument—and that technique perhaps more than any other 'broke open' the field of trace analysis, at least for the organochlorine pesticides but also for PCBs, dioxins and many other classes of food and environmental contaminants. The high maintenance, hands-on equipment of the 1960s gave way to instruments with auto-injectors, integrating recorders, and whole new lines of steadily improving element selective and mass spectrometry (MS)-based detectors in the 1970s and beyond, for both GC and HPLC.

To summarize briefly some of the recent improvements that have spurred the development of residue chemistry, mention should be made of new extraction tools such as the use of Solid Phase Extraction (SPE) (*15*), universal solvent systems, miniaturized systems that cut back on solvent use, and techniques like headspace sampling that can make solvent extraction unnecessary (*13*). Supercritical fluid extraction (SFE) uses primarily CO_2, thus largely replacing organic solvents (*16*). Among the most cited papers in residue chemistry are those describing new cleanup tools, such as gel permeation, which was among the first of the cleanup techniques to be automated (*17*), and the silica gel, Florisil, and other adsorbent based fractionation schemes (*13*). HPLC fractionation, with automation, can be used for difficult separation problems (e.g., isolation of 2,3,7,8-TCDD in a matrix containing all of the other chorodioxins) (*18*). Derivatization has been indispensable, for converting non-volatile or unstable pesticides into gas chromatographable species or non-absorbing, not fluorescing materials such as glyphosate into fluorescing derivatives amenable to HPLC measurement (*19*). And the changeover from packed to capillary columns in GC, occurring roughly in the late 1970s and 1980s, has made a profound improvement in that technique, as has the microparticulate, bonded phases with HPLC (*20*). GC chiral phases are now readily available for resolving enantiomers of biorational, biologically based pesticides as well as some synthetics (*21*). Affinity columns, with high selectivity toward particular analytes, or classes of analytes, are becoming more available, for use in the extraction, cleanup, and resolution steps of trace analysis (*22*).

Quite an evolution has resulted in the past 50-plus years, dating back to the bioassay and gravimetric procedures, colorimetry, and paper- and thin-layer chromatography (TLC)-based methods of the 1940s and 50s, which were quite tedious in terms of hands-on time commitments (Figure 1). The Schecter-Haller

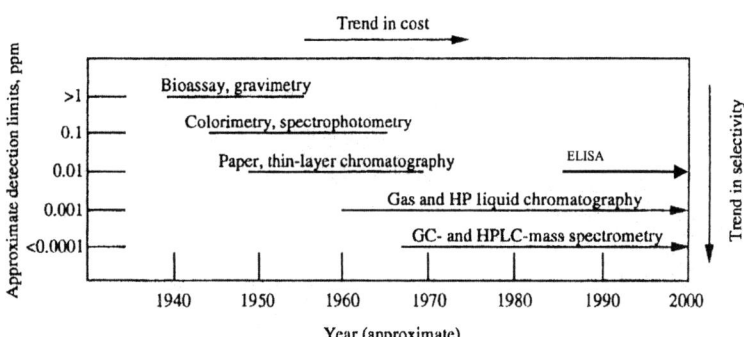

Figure 1. Evolution of analytical techniques used for pesticide residue analysis, 1940 to 2005. (Reproduced from ref. 13.)

and Averill-Norris colorimetric methods (8) have gone by the wayside for routine use, but they are still of instructional use in demonstrating the tenets of residue analysis. Now we are clearly in the era dominated by GC, HPLC, and mass spectrometry, with immunoassays also now included, at least for high-throughput screening. Recent books by George Fong et al. (23) and Philip Lee (24) are among current compendia with more detailed information on the modern methods of trace analysis as applied to pesticides and related environmental contaminants.

Mass spectrometry (MS) has grown to dominate the trace analysis field in general, including for pesticide residues, because of broad applicability and its high degree of selectivity and sensitivity. Trace analysis for chlorodibenzodioxins illustrate the advantages of MS. In the early 1970s, electron-capture (EC)-GC gave limits of detection for the dioxins in the ppb range, such as 50 ppb for TCDD in birds (25).

Just five years later, with the use of GC-MS for final resolution and quantitation, the limit of detection had dropped to 1 ppt, roughly an improvement of 5 orders of magnitude (18). Although a GC-MS costs more than a GC with EC or other selective detectors, the level of investment is quickly recovered by the greater throughput of MS-based methods. MS can be used for virtually any analyte that will go through a GC or HPLC, whether it contains halogen, phosphorous, sulfur, or nitrogen, or not.

ELISA has advantages, too (26). ELISA's sensitivity is comparable to GC and GC-MS-based methods for relatively clean substrates such as water (27) and can give even greater throughput in part because of the automated or semi-automated sample preparation and detection adaptable to the ELISA 96-well sample format. But it is generally less precise than GC or HPLC, and it can show more false positives and false negatives than MS methods. MS is still the "gold standard" as far as legally defensible residue results are concerned.

Improvements in Food Safety

Our food in the U.S. today is much safer than ever in the past, certainly so with respect to pesticide residues, and this record has been compiled during a major transition in food delivery that has included much more fresh food, much more ethnic and/or specialty foods, a larger availability of food products, ranging from staples to whole meals prepared in nearly ready to eat form, and increased international trade in food items. The problems in food safety of the past 10-15 years have largely been from pathogenic microorganisms. When chemical residues from pesticides, PCBs and other industrial materials have been problematic, it has largely been because of trade issues rather than any demonstrable threat to the safety of the consumers.

This improvement in 'chemical safety' of foods has resulted from changes in pesticide availability, use practices, and regulations. The shift away from organochlorine and organophosphate pesticides has reduced the analytical occurrence of residues, and also lowered the probabilistic risk associated with residues of pesticides on foods. Low rate, low residual chemicals, such as glyphosate and synthetic pyrethroids, and the larger inroads of *Bacillus thuriengensis* (B.T.) and other biorational materials have replaced some of the more problematic pesticides of the past. Better education of spray applicators, processors, and others in the production-to-retail chain has occurred, and improved regulatory enforcement is apparent in virtually all aspects of food production, processing, transportation, and retailing. Along with all of these advances, residue chemistry has provided key data showing the impact of changes, such as banning of most of the organochlorine (OC) insecticides and their replacement with 'softer' alternatives, and of better, more even, less drift-prone application methods, and of the reduction in residues with washing, peeling, cooking, and other food handling and processing techniques.

Monitoring of foods consistently shows little to no residue remaining in the final commodity of products for consumption (*28*). For example, 99.3% of randomly selected marketplace samples in California in a recent year were either detectably free (78%), or within established tolerances (21-plus %). Of the less than 1% of samples that were violative, nearly 2/3rds involved a chemical for which no tolerance was established on the crop sampled, although tolerance was very likely established on another crop, perhaps even grown side-by-side with the crop from which the 'illegal' residue originated.

The record for monitoring priority chemicals, and of produce destined for processing was equally as good. When there have been alarms, e.g. in a recent National Academy of Sciences report (*29*), residue chemistry stepped in to help define the parameters of the issue, and established baselines from which the impact of changes in pesticide use patterns could be determined.

Monitoring of our food supply for chemical residues is clearly important, so that the occurrence of hazards from pesticides, other environmental chemicals, or natural chemicals can be detected and dealt with as quickly as possible. A recent foodborne chemical of concern is acrylamide in high temperature prepared carbohydrate-rich foods. An early hypothesis was that the acrylamide, reported in 2002 by Swedish researchers using a new HPLC based MS-MS method (*30*), came from the residues of polyacrylamide used in some herbicide formulations. This turned out not to be the case, and available evidence is that acrylamide is formed from the amino acid asparagine during high temperature processing or cooking, perhaps with some contribution from acrolein residues in cooking oils (*31*). The mechanism of acrylamide formation is still under active investigation, but the confirmed occurrences are fairly dramatic: In excess of 1 ppm acrylamide in potato chips and close to 1 ppm in French Fried potatoes and several other cooked food items. Recent results indicate that the true residues may be even higher, as more rigorous extraction (Soxhlet) (*32*) is used rather than that first reported (extraction with water in a Waring blender) (*30*). The acrylamide issue reinforces the need for vigilance, using the best and latest analytical techniques.

Future challenges in residue analysis can be expected in developing methods for new biorational pesticides with complex chemical structures, such as the avermectins and spinosads, or in determining all of the toxicologically significant breakdown products of the biorationals as well as 'conventional' pesticides such as trifluralin. There will also be challenges associated with developing methods for chemicals with newly recognized toxic effects, such as endocrine disruption. Endosulfan, vinclozalin, and atrazine are a few of the chemicals on the list of suspected endocrine disruptors highlighted by the Food Quality Protection Act (*33*).

Most residue chemists are also good at isolating and identifying, and quantifying bioactive natural products. Low level analysis of aflatoxins in commodities such as corn, peanuts and tree nuts, produced by Aspergillus fungi that enter the produce through insect damage in the field, has been studied in a multiyear research program carried out by USDA Agricultural Research Service scientists (*34*). An area of particular interest is in developing control of damaging insects using natural products, ranging from polyphenols in the hull and nut covering of certain tree nut varieties which inhibit the biosynthesis of aflatoxins, to identification of volatiles from orchard foliage that attract the insects to the orchard. In the latter case, this has resulted in developing lures for use in toxicant-laced traps to reduce pest populations without spraying synthetic pesticides throughout the orchard.

Improvements in Worker Safety

Pesticide residue chemists have played an important role in improving safety for farm workers and those living in and around pesticide treatment areas, working hand-in-hand with occupational toxicologists, chemical company field researchers, and spray applicators. Much of the improvements have resulted from better education (e.g. through Cooperative Extension programs), better use of protective devices and clothing, substitution of reduced-risk pesticides for the older synthetics, and better, smarter spray systems (*35*). Residue chemistry has helped to pinpoint what works and what does not work, and is increasingly used as a monitoring tool by agencies that have farmworker and nearby resident safety and well being within their responsibilities. For example, residue chemistry showed clearly that the primary exposure route for orchard speed spraying (and many other spray scenarios) was dermal rather than respiratory (Table 1). The dermal/respiratory ratio for most chemicals of usual physicochemical properties was found to be a hundred or more (*36, 37*). Exceptions exist for chemicals that release very volatile breakdown products in the environment, or the very volatile fumigants such as Telone, methyl bromide and methyl isothiocyanate (MITC) (*38*).

And residue chemistry is increasingly employed in determining residue distribution patterns in defined areas, such as watersheds or airsheds. Methyl bromide distribution in the air during fumigation treatment of fields in the Salinas Valley, Monterey County, California is an example (*39*). Air samplers were operated at several locations for a week-long period, along the Salinas Valley floor and foothills, focusing particularly on areas proximate to the city of Salinas. The data showed that the highest airborne residues found were one hundredth of the Lowest Observed Adverse Effect Level (LOAEL) (Table 2). This represents a 100-fold or so margin of safety, which is the minimum allowed. The data also helped to confirm the applicability of the Industrial Source Complex Short Term (ISCST) and CalPuff dispersion models for estimating area-wide exposure patterns. Residue chemists need to be involved in more studies of this type, to help pinpoint 'hot spots' of contamination, so appropriate steps can be taken to safeguard health.

California has had a fairly vigorous program, which has involved taking large numbers of air samples around and downwind from fields treated with volatile pesticides that are considered by California law to be potential Toxic Air Contaminants (*40*). However, the issue continues to resurface, most recently in a report issued by Californians for Pesticide Reform (*41*).

Table I. Dermal/Respiratory Exposure for Applicators During Airblast Spraying in Orchard

Chemical	*Dermal (mg/hr)*	*Respiratory (mg/hr)*
azinphosmethyl	12.5	0.26
carbophenothion	41.3	0.11
Dilan	75.1	0.26
Endosulfan	24.7	0.02
Parathion	77.7	0.16
Perthane	59.4	0.14
$\overline{X} =$	50.5	0.16
$\dfrac{\overline{X}\ \text{dermal}}{\overline{X}\ \text{respiratory}} = 320$		

Source: Based on refs. 36 and 37.

Table II. Ambient Concentrations ($\mu g/m^3$) of Methyl Bromide in the Salinas Valley, CA Resulting from Soil Fumigation Sources

	Salinas North	Salinas South	Arroyo Seco	King City
Highest 24 H Average concentration measured	5.68	5.94	1.65	0.51
Annual time weight average concentration (WAC)	2.48	2.51	0.76	0.29
Worst case (ISCST3) = 2x WAC	4.96	5.02	1.52	0.58
Worst case (CALPUFF) = 1.66xWAC	4.12	4.17	1.26	0.48
NOAEL = 81600 $\mu g/m^3$		LOAEL (HEC) = 480 $\mu g/m^3$		

NOAEL = No Observed Adverse Effect Level; LOAEL = Lowest Observed Adverse Effect Level
Source: Based on ref. 39.

Improvements in Environmental Safety

The last 20 years or so have also witnessed significant improvements in environmental safety, accompanied by less environmental contamination by the more long-lived and acutely/chronically toxic pesticides, which has been well demonstrated by residue analysis of environmental samples and indicator organisms (*42*). Better understanding of environmental fate processes, better pre-market screening for potentially adverse effects, and the use of new, reduced risk pesticides are contributing factors. But residue chemists have been very adept at applying their techniques, developed primarily for human foods, to environmental media (soil, air, water, wildlife) and designing monitoring programs which can show more clearly spatial and temporal trends.

For example, the herbicide molinate (Ordram®), has been used extensively on rice grown in the Sacramento Valley. This has led to runoff into the drainage tributaries to the Sacramento River and, in extreme cases, to off flavor and odor of drinking water from treated river water. Through an aggressive program of controlling emissions to the river by, for example, holding the treated waters in the rice paddies long enough for molinate to dissipate, contamination was reduced dramatically to one-fifth of less of residues found in the early 1980s in public waterways.

Unfortunately, one of the pathways of dissipation of molinate is by volatilization from the paddy water, so that air quality remains a concern. The diurnal pattern of molinate flux from a treated paddy heavily instrumented with air samplers and meteorological devices showed significant flux to the air still occurred on the 7th day after application, with about the same intensity as on the first and second days. At least half of the molinate dissipation is by volatilization, the other half by hydrolysis and other breakdown mechanisms. Air sampling conducted in the small towns in and around the rice growing region in the Sacramento Valley confirmed the presence of molinate *(43)*. Since molinate has potential adverse reproductive effects along with its eerie sweet odor, this has been a point of contention in California, and will require additional work.

Future of Pesticide Residue Chemistry

No field can rest on its laurels no matter how productive and dramatic the development of its science and its use by society. Residue chemistry must continue to evolve, in such areas as:

- Real-time in situ methods, that can give results virtually instantaneously, with samples containing the matrix or a very simple extract of the matrix. Methods that can be used in the field are of particular interest.
- Automation and miniaturization of lab methods, to reduce costs, improve reproducibility, and improve throughput.
- New data processing tools that can not only process the data from detection instruments, such as Laboratory Information Monitoring System (LIMS), but also connect that data with computer dispersion models or landscape models, so we can begin to see residue movement in dimensions of space and time, not just as concentrations at a particular space and time.
- Increasing use of biologically based methods, that not only detect and measure but also determine aspects of the biological relevance of residues. This may include further developments of antibody-based methods, including as parts of chips, but also development of tools to detect the products and fragments of biotechnology in genetically altered materials, such as allergenic proteins in corn genetically modified for pest control, or the movement of altered DNA from plants which should contain them to ones nearby that should not.
- We can expect more and new applications of MS, including the Matrix-Assisted Laser Desorption Ionization (MALDI)-Time of Flight (TOF) family that is literally revolutionizing protein analysis.

- We can aim for methods that allow an integrated approach to detecting pathogenic microorganisms as well as chemical residues. MALDI-TOF MS may allow this to occur.
- And we would hope to see more residue distribution studies so that we will know where on or in an apple the residue has accumulated as well as how it is distributed within an orchard, or in a valley, or in a watershed or airshed.
- These areas will increasingly require residue chemists to work as parts of teams that might include engineers, computer specialists, pest control experts, ecologists, and molecular biologists.

If the past fifty years of developments in pesticide residue chemistry are any indication, these and many more developments are sure to become commonplace in the not-too-distant future.

References

1. Seiber, J. N. In *Handbook of Residue Analytical Methods for Agrochemicals;* Lee, P.W., Ed.; John Wiley and Sons, 2003, pp 1-9.
2. Zweig, G. In *"Essays in Toxicology"*, Academic Press, Vol. 2 (**1970**).
3. Washington Report, *J. Agric. Food Chem.* **1953**, 1.
4. Mills, P. A. *J. Assoc. Off. Agric. Chem.* **1959**, *42*, 734.
5. Mitchell, L. C. *J. Assoc. Off. Agric. Chem.* **1958**, *41*, 781.
6. Schechter, M. S.; Soloway, S. B.; Hayes, R. A.; Haller, H. L. *Ind. Eng. Chem. Anal. Ed.* **1945**, *17*, 704.
7. Carson, R. *Silent Spring*, Houghton, **1962**.
8. Gunther, F. A.; Blinn, R. C. *Analysis of Insecticides and Acaricides*, Interscience, **1955**.
9. *Analytical Methods for Pesticides and other Foreign Chemicals in Foods and Feeds.* Academic Press, Vol. 1, up to Vol. 17. Zweig, G.; Ed.
10. *Residue Reviews,* Gunther, F. A., Ed.; Academic Press, Vol. 1, **1962**, through Vol. 171, **2001**.
11. Food and Drug Administration. *Pesticide Analytical Manual,* U.S. Department of Health and Human Services, Washington, DC. **1994**.
12. Gunther, F. A.; Blinn, R. C. *J. Agric. Food Chem.* **1953**, *1*, 325.
13. Seiber, J. N. In *Pesticide Residues in Foods*: *Methods, Techniques and Regulations.* Fong, G. W.; Moye, H. A.; Seiber, J. N.; and Toth, J. P., Eds.; John Wiley, Chapter 2, **1999**.
14. Dimick, K. P.; Hartman, H. *Residue Review*, **1963**, *4*, 150.
15. Moye, H. A. In *Pesticide Residues in Foods: Methods, Techniques, and Regulations.* Fong, G. W.; Moye, H. A.; Seiber, J. N.; Toth, J. P., Eds.; John Wiley, **1999**, Chapter 5.
16. Taylor, L. T. *Supercritical Fluid Extraction*, John Wiley, **1996**.

17. Johnson, L. D. *J. Assoc. Official Anal. Chem.* **1975**, *59*, 174-187.
18. Langhorst, M. L.; Shadoff, L. A. *Anal. Chem.* **1980**, *52*, 2037.
19. Moye, H. A. *Analysis of Pesticide Residues*, John Wiley, **1981**; Chapter 4.
20. Seiber, J. N. In *Pesticide Residues in Foods: Methods, Techniques, and Regulation;* Fong, G. W.; Moye, H. A.; Seiber, J. N.; and Toth, J. P., Eds.; John Wiley, **1999**, Chapter 3.
21. Wong, S. C.; Hoefstra, P. F.; Karlsson, H.; Backos, S. M.; Mabury, S. A.; Muir, D. C. G. *Chemosphere.* **2002**, *49*, 1339-1347.
22. Ferguson, P. L.; Iden, C. R.; McElroy, A. E.; Brownawell, B. J. *Anal Chem.* **2001**, *73*, 3890-3895.
23. Fong, W. G.; Moye, H. A.; Seiber, J. N.; Toth, J. P., Eds.; *Pesticide Residues in Foods: Methods Techniques and Regulations*, John Wiley, **1999**.
24. Lee, P. *Handbook of Residue Analytical Methods for Agrochemicals*, John Wiley, Vol. 1 and 2, **2003**.
25. Woolson, E. A.; Ensor, P. D. J.; Reichel, W. L.; Young, A. L. *Chlorodioxins—Origin and Fate;* Blair, E., Ed.; American Chemical Society, Chapter 12, **1973**.
26. Moye, H. A. In *Pesticide Residues in Foods: Methods, Techniques, and Regulations.* Fong, G. W.; Moye, H. A.; Seiber, J. N.; Toth, J. P, Eds.; John Wiley, **1999**, Chapter 6.
27. Gee, S. J.; Miyamoto, T.; Goodrow, N. H.; Buster, D.; Hammock, B. D. *J. Agric. Food Chem.* **1988**, *36*, 863-870.
28. Fong, W. G. In *Pesticide Residues in Foods: Methods, Techniques, and Regulations.* Fong, G. W.; Moye, H. A.; Seiber, J. N.; Toth, J. P. Eds.; John Wiley, **1999**, Chapter 7.
29. National Research Council, *Pesticides in the Diets of Infants and Children.* National Academy Press, Washington, DC, **1993**.
30. Tareke, E.; Rydberg, P.; Karlsson, P.; Ericksson, S.; Törnqvist, M. *J. Agric. Food Chem.* **2002**, *50*, 4998-5006.
31. Friedman, M. *J. Agric. Food Chem.* **2003** 51, 4504-4526.
32. Pedersen, J. R.; Olsson, J. O. *The Analyst,* **2003**, *128*, 332-334.
33. Keith, L. H.; Jones-Lepp, T. L.; Needham, L. L. *Analysis of Environmental Endocrine Disruptors.* ACS Symposium Series 747, American Chemical Society, Washington, DC, **2000**.
34. Campbell, B. C., Molyneux, R. J. and Schatzki, T. F. *J. Toxicol.-Toxin Rev.* **2003**, *22,* 225-266.
35. Hall, F. R. *Pesticides: Managing Risks and Optimizing Benefits;* Ragsdale, N. N.; Seiber, J. N., Eds.; ACS Symposium Series 734, American Chemical Society, **1999**, pp 96-116.

36. Wolfe, H. R.; Durham, W. F.; Armstrong, J. F. *Arch. Environ. Health.* **1967**, *14*: 622-663.
37. Wolfe, H. R.; Armstrong, J. F.; Staiff, D. C.; Comer, S. W. *Arch Environ. Health.* **1972**, *25*, 29-31.
38. Dowling, K. C.; Seiber, J. N. *Intl. J. Toxicol.* **2002**, *21*, 371-381.
39. Honaganahalli, P.; Seiber, J. N. *Atmos. Environ.* **2000**, *34*, 5311-3523.
40. Baker, L.; Fitzell, D. L.; Seiber, J. N.; Parker, T.R.; Shibamoto, T.; Poore, M. W.; Longley, K. E.; Tomlin, R. P.; Propper, R.; Duncan, D. W. *Environ. Sci. Technol.* **1996**, *30*, 1365-1368.
41. Kegley, S.; Katta, A.; Moses, M. *Secondhand Pesticides: Airborne Pesticide Drift in California*, Californians for Pesticide Reform, 79 pp. **2003**.
42. Seiber, J. N. In *Pesticides in Agriculture and the Environment*, Wheeler, W. B., Ed.; Marcel Dekker, New York, NY **2002,** pp 127-161.
43. Seiber, J. N.; McChesney, M. M.; Woodrow, J. E. *Environ. Toxicol and Chem.* **1989**, *8*, 577-588.

Chapter 3

Multiresidue Analysis of Pesticides in Foods Using Liquid Chromatography/Mass Spectrometry

Futoshi Sato, Hitoshi Iwata, Takakazu Nomura, and Kazuhiro Komatsu

Japan Food Research Laboratories, Tama Laboratory, Tama-si, Tokyo 206-0025, Japan

We developed a multiresidue method (MRM) using LC/MS which enabled to analyze 50 thermolabile or involatile pesticides in foods because of remaining unavailable for the GC mehod. The sample wasa extracted with acetonitrile. After a salting-out with sodium chloride, the acetonitrile extract was purified by gel permeation chromatography. The pesticide fraction was purifeied by ENVI-Carb/NH2 column and determined by LC/MS. Apples, grapefruits, Japanese radishes, brown rice, spinach, and green tea were used for recovery tests in triplicates. The recoveries were more than 70% in 81% (264/324) of the tested items. The complemental use of this method for the conventional GC/MS methods would enable analysis of a wide range of pesticides in a large number of crops.

Introduction

At present, the Ministry of Health, Labor and Welfare (MHLW) in Japan has set the MRL of 229 pesticides in foods and plans a further increase of the number by more than 400 in a few years. The ministry is going to introduce a "positive list system" in which pesticides having no MRLs should be treated as the illegal substance if detected in foods. According to increasing the pesticides listed for MRLs, their physicochemical properties should expand widely. Therefore multiresidue method (MRMs) are required to cover considerably wider range of polarity.

During the recent decades, capillary GC with different sensitive and/or selective detectors has been the most common tool for the multiresidue determination of pesticides in foods. Among these GC systems, the multi-channel mass spectrometer has superseded the element-selective detectors such as ECD and FPD for some years. However, the thermolabile and/or involatile pesticides still remained unavailable for the GC/MS method. On the other hand, it was obvious that the LC/MS suites for such relatively high-polar pesticides. Therefore we developed a MRM using LC/MS which enabled to analyze a total of 50 pesticides consisting of 11 categories (benzoylurea, carbamate, dicarboximide, neonicotinoid, organophosphate, oxim carbamate, pyazole, strobilurin, triazole, urea and the others).

Experimental

Standard preparation

Fifty pesticides were selected and their reference calibrants of garanteed grade or equivalent were obtained from Kanto Kagaku (Tokyo, Japan), Hayashi Pure Chemical (Osaka, Japan), and Wako Pure Chemical (Osaka, Japan). Each stock solution (500mg/L) of the individual pesticide was prepared in acetone. Portions of all the stock solutions were combined in a volumetric flask and made up with acetone to prepare a mixed standard solution. Pesticides concentrations of working solutions for LC/MS analysis were prepared between 0.02 and 1 mg/L using acetnitrile.

Sample preparation

Basically 20g of a homogenized sample (apples, grapefruits, Japanese radishes, and spinach) was weighed, added with 60ml acetnitrile, and blended thoroughly. As for brown rice and green tea, the sample size was scaled down to 10g and 2g, respectively, and 30ml of water was added to these samples before added with acetonitrile. The filtrate of the mixture was combined with 10g of sodium chloride and 10ml of 0.05mol/L phosphate buffer (pH=7), shaken for five minutes. The acetonitrile layer was collected and evaporated near to dryness. Then the residue was

Table I. Instruments conditions of MS

#		Pesticide	M.W.	m/z	mode	F.V.
1		oxamyl	219.3	237	pos	60
2		thiamethoxam	291.7	292	pos	80
3		imidacloprid	255.7	256	pos	100
4		acetamiprid	222.7	223	pos	100
5		aldicarb	190.3	213	pos	120
6		propoxur	209.2	210	pos	120
7		bendiocarb	223.2	224	pos	120
8		carbofuran	221.3	222	pos	120
9		isouron	211.3	212	pos	120
10		carbaryl	201.2	202	pos	80
11	a	ethiofencarb	225.3	226	pos	80
11	b	ethiofencarb sulfoxide	241.3	242	pos	100
11	c	ethiofencarb sulfone	257.3	258	pos	100
12		pirimicarb	238.3	239	pos	120
13		isoprocarb	193.2	194	pos	120
14	a	metominostrobin-E	284.3	285	pos	100
14	b	metominostrobin-Z	284.3	285	pos	80
15		methabenzthiazuron	221.3	222	pos	120
16		diuron	233.1	233	pos	120
17		fenobucarb	207.3	208	pos	120
18		azoxysrtobin	403.4	404	pos	120
19		siduron	232.3	255	pos	120
20	a	methiocarb	225.3	226	pos	120
20	b	methiocarb sulfoxide	241.3	242	pos	100
20	c	methiocarb sulfone	257.3	258	pos	100
21		dimethomorph	387.9	388	pos	120
22		linuron	249.1	249	pos	120
23		daimuron	268.4	269	pos	120
24		cumyluron	302.8	303	pos	120
25	a	triflumizole	345.7	346	pos	80
25	b	triflumizole-M	294.6	295	pos	120
26		procymidone	284.1	284	pos	120
27		tebufenozide	352.5	297	pos	80
28	a	iprodione	330.2	330	pos	80
28	b	iprodione-M	330.2	330	pos	80
29		diflubenzuron	310.7	311	pos	80
30		cyprodinil	225.3	226	pos	80
31		famoxadone	374.4	373	neg	140
32		phoxim	298.3	299	pos	80
33		etobenzanid	340.2	340	pos	80
34		bitertanol	337.4	338	pos	80
35		clofentezine	303.1	303	pos	80
36		difenoconazole	406.6	406	pos	120
37		hexaflumuron	461.1	461	pos	120

Table I. Instruments conditions of MS (continue)

#		Pesticide	M.W.	m/z	Mode	F.V.
38		furathiocarb	382.5	383	pos	120
39		tebufenpyrad	333.9	334	pos	120
40		tolfenpyrad	383.9	384	pos	80
41	a	fenpyroximate-E	421.5	422	pos	120
41	b	fenpyroximate-Z	421.5	422	pos	120
42		piperonyl butoxide	338.4	356	pos	120
43		teflubenzuron	381.1	379	neg	100
44		hexythiazox	352.9	353	pos	120
45		flufenoxuron	488.8	489	pos	120
46		lufenuron	511.2	511	pos	80
47		chlorfluazuron	540.7	540	pos	120
48		methamidophos	141.1	142	pos	120
49		acephate	183.2	184	pos	120
50		monocrotophos	223.2	224	pos	120

#: Peak number assigned by the order of elution from LC/MS column,
M.W.: molecular weight, m/z: mass-to-charge ratio, Mode: mode of API-ES,
F.V.: fragmenter voltage, -M: metabolite

dissolved with 30mL of ethyl acetate and the solution was dehydrated with anhydrous sodium sulfate.

After evaporating the solvent, the residue was dissolved with ethyl acetate/cyclohexane (1:1), and a 2.5g portion of the solution was loaded on a gel permeation chromatography (GPC) column. As for green tea, the volume loaded to GPC was changed to a 1g portion to avoid interfering effects from the matrix. The corresponding fraction from the GPC was collected and evaporated to dryness.

The residue was dissolved in toluene/acetonitrile (1:3) and loaded on an ENVI-Carb/NH$_2$™ column, and then eluted with 30ml toluene/acetonitrile (1:3). The eluate was evaporated to dryness and the residue was dissolved in 1ml acetonitrile. To verify the accuracy and precision of the method we conducted recovery tests in triplicates at 0.2mg/kg as the spiking level.

Instruments and conditions

The GPC was performed using a LC-10 Series GPC system (Shimadzu, Kyoto, Japan). A glass column (φ25 x 600mm) packed with 60g of Envirobeads S-X3 Select (O.I.Analytical, Houston, USA) was operated at room temperature and at a flow-rate of 5ml/min. The mobile phase was ethyl acetate/cyclohexane (1:1), and the sample injection volume was 5ml. LC/MS determination was achieved using "1100 series" LC system equipped with "G1946D" MS instrument (Agilent Technologies, CA, USA). The chromatographic separations were carried out in gradient modes using methanol and 2mmol/L ammonium acetate solution as the solvent. The gradient was started from 20% methanol, increased linearly to 95% methanol in 45 min, and held

Table II. Elution profiles on GPC and ENVI-Carb/NH$_2$™

Pesticide	GPC Elution volume (ml)	Rec(%)	ENVI-Carb/NH$_2$™ Elution volume (ml)	Rec(%)
benzoylurea				
chlorfluazuron	105-135	76	10-70	108
diflubenzuron	130-160	97	0-15	95
flufenoxuron	105-135	83	0-15	100
hexaflumuron	110-130	57	0-15	90
teflubenzuron	120-150	67	5-30	106
carbamate				
bendiocarb	140-180	100	0-10	91
carbaryl	150-190	101	0-15	99
carbofuran	140-180	96	0-15	100
ethiofencarb	150-170	14	0-10	93
ethiofencarb sulfone	140-180	75	0-15	103
ethiofencarb	145-190	146	0-15	100
fenobucarb	135-170	83	0-10	92
furathiocarb	145-175	103	0-10	102
isoprocarb	135-170	91	0-15	92
methiocarb	145-180	110	0-15	100
methiocarb sulfone	150-180	17	0-10	94
methiocarb sulfoxide	150-200	82	0-10	100
pirimicarb	160-200	84	0-15	97
propoxur	140-170	91	0-10	95
dicarboximide				
iprodione	135-170	90	0-20	96
iprodione-M	140-170	64	0-10	102
procymidone	145-175	98	0-10	105
neonicotinoid				
acetamiprid	150-200	105	0-10	94
imidacloprid	150-200	103	0-15	99
thiamethoxam	165-195	79	0-10	97
organophosphosphate				
acephate	155-185	89	10-70	106
methamidophos	155-185	83	0-10	100
monocrotophos	145-185	98	0-10	115
phoxim	140-180	89	0-10	99
oxime carbamate				
aldicarb	150-180	76	0-10	89
oxamyl	160-190	87	0-15	99
pyrazole				
fenpyroximate-E	130-180	97	0-25	99
fenpyroximate-Z	130-180	118	0-15	109
tebufenpyrad	135-165	131	0-10	128
tolfenpyrad	145-175	116	0-10	119

Table II. Elution profiles on GPC and ENVI-Carb/NH$_2$™ (continue)

Pesticide	GPC Elution volume (ml)	Rec(%)	ENVI-Carb/NH$_2$™ Elution volume (ml)	Rec(%)
strobilurin				
azoxysrtobin	150-200	102	0-10	103
famoxadone	145-165	90	0-10	103
metominostrobin	165-185	87	0-10	102
triazole				
bitertanol	125-170	90	0-15	99
difenoconazole	155-185	109	0-10	104
triflumizole	125-155	98	0-10	102
triflumizole-M	125-155	88	0-10	96
urea				
diuron	135-180	102	0-15	97
isouron	145-175	100	0-10	100
linuron	140-170	135	0-10	108
lufenuron	105-140	119	0-15	118
methabenzthiazuron	180-220	92	0-25	91
siduron	135-170	117	0-15	107
others				
clofentezine	170-210	98	0-20	96
cumyluron	135-170	117	0-10	106
cyprodinil	150-190	89	0-20	94
daimuron	130-160	117	0-10	106
dimethomorph	170-210	97	0-15	101
etobenzanid	145-190	97	10-70	114
hexythiazox	145-200	110	0-10	114
piperonyl butoxide	130-170	108	0-15	104
tebufenozide	120-160	110	0-10	101

Rec.: Recovery (%) of elution test, -M: metabolite

at 90% for 10min except for acephate, methamidophos and monocrotophos. A Develosil C30™ (ϕ2 x 250mm, Nomura Chemical, Aichi, Japan) was used as the column. For acephate, methamidophos and monocrotophos, the gradient was from methanol 5% to 50% linearly in 15 min and held at 50% for 10 min, and the column was a Hydrosphere C18 (ϕ2 x 150mm, YMC, Kyoto, Japan). The temperature, flow rate, and injection volume were at 40°C, 0.2ml/min, and 2μL, respectively, in all cases. The API-ES interface was operated in both positive and negative modes. The drying gas was operated at 350°C under 10 L/min of flow rate.

The nebulizer gas pressure was 35 psi. The capillary voltages were 4000V and 3500V for the positive and negative modes, respectively. SIM mode was used for the determination. The details of the fragmenter voltage and *m/z* were shown in Table-1.

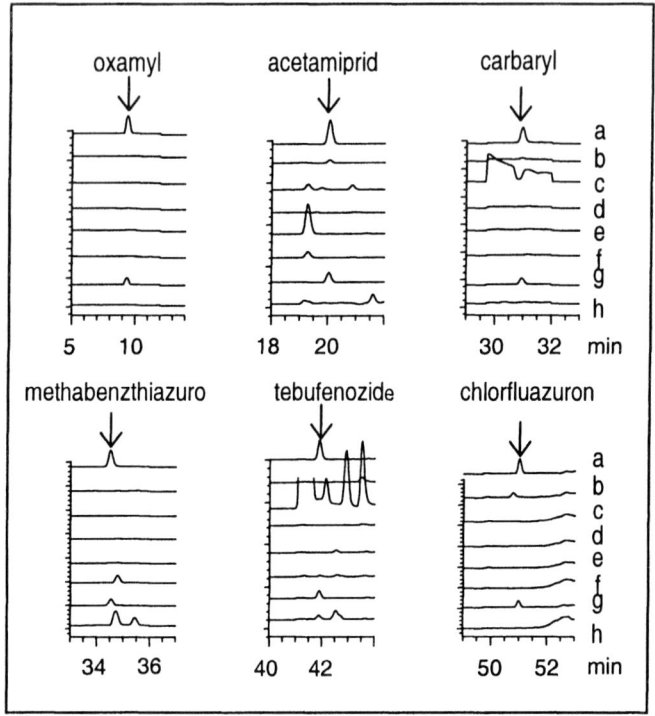

a: standard (0.1ng), b: apples, c: grapefruits, d: J. radishes, e: brown rice, f: spinach, g:standard (0.04ng), h: green tea

Figure 1. Chromatograms for blank extracts.

Results and discussion

Shibata et. al. (1) have reported that ENVI-Carb/NH$_2$™ was efficient for purification and enough recoveries of pesticides in foods by MRMs. Table-2 shows typical elution profiles of the pesticides on the GPC and ENVI-Carb/NH$_2$™.

We have chosen to collect the fraction of the GPC from 105ml to 215ml based on the result shown in Table II. As for ENVI-Carb/NH$_2$™, all pesticides were eluted with 30ml, except for chlorfluazuron, acephate, and etobenzanid. To separate interfering materials which elutes from 30ml to 70ml in the eluate, as well as to save the analytical time, we chose to collect first 30ml for the elution volume despite incomplete elution of 3 pesticides.

For most of the pesticides, the highest response was obtained in the positive mode, where [M+H]$^+$ was the highest abundant ion (Table I). In contrast, the highest abundant ion for famoxadone and teflubenzuron was [M-H]$^-$ in negative mode.

Table III. Recoveries of pesticides for spiked test

#		Pesticide	apples	grape fruits	Japanese radishes	brown rice	spinach	green tea
		benzoylurea						
47		chlorfluazuron	78	58	88	85	73	85
29		diflubenzuron	94	61	96	82	94	82
45		flufenoxuron	60	78	87	84	83	55
37		hexaflumuron	95	82	105	76	93	81
43		teflubenzuron	95	74	103	76	74	69
		carbamate						
7		bendiocarb	92	84	88	91	96	87
10		carbaryl	93	26	88	91	90	83
8		carbofuran	91	79	88	91	90	85
11	a	ethiofencarb	76	70	84	91	87	37
17		fenobucarb	87	73	79	89	85	87
38		furathiocarb	100	92	65	97	90	90
13		isoprocarb	82	74	72	86	80	87
20	a	methiocarb	87	74	94	92	92	65
12		pirimicarb	56	46	83	89	87	43
6		propoxur	91	77	84	90	92	89
		dicarboximide						
28	a	iprodione	89	82	100	81	98	68
28	b	iprodione-M	89	62	60	62	52	70
26		procymidone	94	84	92	102	92	90
		neonicotinoid						
4		acetamiprid	71	60	82	87	87	32
3		imidacloprid	68	59	79	84	89	28
2		thiamethoxam	70	53	65	84	79	17
		organophosphorus						
49		acephate	67	58	84	98	98	40
48		methamidophos	58	49	76	85	80	36
50		monocrotophos	90	76	114	126	125	20
32		phoxim	90	60	72	89	58	82
		oxime carbamate						
5		aldicarb	27	73	72	84	72	36
1		oxamyl	66	59	87	90	88	30
		pyrazole						
41	a	fenpyroxim-E	100	81	107	98	95	98
41	b	fenpyroxim-Z	88	81	79	85	85	95
39		tebufenpyrad	99	89	101	100	91	96
40		tolfenpyrad	98	85	96	93	90	80

Continued on next page

Table III. Recoveries of pesticides for spiked test (continue)

#		Pesticide	Recovery (%)					
			apples	grape fruits	Japanese radishes	brown rice	spinach	green tea
		strobilurin						
18		azoxystrobin	93	84	88	90	89	85
31		famoxadone	92	86	91	97	95	86
14	a	metominostrobin-E	97	79	89	96	86	81
14	b	metominostrobin-Z	95	68	85	95	88	82
		triazole						
34		bitertanol	94	75	94	88	94	98
36		difenoconazole	94	74	97	98	97	87
25	a	triflumizole	88	37	76	91	87	64
25	b	triflumizole-M	78	59	77	90	84	76
		urea						
16		diuron	91	75	88	90	91	84
9		isouron	92	77	85	94	90	96
22		linuron	96	85	92	93	93	88
46		lufenuron	64	81	85	82	87	60
15		methabenzthiazuron	83	61	87	90	92	83
19		siduron	93	82	89	92	90	91
		others						
35		clofentezine	73	73	74	81	86	92
24		cumyluron	90	44	86	88	86	90
30		cyprodinil	82	50	89	90	89	83
23		daimuron	89	76	86	89	87	89
21		dimethomorph	89	60	90	88	88	61
33		etobenzanid	40	26	46	46	48	79
44		hexythiazox	94	91	86	68	84	97
42		piperonyl butoxide	94	87	88	94	92	98
27		tebufenozide	94	70	89	91	90	107

#: Peak number assigned by the order of elution from LC/MS column, -M: metabolite,

Figure I shows ion chromatograms for blank extracts of some crops compared with six representative pesticides at two different doses. Using this method, the substantial responses of the standards were obtained at around 0.04ng with over 3 signal-noise ratio, except for iprodione and teflubenzuron. Matrix-derived interfering peaks observed in some samples such as grapefruits and green tea, make it difficult to determine at the level of 0.02mg/kg for carbaryl, triflumizole, methabenzthiazuron and tebufenozide. Except for these samples, no significant interfering peaks were observed for target pesticides. Table III shows the results of recovery tests. For Japanese radishes, brown rice, and spinach, the recoveries were above 70% for most pesticides. For green tea, high-polar pesticides such as neonicotinoid, oxime

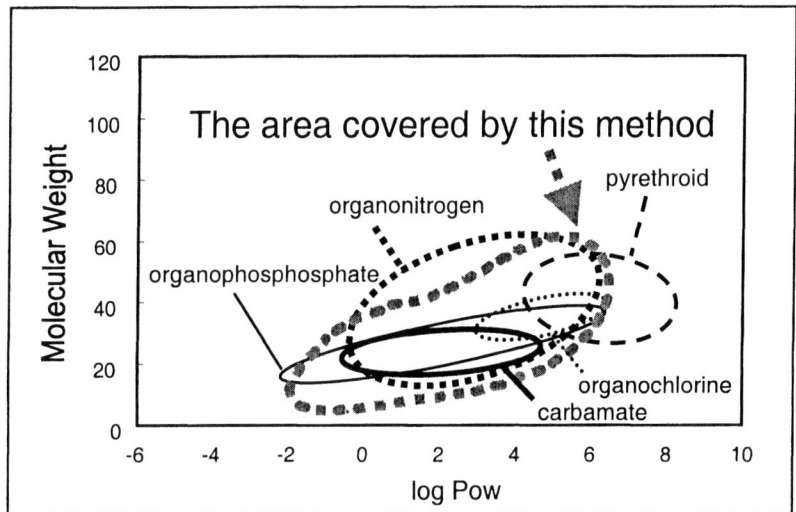

Pow: octanol-water partition coefficient

Figure 2. Distribution map of pesticides determined by this method and major pesticide classes with MRLs in Japan

carbamate and a part of organophosphate showed less than 50% recoveries. The relatively lower recoveries for grapefruit and green tea can be accounted for the suppressed ionization due to the abundant interfering matrices in these samples. The recoveries of etobenzanid were less than 50% in all crops, because of their poor recoveries from ENVI-Carb/NH_2.

However, chlorfluazuron and acephate showed enough recoveries. Figure 2 shows the distribution map of pesticides on the bases of their polarities and molecular weights. As the results, most pesticide classes were covered by this method with the exception of nonpolar pesticides like pyrethroid.

In conclusion, our proposed method enabled to determine about 50 pesticides at LOD of 0.02mg/kg or below. The recoveries were more than 70% in 81% (264/324) of the tested items. The complemental use of this method for the conventional GC/MS methods would enable analysis of a wide range of pesticides in a large number of crops.

References

1. Shibata, Y.; Oyama, M.; Sato, H.; Nakao, K,; Tsuda, M.; Sonoda, M.; Tanaka, F.: J.Food Hyg. Soc. Jpn. 1998, 39, 241-250.

Chapter 4

Multiresidue Monitoring of Pesticides in Domestic and Imported Foods Collected in Hyogo Prefecture, Japan

Yumi Akiyama, Naoki Yoshioka, and Kiyoshi Teranishi

Hyogo Prefectural Institute of Public Health and Environmental Sciences, Kobe City, Hyogo Prefecture 652–0032, Japan

During 7 years' monitoring survey (April 1995 — March 2002) of pesticide residues in agricultural products, 1092 samples (701 domestic; 391 imported) collected in Hyogo prefecture, Japan were analyzed. The number of pesticide tested increased from 107 in fiscal year (FY) 1995 to 228 in FY 2001. The purpose of this study was to promote consumer safety by excluding the food illegally containing pesticide residues from the markets. Overall, 49% of domestic and 31% of imported samples contained no detectable residues. Multiple residues were detected in 31% of domestic and 53% of imported samples. Of the detectable residues above 0.01μg/g, 53% in domestic and 37% in imported samples were <10% of the maximum residue limit (MRL). Violations of the MRL were observed in 4 samples. Of all the samples, 2.4% contained more than 5 different pesticides. Tomatoes, strawberries, apples, and citrus fruits tended to have more multiple residues with the combination of insecticides and fungicides. The imported frozen vegetables from China, such as spinach and baby kidney bean, contained multiple insecticides.

Introduction

Pesticide residue in agricultural products is one of the most important issues in food sanitation. The number of pesticides regulated under the Japanese Food Sanitation Law (FSL) had remarkably increased from 26 to 229 during 1992-2002 (1,2). However, most Japanese official analytical methods are for single-residue analysis. Because a multiple analysis of residual pesticides was needed for regulatory monitoring, we developed the simultaneous screening method for multi-pesticide residues by gas chromatography/mass spectrometry (GC/MS; 3-5). By this method, we have continued the monitoring and expanded the coverage annually by incorporating the newly regulated pesticides.

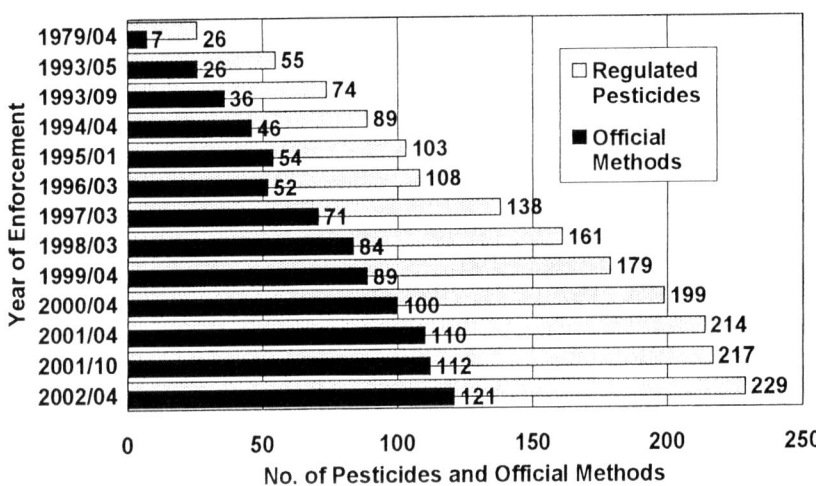

Figure 1. Number of Pesticides Regulated and Official Methods under the Japanese Food Sanitation Law

This study reports on 7-fiscal year (FY) period (1995-2001) monitoring program conducted on about 150 agricultural products per year (6-8). In Japan, agricultural products containing pesticide residues exceeding the maximum residue limit (MRL) established under the FSL are prohibited from the market. However, they are not restricted if there is no specified MRL for the pesticide-commodity combination. In FY2002 we also investigated pesticide residues in the imported frozen vegetables. Here we describe residual pesticide levels and the distribution of multiple residues on agricultural products and frozen vegetables.

Multiple Residue Analysis

Determination and confirmation could be performed simultaneously by GC/MS. A representative total ion chromatogram of 258 mixtures of pesticides and their metabolites is shown in Figure 2. We adopted high-pressure injection system to increase injection volume, and this enabled more sensitive determinations and smaller-scale sample preparation.

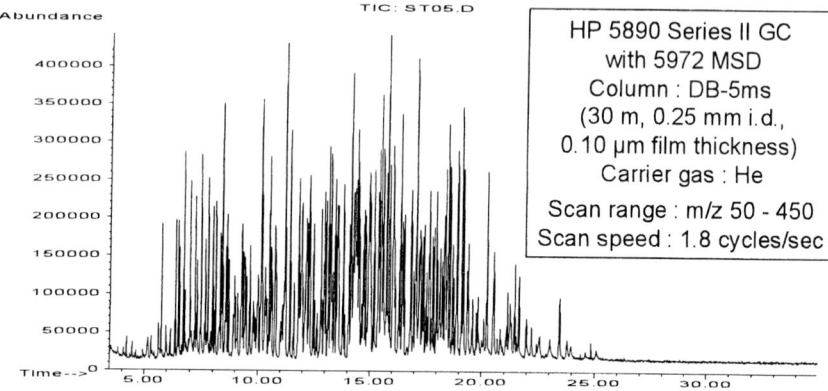

HP 5890 Series II GC
with 5972 MSD
Column : DB-5ms
(30 m, 0.25 mm i.d.,
0.10 μm film thickness)
Carrier gas : He
Scan range : m/z 50 - 450
Scan speed : 1.8 cycles/sec

Oven temp. : $80°$ (3 min) →$30°$ /min →$170°$ (4 min) →$10°$ /min →$270°$ (15 min)
Injector temp. : $250°$; Interface temp. : $280°$; Ionization energy : 70 eV
Inlet pres. : 30 psi (1 min) →80 psi/min →8 psi (0.2 min) →const. flow (0.9 mL/min)
Initial flow : 4 mL/min ; Injection volume : 4 μL(splitless) ; Purge off time : 1 min

Figure 2. Total Ion Chromatogram (TIC) of Pesticides Monitored by GC/MS with Scan Mode

We mainly used SCAN mode and utilized selective-ion monitoring (SIM) mode only for less sensitive compounds. Retention times and monitor ions for GC/MS analysis are listed in Table I. Duplicate injections per sample for both SCAN and SIM mode enabled to ensure the limits of quantitation 0.01 μg/g (S/N≥10) for all compounds. The limits of detection were set at 0.001 μg/g (S/N≥3) and Trace level (Tr.) indicated that remaining could be confirmed at the range from 0.001 to 0.01 μg/g.

Positive analytes were confirmed by retention time and relative response ratio of 2 fragment ions. Rapid and reliable confirmation was available by printing reports containing the extracted ion chromatograms automatically according to our original macro program.

CH₃CN Extraction
| Sample 25g
| CH₃CN 60mL
| homogenize for 3min
| filter with filter paper (No.5A)

ODS Purification
| pass through ODS(1g) cartridge

Salting-out
| shake eluate with NaCl 6g
| and 2M phosphate buffer/
| sat. brine soln. (pH7.0) 10mL
| collect CH₃CN layer 36mL
| (equivalent to sample 15g)
| add internal standard soln.
| (Triphenylphosphate 5ppm 0.15mL)

| dehydrate and evaporate
| adj. to 3mL with n-hexane-acetone(1:1)
| centrifuge for 5min

PSA Purification
| load 2mL aliquot on
| PSA(200mg) cartridge
| add n-hexane-acetone(1:1) 2mL×3
| collect all eluate and
| concentrate to near dryness
| adj. to 2mL with n-hexane-acetone(4:1)
| (equivalent to sample 5g/mL)

GC/MS analysis

Figure 3. Sample Preparation Method for Multi-pesticide Residue Analysis

A simple and rapid sample preparation method (Figure 3) was developed by modification of both CDFA and Luke II methods (9,10). The acetonitrile extract of sample was purified with an ODS cartridge, and then the acetonitrile was separated by salting-out. After evaporation, fatty acids and chlorophylls were eliminated by purification with a PSA cartridge. The final sample solution was concentrated 5 times and determined by GC/MS.

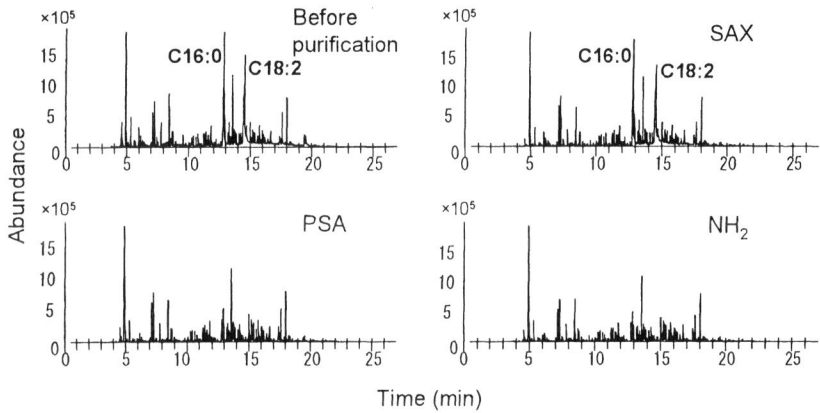

Figure 4. TIC for Sample Extract before and after Purification with Anion-exchange Cartridge SAX, PSA, NH₂
(Reproduced with permission from ref. 3. Copyright 1996 Food Hygienics Society of Japan.)

Table I. Qualitative Parameters for Pesticides Monitored by GC/MS

Pesticide	Retention Time (min) SCAN	SIM	Monitor Ion (m/z) Target	Qualifier	Pesticide	Retention Time (min) SCAN	SIM	Monitor Ion (m/z) Target	Qualifier
Aldicarb (deg.)	3.58		115.10	100.00	Chlorothalonil	10.90		265.85	263.80
Oxamyl (deg.)	3.95		98.10	72.10	Pyrimethanil	10.94		198.10	77.00
Pencycuron (deg.)	4.25		119.00	91.05	Parathionmethyl oxon	11.15		108.95	229.95
Methomyl (deg.)	3.90	5.03	105.05	88.05	Terbacil	11.26		161.00	117.00
Metolcarb (deg.)	4.06		108.00	79.05	d-BHC	11.29		180.90	216.90
Phoxim (deg.)	5.11	5.69	103.05	130.05	Tetluthrin	11.30		177.00	197.00
XMC (deg.)	5.13		122.05	107.10	Etrimfos	11.30		292.05	153.10
Clofentezine (deg.)	5.15		137.00	102.05	Phenothiol	11.34		244.00	155.00
Propoxur (deg.)	5.18		109.95	152.05	Iprobenfos	11.53		203.95	91.05
MPMC (deg.)	5.35		122.05	107.10	Pirimicarb	11.53		166.05	72.10
Isoprocarb (deg.)	5.40		121.00	136.00	Ethiofencarb	11.72		107.05	168.05
Methamidophos	5.67		94.00	141.00	Benfuresate	11.91		163.05	256.10
Dichlorvos	5.78		108.95	184.95	Dimethenamid	11.92		154.10	230.05
Iprodione Met. (deg.)	5.90		186.85	123.95	Phosphamidon Z	11.93		127.05	264.00
Bendiocarb (deg.)	5.97		150.95	125.95	Fenitrothion oxon	11.99		244.00	108.95
Fenobucarb (deg.)	6.01		121.00	149.95	Propanil	12.04		160.90	217.00
Carbofuran (deg.)	6.16		164.05	148.95	Chlorpyrifosmethyl	12.05		285.95	124.95
Swep (deg.)	6.25		186.85	123.95	Terbucarb(MBPMC)	12.05		205.15	220.20
Imibenconazole Met.	6.39		160.90	162.95	Metribuzin	12.12		198.00	144.05
Iprodione Met. (deg.)	6.55		126.95	100.90	Vinclozolin	12.20		211.95	284.90
EPTC	6.56		128.05	189.05	Parathionmethyl	12.26		263.00	124.95
Diphenyl	6.76		154.10	76.10	Tolclofosmethyl	12.27		264.90	124.95
Propamocarb	6.82		58.10	188.15	Alachlor	12.28		160.10	188.05
Iprodione Met. (deg.)	6.90		160.90	126.00	Malathion oxon	12.30		127.05	195.00
Ethiofencarb (deg.)	6.99		107.05	167.95	Heptachlor	12.36		271.75	100.00
Butylate	7.04		146.15	156.10	Acibenzolar-S-methyl	12.38		181.90	135.00
Acephate	7.20		136.00	94.00	Simetryn	12.40		213.05	170.05
Metolcarb	7.28		108.05	107.05	Carbaryl	12.42		144.05	115.10
Trichlorfon	7.30		79.00	109.05	Cinmethylin	12.57		105.05	123.15
Folpet (deg.)	7.36		147.05	104.05	Parathion oxon	12.65		109.05	274.95
Methacrifos	7.53		124.95	180.00	Demeton-S-methyl sulfone	12.78		168.95	108.95
Captafol,Captan (deg.)	7.54		151.00	79.05	Pirimiphosmethyl	12.91		290.05	276.00
Terbucarb(MBPMC) (deg.)	7.59		205.20	220.15	Fenitrothion	12.93		277.00	124.95
o-Phenylphenol	7.72		170.05	141.00	Methiocarb	12.97		168.05	153.00
Carbaryl (deg.)	7.72		115.00	144.05	Dichlofluanid	13.05		123.05	223.90
Isoprocarb	7.84		121.00	136.10	Dimethylvinphos E	13.06		294.95	108.95
Methiocarb (deg.)	7.98		167.95	153.00	Esprocarb	13.13		222.00	162.10
XMC	8.06		122.05	107.05	Bromacil	13.14		204.95	206.95
Diethyltoluamide	8.17		119.00	190.05	Malathion	13.23		173.10	124.95
Methomyl	8.19	8.14	105.05	87.95	Aldrin	13.25		262.80	292.85
Xylylcarb(MPMC)	8.40		122.05	107.05	Metolachlor	13.25		162.10	238.10
Omethoate	8.43		156.00	109.95	Chlorpyrifos	13.33		196.90	313.85
Fenobucarb	8.47		121.00	150.05	Thiobencarb	13.34		100.10	257.00
Propoxur	8.50		110.05	152.05	Dimethylvinphos Z	13.39		294.95	108.95
Demeton-S-methyl	8.63		88.05	60.00	Fenthion	13.44		278.00	124.95
Diphenylamine	8.68		169.05	168.05	Dichofencarb	13.51		225.05	267.15
Ethoprophos	8.77		157.90	200.00	Parathion	13.52		290.95	108.95
Naled	8.96		144.95	108.95	Cyanazine	13.57		225.95	198.00
Chlorpropham	9.01		127.05	213.05	Triadimefon	13.62		208.05	181.00
Trifluralin	9.08		306.05	264.00	Dicofol (deg.)	13.65		139.00	249.95
Dichlofluanid (deg.)	9.10		200.00	92.10	Phthalide	13.65		242.80	271.85
Bendiocarb	9.18		151.05	166.05	Isofenphos oxon	13.68		229.05	201.00
Methabenzthiazuron	9.20		164.00	135.95	Tetraconazole	13.69		335.95	159.00
Cadusafos	9.36		158.90	157.90	Diphenamid	13.90		167.05	239.10
Pencycuron	9.37	9.30	125.05	180.10	Fosthiazate1	13.92		195.00	283.00
Monocrotophos	9.41		127.05	191.95	Fosthiazate2	13.97		195.00	283.00
Phorate	9.43		75.00	260.00	Pendimethalin	14.13		252.05	162.00
a-BHC	9.52		180.90	216.90	Cyprodinil	14.15		224.10	225.15
Hexachlorobenzene	9.57		283.80	141.90	Oxychlordane	14.23		115.00	184.95
Thiometon	9.73		88.05	124.95	Chlorfenvinphos1	14.23		266.95	322.90
Dimethoate	9.97		87.05	124.95	Heptachlorepoxide(endo)	14.24		182.90	81.05
Ethoxyquin	9.99		202.10	174.10	Heptachlorepoxide(exo)	14.24		352.85	81.05
Desmedipham deg.	10.00	9.90	181.00	122.05	Penconazole	14.27		159.00	248.05
Carbofuran	10.14		164.05	149.05	Fipronil	14.28		366.90	212.85
Simazine	10.17		201.00	186.05	Tolylfluanid	14.34		137.00	237.90
Quintozene	10.25		141.90	236.80	Pyrifenox1	14.36		262.00	186.95
Swep	10.29		186.95	220.90	Isofenphos	14.40		212.95	121.00
Atrazine	10.30		215.10	200.00	Chlorfenvinphos2	14.43		266.95	322.90
Triflumizole Met.	10.31		167.05	201.00	Allethrin1	14.44		123.15	136.10
b-BHC	10.31		180.90	216.90	Ethychlozate	14.46		165.00	237.95
Dimethipin	10.31		54.10	118.00	Allethrin2	14.48		123.15	136.10
g-BHC	10.42		180.90	216.90	Prothiofos oxon	14.48		139.00	292.95
Diazinon oxon	10.51		137.10	273.05	Captafol	14.50		79.00	149.05
Tolylfluanid (deg.)	10.53		106.05	214.10	Thiabendazole	14.53		201.00	173.95
Terbufos	10.61		230.95	153.00	Quinalphos	14.54		146.05	157.10
Cyanophos	10.61		108.95	243.00	Phenthoate	14.54		273.95	124.95
Fonofos	10.68		108.95	246.05	Allethrin3	14.56		123.15	136.10
Propyzamide	10.76		172.95	255.00	Procymidone	14.59		96.10	283.00
Diazinon	10.85		179.10	137.10	Allethrin4	14.60		123.15	136.10
Phosphamidon E	10.87		127.05	264.00	Triadimenol1	14.62		112.10	168.05

Table I. Continued

Pesticide	Retention Time (min)		Monitor Ion (m/z)		Pesticide	Retention Time (min)		Monitor Ion (m/z)	
	SCAN	SIM	Target	Qualifier		SCAN	SIM	Target	Qualifier
Folpet	14.63		259.90	294.85	Diflufenican	17.75		266.05	101.00
Triflumizole	14.72		205.95	278.00	Piperonyl butoxide	17.85		176.10	149.05
trans-Chlordane	14.79		372.75	374.75	Bioresmethrin	17.92		123.15	171.05
Triadimenol2	14.81		112.10	168.05	Pyributicarb	18.04		165.05	108.05
Methidathion	14.82		144.95	85.05	Chlomethoxyfen	18.11		265.95	312.95
Methoprene1	14.87		73.10	111.05	Acetamiprid	18.16	18.00	151.95	126.05
Methoprene2	14.87		73.10	111.05	Iprodione	18.18		313.95	186.95
Pyrifenox2	14.96		262.00	186.95	Phosmet	18.23		160.00	317.00
Tetrachlorvinphos	15.01		328.95	108.95	Tetramethrin1	18.24		164.05	123.15
Paclobutrazol	15.01		236.00	125.05	EPN	18.31		185.05	323.00
Trichlamide	15.04		148.05	121.00	Bifenthrin	18.33		181.10	165.05
Vamidothion	15.05		87.05	145.05	Bromopropylate	18.34		340.90	182.90
a-Endosulfan	15.07		194.90	158.90	Tetramethrin2	18.42		164.05	123.15
cis-Chlordane	15.08		372.75	374.75	Methoxychlor	18.46		227.05	113.60
Butachlor	15.10		176.10	160.10	Dicofol	18.48		139.00	250.95
trans-Nonachlor	15.14		408.75	406.70	Etoxazole	18.52		204.15	141.00
Mepanipyrim Met.	15.24		198.95	243.00	Indanofan	18.55		139.00	310.05
Mepanipyrim	15.26		222.10	77.00	Fenpropathrin	18.56		97.10	181.00
Butamifos	15.27		286.05	200.00	Tebufenpyrad	18.65		170.95	318.15
Hexaconazole	15.44		214.00	83.05	Bifenox	18.65		340.90	75.10
Prothiofos	15.48		266.95	308.95	Furametpyr	18.78		157.00	298.10
Pymetrozine	15.48	15.35	113.00	98.00	Phenothrin1	18.82		123.15	183.00
Tricyclazole	15.48	15.35	189.05	162.00	Iprodione Met.	18.86		328.95	186.95
Imazalil	15.49		214.95	172.95	Phenothrin2	18.95		123.15	183.00
Flutolanil	15.52		172.95	145.05	Phosalone	18.96		182.00	366.90
Pretilachlor	15.57		162.10	238.10	Azinphosmethyl	18.99		160.00	132.05
Isoprothiolane	15.58		118.00	290.05	Pentoxazone	19.03		285.00	70.05
Fludioxonil	15.59		248.05	127.05	Clofentezine	19.15		136.90	101.90
Profenofos	15.60		139.00	207.85	Pyriproxyfen	19.20		136.10	78.00
Isoxathion oxon	15.61		105.05	161.00	Mefenacet	19.22		191.95	120.10
p,p'-DDE	15.65		245.95	317.90	Cyhalothrin1	19.25		181.00	197.00
Dieldrin	15.66		79.10	262.80	Mirex	19.25		271.75	236.80
Uniconazole P	15.69		233.95	70.05	Furametpyr Met.	19.26		296.10	159.00
Imazalil Met.	15.80	15.68	82.10	174.95	Cyhalofopbuthyl	19.26		256.10	229.05
Myclobutanil	15.80		179.00	149.95	Cyhalothrin2	19.41		181.00	197.00
Thifluzamide	15.81		194.00	165.95	Fenarimol	19.49		139.00	329.95
Flusilazole	15.86		233.05	206.05	Tebufenozide	19.49		133.10	296.10
Buprofezin	15.87		105.05	172.05	Acrinathrin1	19.50		181.00	208.05
Imibenconazole Met.	15.87	15.74	83.05	235.00	Acrinathrin2	19.66		181.00	208.05
Kresoximmethyl	15.95		116.10	206.05	Pyraclofos	19.82		360.00	194.00
Chlorfenapyr	16.08		59.10	247.05	Fenoxapropethyl	19.91		288.05	361.00
Cyproconazole1	16.09		222.00	139.00	Bitertanol1	20.08		170.00	112.05
Endrin	16.10		262.80	81.05	Permethrin1	20.17		183.00	163.05
Cyproconazole2	16.12		222.00	139.00	Bitertanol2	20.19		170.00	112.05
Isoxathion	16.14		105.05	177.00	Pyridaben	20.30		147.15	117.10
Nitrofen	16.16		282.90	202.00	Permethrin2	20.31		183.00	163.05
Carpropamid	16.20	16.09	139.00	103.05	Prochloraz	20.32	20.19	180.10	307.95
b-Endosulfan	16.35		194.90	158.90	Etobenzanide	20.58		179.00	121.00
Chlorobenzilate	16.39		250.95	139.00	Cafenstrole	20.62		100.10	188.05
Pyriminobacmethyl Z	16.46		302.10	256.10	Cyfluthrin1	20.78	20.63	163.05	206.05
Fensulfothion	16.46		292.95	141.00	Cyfluthrin2	20.90	20.76	163.05	206.05
cis-Nonachlor	16.47		408.75	406.70	Cyfluthrin3	20.97	20.81	163.05	206.05
o,p'-DDT	16.54		235.00	165.05	Cyfluthrin4	21.03	20.88	163.05	206.05
p,p'-DDD	16.55		235.00	165.05	Cypermethrin1	21.16	21.00	181.00	163.05
Ethion	16.61		230.95	153.00	Halfenprox	21.22		262.90	183.00
Clethodim	16.84	16.70	164.05	205.05	Cypermethrin2	21.29	21.07	181.00	163.05
Mepronil	16.93		119.00	269.15	Quizalofopethyl	21.33		372.05	299.00
Triazophos	16.95		77.10	161.00	Cypermethrin3	21.34	21.13	181.00	163.05
Chlornitrofen	17.03		316.90	236.00	Flucythrinate1	21.38	21.23	199.10	157.00
Pyrethrin I	17.05	16.89	123.10	162.00	Cypermethrin4	21.39	21.19	181.00	163.05
Carbophenothion	17.11		157.00	341.90	Ethofenprox	21.56		163.15	135.10
Edifenphos	17.15		172.95	309.95	Flucythrinate2	21.65	21.48	199.10	157.00
Cyanofenphos	17.15		157.00	169.05	Silafluofen	21.77		179.10	286.15
Endosulfan sulfate	17.17		271.75	228.85	Pyrimidifen	22.09		184.05	186.05
Propiconazole1	17.21		172.95	259.00	Fenvalerate1	22.33	22.14	125.05	167.05
Lenacil	17.26		153.10	136.00	Fluvalinate1	22.57	22.37	250.05	181.00
p,p'-DDT	17.30		235.00	165.05	Fenvalerate2	22.66	22.45	125.05	167.05
Propiconazole2	17.32		172.95	259.00	Fluvalinate2	22.69	22.50	250.05	181.00
Pyriminobacmethyl E	17.34		302.10	256.10	Difenoconazole1	23.00	22.78	323.00	264.90
Pyraflufenethyl	17.42		411.95	348.95	Difenoconazole2	23.11	22.89	323.00	264.90
EPN oxon	17.45		306.05	141.00	Pyrazoxyfen	23.15	22.94	105.05	172.95
Thenylchlor	17.50		127.05	288.05	Deltamethrin1	23.30	23.01	181.00	252.85
Tebuconazole	17.61		125.05	250.05	Deltamethrin1(Tralomethrin)	23.59	23.37	181.00	252.85
Propargite1	17.70		135.10	173.10	Azoxystrobin	23.93	23.71	344.00	171.95
Propargite2	17.72		135.10	173.10	Dimethomorph1	24.10	23.86	301.00	165.05
IS (Triphenylphosphate)	17.72	17.62	326.00	170.05	Dimethomorph2	24.75	24.50	301.00	165.05
Captafol	17.72		79.05	183.00	Imibenconazole	25.25	24.98	125.05	252.95

Column:DB-5ms (30m, 0.25mm I.d.,0.10μm film thickness)
Oven temp. 80°C(3min)→30°C/min→170°C(4min)→10°C/min→270°C(15min) for SCAN mode
Oven temp. 50°C(2min)→30°C/min→170°C(4min)→10°C/min→270°C(15min) for SIM mode

The effectiveness of purification procedure with 3 kinds of anion-exchange cartridges was compared. Peaks of palmitic acid and linoleic acid were observed in the chromatograms obtained before purification and after purification with SAX cartridge. As shown in Figure 4, these fatty acids could be removed by purification with PSA or NH_2 cartridge. Chlorophylls were also removed by PSA or NH_2 cartridge. Most of the pesticides were eluted rapidly from PSA or NH_2 cartridge. However, elution of acephate and trichlamide were delayed, and the recovery of trichlamide from cartridge was low. Therefore, we preferred PSA to NH_2 cartridge (3).

In order to apply this multiresidue screening method to pesticides which were easily degraded during sample preparation or GC analysis, we monitored degradation products together with their parent pesticides. And we also investigated the recoveries of degradation products and metabolites in plants by using their standard materials. We accepted recoveries 50-140% available for screening analysis. Therefore 229 pesticides and 26 their degradation products and metabolites could be determined simultaneously in FY 2002.

The degradation ratios of 5 N-trihalomethylthio fungicides (captan, captafol, folpet, dichlofluanid, and tolylfluanid) during sample preparation were examined (Figure 5). For cabbage, all of the parent pesticides disappeared completely and more than 80% were recovered as degradation products. For pumpkin, 39-81% of pesticides were recovered as degradation products and total recoveries of parent pesticides and degradation products were 94-110%. Decreases of the parent pesticides corresponded closely to the formation of the degradation products (5).

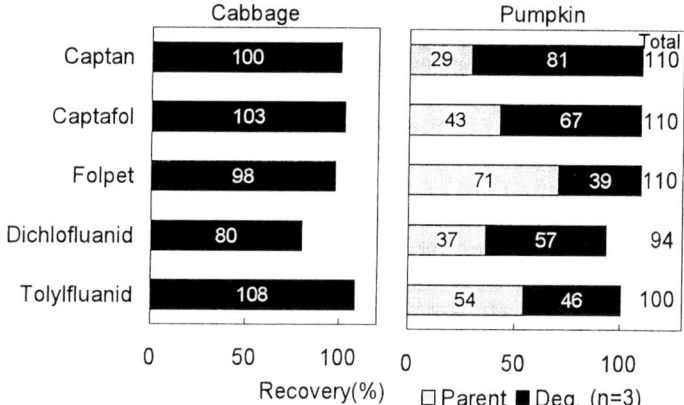

Figure 5. Recoveries and Degradation Ratios of Pesticides Added to Agricultural Pesticides

Pesticide Residues

During 7 years' monitoring survey (April 1995 — March 2002) of pesticide residues in agricultural products, we analyzed 1092 samples collected in Hyogo prefecture, Japan. Overall, 49% of domestic and 31% of imported samples contained no detectable residues. Multiple residues were detected in 31% of domestic and 53% of imported samples (Figure 6).

Detection rates in fruits were higher than those in vegetables and cereals. Imported citrus fruits showed high detection rates, however, for the other fruits and vegetables there were no remarkable differences between domestic and imported samples.

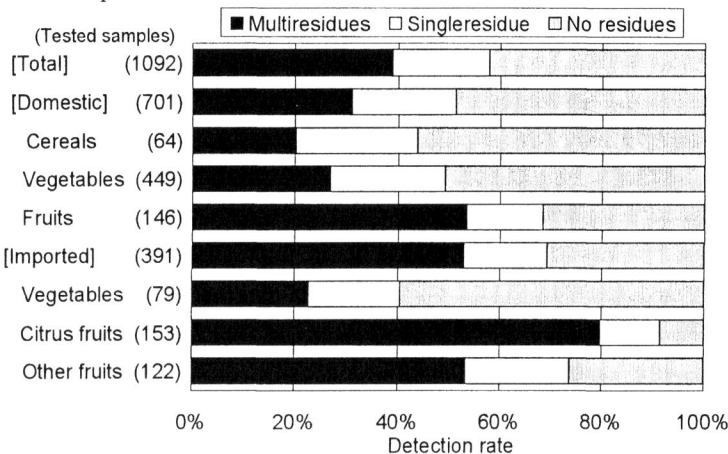

Figure 6. Pesticide Residues in Agricultural Products (FYs1995-2001)

Table II shows the distribution of pesticide residues detected above 0.01 μg/g by percent of MRL (8). In domestic samples, the concentrations of 53% of the detectable residues were less than 10% of the MRL, and 9% were within the range of 10–50% of the MRL. There was no MRL for 37% of the total residues detected. In imported samples, the levels of 37% of the detectable residues were ≤10% of the MRL, and 39% were 10–50% of the MRL. Most of the residues distributed in the range of 10 to 50% were antifungal agents used for citrus fruits. No MRLs were set for 20% of the total residues detected and the ratio was high especially for imported vegetables.

We found violations of the MRL in 4 samples. They were diazinon in garland chrysanthemum (0.94 μg/g, MRL 0.1 μg/g), dieldrin in cucumber (0.03 μg/g, MRL 0.02 μg/g), EPN in eggplant (0.19 μg/g, MRL 0.1 μg/g), and bitertanol in banana (0.65 μg/g, MRL 0.5 μg/g).

Table II. Distribution of Pesticide Residues (≥0.01ppm) by percent of MRL

Sample group	Total residues (=0.01ppm)	% of MRL*				No MRLs
		<10%	<50%	<100%	100%=	
Domestic						
Cereals	17	12	2	0	0	3
Beans	5	0	0	0	0	5
Nuts & seeds	0	0	0	0	0	0
Vegetables	284	130	23	3	3	125
Teas	7	7	0	0	0	0
Fruits	191	117	20	1	0	53
Total	504	266	45	4	3	186
Rate(%)		52.8	8.9	0.8	0.6	36.9
Imported						
Cereals	0	0	0	0	0	0
Beans	1	0	1	0	0	0
Nuts & seeds	0	0	0	0	0	0
Vegetables	38	13	1	1	0	23
Citrus fruits	359	121	177	15	0	46
Other fruits	133	61	28	4	1	39
Total	531	195	207	20	1	108
Rate(%)		36.7	39.0	3.8	0.2	20.3

* Maximum residue limits established under the Food Sanitation Law
Source: Reproduced with permission from ref. 8. Copyright 2003 Taylor & Francis.)

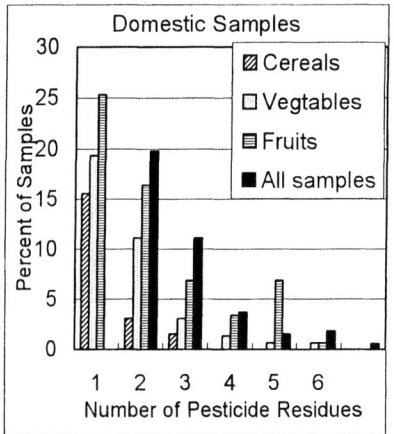
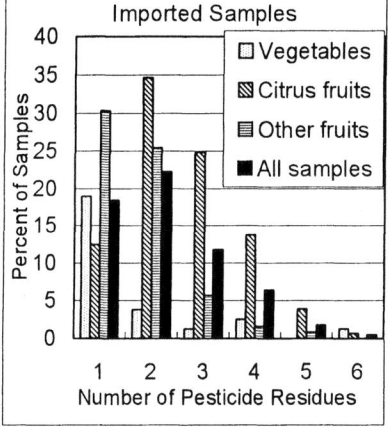

Figure 7 Distribution of Multi-Pesticide Residues for Each Sample Group

The distribution of multiple pesticide residues detected above 0.01 μg/g is shown in Figure 7 for each sample group. Imported citrus fruits contained 2 or more residues with high frequency. Of all the samples, 2.4% contained more than 5 different pesticides. In domestic samples, 7.5% of fruits and 1.4% of vegetables were contaminated with 5 or 6 different pesticides. As for imported samples, 4.6% of citrus fruits, 0.8% of other fruits, and 1.3% of vegetables were contaminated with 5 or 6 different pesticides. Tomatoes, strawberries, apples, and citrus fruits tended to have more multiple residues.

Details about pesticides detected and their levels for typical samples containing multiple residues are shown in Table III. Domestic garland chrysanthemum contained diazinon, dimethoate, isoxathion above MRLs. Tomato was contaminated with phenthoate above MRL. It was found this pesticide was sprayed on the companion crop "soybean". In domestic fruits, residues above MRL were not detected. Most of the samples were contaminated with both insecticides and fungicides. In imported products, residues with no MRLs, such as methamidophos and omethoate, were detected. Residue level of cypermethrin in baby pea exceeded 50% of MRL, which is set at low level comparing with other agricultural products

Table III. Multiple Pesticide Residues in Agricultural Products

(Residue / MRL* (ppm))

Chrysanthemum	(1997.12)	Apple	(2000.11)	Baby pea (China	(2000.12)
Diazinon	0.94 / 0.1	Chlorpyrifos	0.01 / 1.0	Chlorpyrifos	Tr. / 0.1
Dimethoate	1.26 / 1	Fenitrothion	Tr. / 0.2	Dimethoate	Tr. / 1
Omethoate	0.25	Dicofol	0.02 / 3.0	Omethoate	0.06
Isoxathion	4.64 / 0.1	Cyfluthrin	Tr. / 1.0	Methamidophos	0.12
Phenthoate	Tr. / 0.1	Tebufenpyrad	Tr. / 0.5	Dicofol	Tr. / 2
Permethrin	0.52 / 3.0	Bromopropylate	0.04 / 2	Chlorfenapyr	0.01
		Iprodione	0.55 / 10	Cypermethrin	0.03 / 0.05
		Triflumizole	0.02 / 2.0	Fenvalerate	0.01 / 0.10
				Fenpropathrin	Tr.
				Triadimenol	0.18 / 1
Tomato	(1999.9)	Strawberry	(1998.5)	Orange (USA)	(1995.11)
Phenthoate	0.30 / 0.1	Acetamiprid	0.02 / 5	Chlorpyrifos	0.03 / 0.3
Cypermethrin	0.02 / 2.0	Tebufenpyrad	0.05 / 1	Dicofol	0.35 / 3
Etofenprox	0.02 / 2	Pyridaben	0.08 / 2.0	Imazalil	0.64 / 5.0
Procymidone	0.33 / 5	Iprodione	0.26 / 20	Thiabendazole	1.74 / 10
Diethofencarb	0.06 / 5.0	Triflumizole	0.02 / 2.0	o-Phenylphenol	0.27 / 10
		Procymidone	0.02 / 10		

* "Pesticide Residues Standards" or "Standards for Withholding Registration of Pesticides on Crop Residues"

In FY2002 we investigated the imported frozen vegetables. Spinach, baby kidney bean, chingensai, leek, and Welsh onion from China were contaminated with multiple insecticides, such as cypermethrin, omethoate, chlorpyrifos, methamidophos.

Violative samples were not found in our survey, however, 1 spinach sample contained chlorpyrifos and cypermethrin at the level of MRL, and 1 baby kidney bean sample contained more than 10 kinds of pesticides. Since ADI for monocrotophos is relatively low (0.6 μg/kg/day), its residue level in baby kidney bean easily reached 80% of MRL. To add the effects of the other pesticides, the sum of the ratios of residues to MRLs were calculated. As shown in Table IV, we found they exceeded 100%. These values are considered to be one of the indexes to represent the risk of multiple residues.

Table IV. Multiple Pesticide Residues in Imported Frozen Vegetables

(Residue (ppm) / MRL (ppm), %)

Baby kidney bean (China)		(2002.7)		Spinach (China)		(2002.9)	
	Residue	MRL*	(%)		Residue	MRL*	(%)
Chlorpyrifos	0.01 /	0.2	5.0	Chlorpyrifos	0.01 /	0.01	100.0
Dimethoate	0.03 /	1	3.0	Methamidophos	0.01		
Omethoate	0.11			Parathion-methyl	Tr. /	1.0	
Methamidophos	0.06			DDT	Tr. /	0.2	
Monocrotophos	0.04 /	0.05	80.0	Cypermethrin	1.93 /	2.0	96.5
DDT	0.01 /	0.2	5.0	Fenvalerate	0.04 /	0.5	8.0
Dicofol	0.07 /	2	3.5				
Buprofezin	Tr.						
Pyridaben	Tr. /	2.0					
Methomyl	0.03 /	0.5	6.0				
Isoprocarb	Tr.						
Cypermethrin	0.04 /	0.5	8.0				
Fenvalerate	0.03 /	1.0	3.0				
Fenpropathrin	0.02						
Propargite	0.01						
Triadimenol	0.01 /	1	1.0				
Total (Residue / MRL) %			114.5	Total (Residue / MRL) %			204.5

* "Pesticide Residues Standards" or "*Standards for Withholding Registration of Pesticides on Crop Residues*"

Conclusion

The results from our monitoring survey are summarized in Table V comparing with those from the other surveys (*11-14*). From our data the detection rate of residues was 58% in all and the violation rate was 0.4%. The detection limit of the analytical method as well as the number and kind of pesticides monitored easily affect the detection rates. The violation rates are also affected by the kinds of pesticide monitored and by the MRL allowed for a particular pesticide–commodity combination in each country.

Table V. Comparison of Monitoring Data

	Hyogo (Japan) 1995-2001	USFDA 2000	USDA 2000	Ontario (Canada) 1991-1995	EU 1996-2001
Detection Rate (%)	58	40	57	69	30
Violation Rate (%)	0.4	2.6	1.4	8.0	2 - 7
above MRL	0.4	0.1	0.2	3.2	
no MRL	not violative	2.5	1.2	4.8	

In Japan, although detection of pesticide residues with no MRLs is not violationve now, Ministry of Health, Labor, and Welfare will adopt a "Positive List" with MRLs for the combination of 400 pesticides and all agricultural products. The new legislation will be fully implemented in 2006.

References

1. Ministry of Health and Welfare, Japan: *Notification No.239* (27 Oct 1992).
2. Ministry of Health, Labour and Welfare, Japan: *Notification No.94* (13 Mar. 2002).
3. Akiyama, Y.; Yano, M.; Mitsuhashi, T.; Takeda, N.; Tsuji, M.: *J. Food Hyg. Soc. Japan* **1996,** 37, 351-362.
4. Akiyama, Y.; Yoshioka, N.; Yano, M.; Tsuji, M.: *Bull. Hyogo Pref. Inst. Publ. Hlth.* **1996,** 31, 117-125.
5. Akiyama, Y.; Yoshioka, N.; Tsuji, M.: *J. Food Hyg. Soc. Japan* **1998,** 39, 303-309.
6. Akiyama, Y.; Yoshioka, N.; Yano, M.; Mitsuhashi, T.; Takeda, N.; Tsuji, M.; Matsushita S.: *J. Food Hyg. Soc. Japan* **1997,** 38, 381-389.
7. Akiyama, Y.; Yoshioka, N.; Tsuji, M.: *J. AOAC Int.* **2002,** 85, 692-703.
8. Akiyama, Y.; Yoshioka, N.: *Reviews in Food and Nutrition Toxicity* **2003,** 1, 400-444.
9. Lee, S.M.; Papathakis, M.L.; Feng, H.-M.C.; Hunter, G.F.; Carr, J.E.: *Fresenius J. Anal. Chem.* **1991,** 339, 376-383.
10. Cairns, T.; Luke, M.A.; Chiu, K.S.; Navarro, D.; Siegmund, E.G.: *Rapid Commun. Mass Spectrom.* **1993,** 7, 1070-1076.
11. U.S. Food and Drug Administration: *Pesticide Program: Residue Monitoring 2000,* <http://www.cfsan.fda.gov/~dms/pes00rep.html> (2002).
12. U.S. Department of Agriculture: *Pesticide Data Program-Annual Summary Calendar Year 2000,* <http://www.ams.usda.gov/science/pdp/00summ.pdf> (2002).
13. Saito, I.: *Food Sanit.Res.* **2002,** 52, 29-45.
14. Ripley, B.D.; Lissemore, L.I.;, Leishman, P.D.; Denomme, M.A.; Ritter L.: *J. AOAC Int.* **2000,** 83, 196-213.

Chapter 5

Long-Term Fate of the Herbicide Azimsulfuron in a Rice-Grown Lysimeter over Six Consecutive Years

J. K. Lee[1], F. Führ[2], S. H. Park[1], E. Y. Lee[1], Y. J. Kim[1], and K. S. Kyung[3]

[1]Department of Agricultural Chemistry, Chungbuk National University, Cheongju 361–763, South Korea
[2]Institute of Chemistry and Dynamics of the Geosphere 5, Radioagronomy, Forschungszentrum Jülich GmbH, D–52425 Jülich, Germany
[3]Pesticide Safety Division, National Institute of Agricultural Science and Technology, Suwon 441–707, South Korea

The long-term fate of the sulfonylurea herbicide azimsulfuron used in rice paddies was investigated using a lysimeter (0.564 m ID x 1m soil depth) which simulates flooded rice paddies and had been treated with [pyrazole-4-^{14}C]azimsulfuron (specific activity, 2.05 MBq/mg) 28 days after rice transplanting. The treated amount (60 g a.i./ha) was four times larger than the ordinary dosage for analytical convenience. The amount of $^{14}CO_2$ evolved from the flooded soil surface during the rice growing period of 12 weeks and that from the non-flooded soil surface for 9 weeks after harvest were 0.09 and 4.73% of the applied ^{14}C-radioactivity, respectively, indicating that the degradation of azimsulfuron in soil is much faster under the aerobic conditions than under the flooded anaerobic ones. The loss by volatilization from the soil surface was negligible. The total ^{14}C-radioactivities absorbed and translocated by rice plants, leached from the lysimeter, and remaining in the soil throughout six years amounted to 4.01, 9.98, and 47.5% of the originally applied amount, respectively. The brown rice grain contained the residue far less than the MRL of 0.1 mg/kg. It can be concluded that azimsulfuron will not reach environmentally significant concentrations in groundwater in practical use, based on the fact that the originally treated amount was four times larger than the ordinary dosage.

Herbicides are an important tool in modern agriculture for the control of various weeds. In recent years, however, environmental and ecotoxicological concerns have been increased over the contamination of groundwater and the potential impact of soil-applied herbicides on non-target aquatic organisms due to leaching through soil profile and run-off on the soil surface (*1*). To prevent or minimize their risk, it will be helpful to elucidate the temporal and spatial fate and behavior of the pesticides in the environment using lysimeters with the aid of radiotracers. A number of lysimeter studies have been conducted for this purpose, in terms of their leaching, volatilization, mineralization, chemical and microbial degradation, and bioavailability to the target plant, typically by using ^{14}C-radiolabelled compounds (*2-9*).

Azimsulfuron (1-(4,6-dimethoxypyrimidin-2-yl)-3-[1-methyl-4-(2-methyl-2H-tetrazol-5-yl)pyrazol-5-ylsulfonyl]urea) is a selective post-emergence sulfonylurea herbicide used for the control of *Echinochloa* species and most of the annual and perennial broad-leaved and sedge weeds in rice (*Oryza sativa*) paddies in Korea (*10*). The initial herbicidal symptoms of this herbicide were known to be cessation of growth and chlorosis, followed by reddish colouration, then necrosis (*11*).

Sulfonylurea degradation in soil is mainly due to chemical hydrolysis and microbial breakdown, the latter being an important factor for field dissipation (*12-16*). Photodegradation and volatilization would be relatively minor dissipation processes (*17*). Nevertheless, the most significant degradation mechanisms for azimsulfuron in rice paddies were reported to be indirect photolysis and metabolism in soil (*15*). Despite their relatively great mobility in soil, sulfonylurea herbicides do not pose a potential problem in groundwater by virtue of their low application rates, low toxicities, and the relatively rapid degradation and/or disappearance characteristics in soil (*13*). Relatively little information has been available on the environmental fate of azimsulfuron, including leaching, metabolism in soil and plants, and bioavailability to the target crop. The purpose of this investigation is to elucidate the long-term fate of azimsulfuron which was treated to the rice paddy soil using a ^{14}C-radiotracer treated once to the rice-grown lysimeter simulating the rice paddy conditions for six consecutive years.

Materials and Methods

Lysimeter and Preparation of Soil Core

A cylindrical lysimeter was manufactured with stainless steel of 8 mm in thickness. Its surface area was 0.254 m^2, the height 1.1 m, and the inner diameter 0.564 m. An undisturbed monolithic soil core of 1.0 m was obtained by pressing the lysimeter down into a rice paddy soil with the aid of a fork-crane as described elsewhere(*3,4,7,9*). It was carried to the lysimeter station on campus and installed on the ground to collect the leachate through the soil profile and to trap the $^{14}CO_2$ and ^{14}C-volatile organic compounds (^{14}C-VOCs) evolved from the soil surface. The lysimeter for collecting leachates and the device for trapping $^{14}CO_2$ and ^{14}C-VOCs were shown previously(*3*). Water was added to the lysimeter to maintain 5 cm of water-depth over the soil surface and the lysimeter was equilibrated for a month. The physicochemical properties of the lysimeter soil layers are presented in Table I.

Growing of Rice Plants

Prior to transplanting rice plants, the lysimeter soil was fertilized with N-P-K at a ratio of 150-90-110 kg/ha, except that in the case of the nitrogen fertilizer 80% of the total was applied at the beginning and the rest 20% at the earing stage. The rice seedlings (*Oryza sativa* cv Akibare, Japan) grown for 35 days were transplanted onto the lysimeter soil in June each year for six consecutive years. Nine hills were transplanted with three seedlings per hill. To simulate the rice paddy, the soil was flooded throughout the cultivation period. After rice harvest, the lysimeter was left uncovered.

Application of [^{14}C]Azimsulfuron and Cultivation

[Pyrazole-4-^{14}C]azimsulfuron (specific activity, 2.05 MBq/mg; purity, >99%, Figure 1) was supplied by DuPont. The commercial granular formulation, Gulriver (0.05% azimsulfuron), in which azimsulfuron was replaced by [^{14}C]azimsulfuron as a tracer, was applied onto the lysimeter soil at a rate of 60 g ai/ha 28 days after rice transplanting only once in the first year. For investigating the long-term behavior of azimsulfuron in terms of leaching, degradation in soil, and absorption and translocation by rice plants, the applied amount was four times larger than the real application rate of 15 g a.i. /ha recommended in Korea. The total amount and ^{14}C-radioactivity of azimsulfuron applied were 1.51 mg and 3.09 MBq, respectively. The [^{14}C]azimsulfuron dissolved in about 10 mL of methanol and the formulation without azimsulfuron were added to about 200 g of soil. After the methanol had evaporated, the treated soil was uniformly spread on the submerged lysimeter soil planted with rice seedlings. Rice plants were grown for six consecutive years according to the same methods as described above.

Table I. Physicochemical properties of each soil layer of the lysimeter

Soil depth (cm)	pH (H$_2$O, 1:5)	Organic matter (%)	CEC^a (cmol(+)/ kg soil)	Texture
0-10	6.4	2.3	12.8	L^b
10-20	5.9	2.5	13.1	L
20-30	6.3	1.9	12.7	L
30-40	7.2	1.1	12.2	SiL^c
40-50	7.4	0.8	9.2	L
50-60	7.4	0.8	13.0	SiCL^d
60-70	7.2	0.8	14.1	L
70-80	6.9	0.6	11.9	L
80-90	6.9	0.5	9.6	L
90-100	6.5	0.5	11.1	SiL

^{a}Cation exchange capacity, ^{b}Loam, ^{c}Silt loam, ^{d}Silty clay loam
(Reproduced from reference 9.)

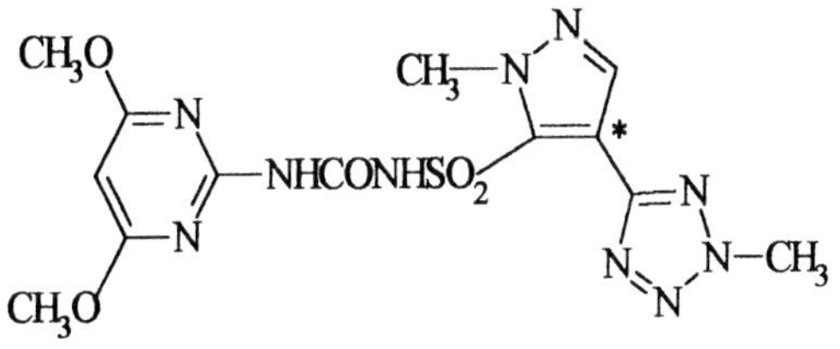

Figure 1. Structural formula and labelled position (*) of [pyrazole-4-^{14}C]azimsulfuron supplied by DuPont. Specific activity: 2.05MBq/mg, Purity: >99%.

Leachate

After the treatment of the lysimeter soil with azimsulfuron, the leachate percolated through the lysimeter was collected weekly in a 20-L polyethylene container for six years, and the volume and radioactivity were measured on a weekly basis. The ^{14}C-radioactivity in the leachate was measured with a liquid scintillation counter (Tri-Carb 1500, Packard Instrument Co., Downers Grove, Ill., U.S.A.). In addition, the ^{14}C-radioactivities of the leachates were partitioned between methylene chloride and distilled water to examine the changes in their polarities with time. That is, 5 mL of the leachates which were acidified to pH 2 with hydrochloric acid was put in the screw-capped tubes and 5 mL of methylene chloride was added.

The tubes are agitated vigorously for a few minutes. After phase separation, 3 mL of organic and aqueous phase, each, was taken and the ^{14}C-radioactivity measured.

Mineralization and Volatilization of Azimsulfuron

Pyrex^R glass devices (8 cm i.d. x 25 cm long) with an inlet and an outlet were used to trap $^{14}CO_2$ and ^{14}C-VOCs evolved from the lysimeter soil treated with [^{14}C]azimsulfuron as reported by Lee et al. (3). Four of them were placed on the surface of the lysimeter soil and CO_2-free air was supplied to them through the inlet at a rate of 1.2-1.5 mL min^{-1} using a vacuum pump connected with the outlet via the serial traps of $^{14}CO_2$ and ^{14}C-VOCs. Each glass device covered an area of 50.24 cm^2, representing about 1/50 of the area of the lysimeter soil surface. $^{14}CO_2$ and ^{14}C-VOCs were trapped in 1 N NaOH and 0.1 N H_2SO_4, respectively, only for 21 weeks of the first year and the radioactivities were measured weekly or biweekly. The total amounts of $^{14}CO_2$ and ^{14}C-VOCs evolved during that period were calculated by multiplying the respective average value obtained with one of the four devices by 50.

Results and Discussion

Mineralization and Volatilization

As can be seen in Figure 2, the amount of $^{14}CO_2$ evolved from the flooded soil surface of the lysimeter during 12 weeks of the rice cultivating period after application of [^{14}C]azimsulfuron in the first year was 0.09% of the originally applied ^{14}C-radioactivity. Meanwhile, the amount of $^{14}CO_2$ evolved from the non-flooded soil surface for 9 weeks after harvest increased remarkably up to 4.73% of the treated ^{14}C. This result strongly indicated that the microbial degradation of azimsulfuron in rice paddy soil was much faster under the aerobic conditions than under the flooded anaerobic ones.

Lee et al. (9) reported that the microbial degradation of the insecticide imidacloprid treated to rice-growing lysimeters was predominant in soil and the larger amount of $^{14}CO_2$ was evolved from the dry soil surface of the lysimeter than from the flooded one. They also suggested that the abrupt increase in $^{14}CO_2$ evolution from the dry soil surface would be due to the fact that many intermediate degradation products formed photochemically and/or microbiologically during the cultivation period were able to be readily mineralized to $^{14}CO_2$ under the dry aerobic conditions. Volatilization of the chemical from the lysimeter soil surface during that period was at the background levels.

Carter (18) reported that volatilization losses are most significant for pesticide residues remaining on the surface of bare soil but their losses by volatilization are low for herbicides. Especially, incorporation into the soil profile can significantly reduce losses.

Soil and Plant Analysis

To measure the ^{14}C-radioactivity distributed in the lysimeter soil profile after harvest, the soil sample of each 5-cm layer was collected down to the 40-cm depth in the first year (21 weeks after application), the 50-cm depth in the second year, the 70-cm depth in the fourth year, and the 90-cm depth in the sixth year with a soil core sampler attached to a stainless steel core of 5.05-cm diameter and 100 cm^3 volume. Soil samples were taken vertically from three random spots of the soil, and those corresponding to the same soil layer were combined after air-drying and ground in a mortar for analysis. The amount of ^{14}C in each soil layer was determined by combusting aliquots of the mixed soils. The rice plants were separated into straw, ear without grains, chaff, and brown rice grains, freeze-dried, and pulverized with a cutting mill to measure the radioactivity by combustion. 0.3 g of each plant or soil sample was combusted with a Biological Oxidizer (R. J. Harvey Instrument Corporation, U.S.A., OX-400) to give $^{14}CO_2$ which was absorbed in the ^{14}C-cocktail (CARBO MAX PLUS LUMAC*LSC B.V., The Netherlands). The radioactivity was measured by LSC. The toluene cocktail was used for measuring the ^{14}C-radioactivity in organic solvents that were evaporated before addition of the cocktail. The radioactivity of $^{14}CO_2$ trapped in 1 N NaOH and ^{14}C-VOCs absorbed in 0.1 N H_2SO_4 was measured using Aquasol (Packard, U.S.A.) as a cocktail.

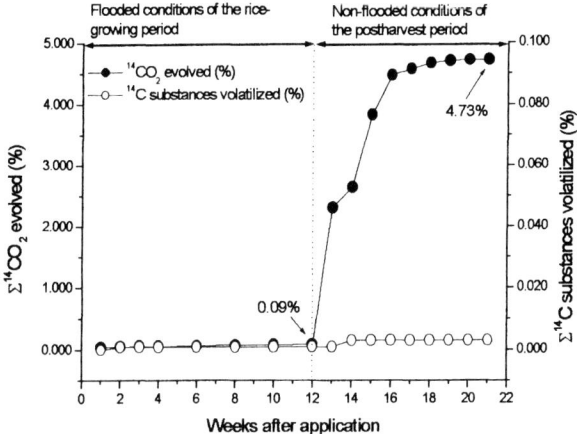

Figure 2. Mineralization and volatilization of [^{14}C]azimsulfuron from the soil surface to the air during the experimental period of 21 weeks in the first year. ^{14}C-Radioactivity applied = 100%.

Leachates from the lysimeter

As can be seen in Table II, the total radioactivities of leachates through the lysimeter soil profile for six consecutive years were 9.98% of the original ^{14}C-radioactivity applied, corresponding to an average of 0.55 μg/L as the azimsulfuron equivalent calculated on the basis of the specific ^{14}C-radioactivity of the azimsulfuron applied. Comparing the annual amounts of ^{14}C percolated through the lysimeter for six consecutive years, the amount increased remarkably in the third year after application and showed a tendency to decrease gradually thereafter, indicating the slow leaching of the azimsulfuron-derived ^{14}C in soil. Peak concentration appeared in the third year, reaching the level of 0.79 μg/L. Lee *et al.* (*7*) reported that the amounts of ^{14}C percolated through the two rice-growing lysimeters treated with the sulfonylurea herbicide cinosulfuron during four consecutive years were less than 3% of the ^{14}C applied.

Partition of the ^{14}C-radioactivity of the leachates between aqueous phase and organic phase

Since an increase in the radioactivity of the aqueous phase with time indicates the gradual formation of polar degradation products of azimsulfuron in soil, the leachates collected during the first three years were acidified to pH 2 with hydrochloric acid and the ^{14}C-radioactivity therein was partitioned between aqueous phase and organic phase. As can be seen in Figure 3, more than 93% of the radioactivity in the first year was distributed in the aqueous phase and less than 7% was distributed in the organic phase.

Table II. Amounts of ^{14}C-radioactivity leached through the lysimeter soil treated once with [^{14}C]azimsulfuron throughout six consecutive years. ^{14}C-Radioactivity applied = 100%

Period (Year)	Volume of leachate (L)	^{14}C-Radioactivity leached (%)	Original amount applied (mg)	Amount leached (mg)	Concentration of azimsulfuron equivalent (μg/L)
1st	12.546	0.106	1.51	0.002	0.159
1st-2nd	36.066	0.224	1.51	0.003	0.083
1st-3rd	109.766	4.161	1.51	0.063	0.574
1st-4th	158.453	6.627	1.51	0.100	0.631
1st-5th	190.583	7.980	1.51	0.120	0.630
1st-6th	275.380	9.980	1.51	0.151	0.548

Meanwhile, in the second year, the ^{14}C-radioactivity distributed in the aqueous phase decreased down to 58.3%, and that in the organic phase increased up to 41.7%. However, in the third year, the ^{14}C-radioactivity of the aqueous phase increased again up to more than 76.4% and that of the organic phase was less than 41.7%. The increased ^{14}C-radioactivity in the organic phase in the second year would be due to the slow and steady leaching of the unchanged azimsulfuron and/or less polar degradation products. In addition, in order to elucidate the ^{14}C-radioactivity contained in the leachates, they were acidified to pH 2 and extracted with methylene chloride. When the concentrated extract was subjected to autoradiography, no parent azimsulfuron could be detected (the autoradiogram is not presented). This fact indicates that after 5 years or so, the parent chemical is no longer present in the leachates.

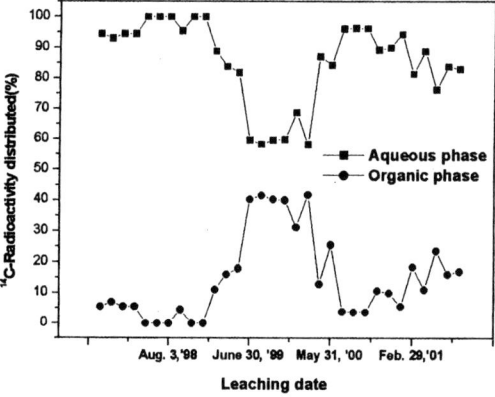

Figure 3. Partition of the ^{14}C-radioactivity of the leachates from the lysimeter treated with [^{14}C]azimsulfuron between aqueous and organic phase. ^{14}C-Radioactivity of aqueous phase + ^{14}C-radioactivity of organic phase = 100%.

Distribution of ^{14}C-radioactivity in rice plants

The ^{14}C-radioactivity absorbed and translocated into each part of rice plants for six consecutive years is presented in Table III. ^{14}C-Radioactivity was detected in all parts such as straw, ears without grains, chaff, and brown rice grains until the sixth cultivation year and the total amount of ^{14}C-radioactivity distributed in rice plants except for roots was 4.01% of the original ^{14}C-radioactivity applied. Larger amounts of azimsulfuron equivalents (0.10-1.50%) were distributed in the straw than in any other parts. More than 90% of the total ^{14}C-radioactivity in rice plants excluding roots remained in the straw except for 80.47% in the sixth year, while 0.15-7.81% in brown rice grain. However, the amounts in the straw decreased gradually year after year. The maximum residue limit (MRL) of azimsulfuron for brown rice grain set by both Korea and Japan is 0.1 mg/kg. From the viewpoint of its safe use, each brown rice grain harvested every year contained the residue far below the MRL, irrespective of the applied amount four times larger than the recommended one in Korea.

Table III. Amounts (%) of the ^{14}C-radioactivity absorbed and translocated in each part of rice plants grown on the lysimeter soil treated with [^{14}C]azimsulfuron. ^{14}C-Radioactivity applied = 100%

Part of rice plants	Year						Total
	1st	2nd	3rd	4th	5th	6th	
Straw	1.502 (0.122)^a	0.613 (0.022)	0.612 (0.010)	0.459 (0.009)	0.429 (0.029)	0.103 (0.003)	3.718
Ears without rice grain	0.007 (0.017)	0.006 (0.020)	0.028 (0.010)	0.008 (0.015)	0.002 (<0.001)	0.003 (0.001)	0.054
Chaff	0.048 (0.017)	0.021 (0.008)	0.029 (0.004)	0.010 (0.004)	0.003 (0.006)	0.012 (<0.001)	0.123
Brown rice grain	0.044 (0.004)	0.027 (0.002)	0.001 (0.002)	0.017 (0.003)	0.011 (0.001)	0.010 (<0.001)	0.110
Total	1.601	0.667	0.670	0.494	0.445	0.128	4.005

^{a}Concentration of the azimsulfuron equivalent ($\mu g/g$) calculated on the basis of the specific ^{14}C-radioactivity (2.05 MBq/mg) of azimsulfuron applied.

Distribution of ^{14}C-radioactivity in soil profile

The distribution of the ^{14}C-radioactivities calculated as azimsulfuron equivalents remaining in the different layers of the lysimeter soil for six consecutive years is shown in Figure 4. A significantly high ^{14}C-radioactivity was distributed in the top 10-cm layer of the lysimeter soil. The soil layer contained a high organic matter content of 2.3% and thus larger amounts of ^{14}C-radioactivities distributed in this soil layer were thought to be bound to organic matter. The amounts of ^{14}C-radioactivities remaining in the different layers of the lysimeter soil decreased gradually year after year, indicating the continuous downward movement and degradation of azimsulfuron in soil.

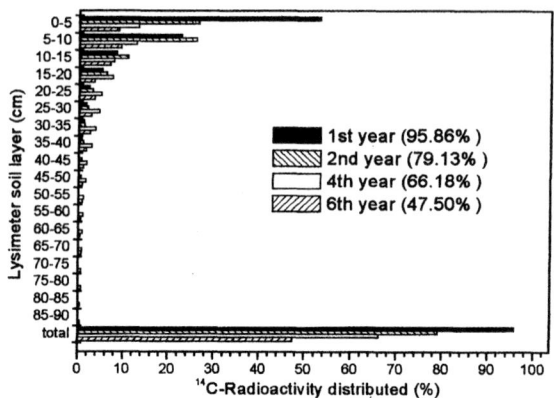

Figure 4. Amounts of the ^{14}C-radioactivity remaining in the different layers down to the 90-cm soil depth of the lysimeter soil for six consecutive years. ^{14}C-Radioactivity applied=100%.

Degradation of azimsulfuron in the lysimeter soil

Figure 5 is the autoradiograms which show the degradation products of azimsulfuron remaining in the top soil (0-10cm) through 4 years.

As can be seen, the authentic azimsulfuron contained two impurities which are thought to be degradation products formed in storage.

In the first two years, [^{14}C]azimsulfuron was degraded in soil to increase the amounts of the two substances. According to Barefoot et al. (15), the hydrolysis of azimsulfuron in sterile buffers of pH 5-9 yielded two principal products formed through cleavage of the sulfonylurea bridge(15-22%), tetrazolylpyrazole sulfonamide and aminodimethoxypyrimidine. They also reported some more degradation products in soil, besides the two products. Similarly, in our study, tetrazolylpyrazole sulfonamide (m/z: 243, the second from the bottom) was identified by GC-MS(chemical ionization) as one of the major degradation products as shown in Figure 5.

Interestingly, in the 4th year, the degradation products could not be detected in the top soil, presumably because they moved downwards and/or disappeared through mineralization. In addition, in Figure 5(A), when the soil sample was extracted with methanol by refluxing, another degradation product was detected (A-4).

Fate of [^{14}C]Azimsulfuron in a Rice Plant-Grown Lysimeter

Table IV summarizes the fate of [^{14}C]azimsulfuron treated once onto the rice plant-grown lysimeter soil over six consecutive years of cultivation. The total ^{14}C absorbed and translocated by rice plants during this period is 4.01% of the originally applied ^{14}C

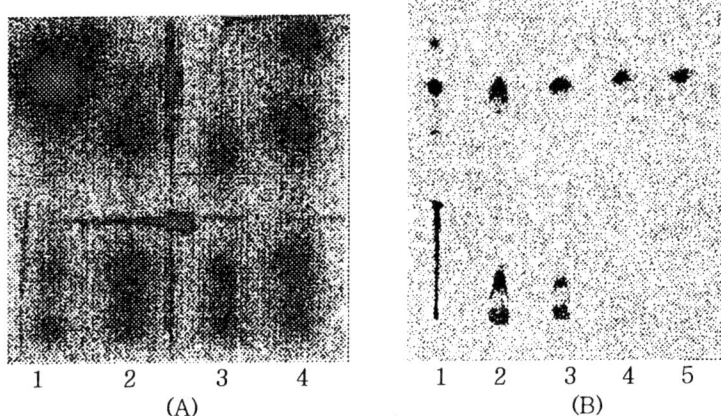

Figure 5. Autoradiogram of the methanol extracts from the lysimeter soil treated with [^{14}C]azimsulfuron.
TLC developing solvent:n-hexane-acetone-acetic acid (10:10:0.1, v/v/v)

A: In the 1st year
1. Authentic [^{14}C]azimsulfuron
2. 0-5 cm soil layer
3. 5-10 cm soil layer
4. 0-5 cm soil layer extracted by refluxing

B: In the 2nd (2,3) and 4th (4,5) year
1. Authentic [^{14}C]azimsulfuron
2. 0-5 cm soil layer
3. 5-10 cm soil layer
4. 0-5 cm soil layer
5. 5-10 cm soil layer

Table IV. Fate of [^{14}C]azimsulfuron treated onto the rice plant-grown lysimeter soil during the experimental period of six consecutive years.
^{14}C-Radioactivity Applied = 100%

Period (Year)	$^{14}CO_2$ evolved^a	^{14}C volatilized^b	^{14}C in rice plants (except for root)	^{14}C leached	^{14}C in soil	Recovery
1st	4.73	BG^b	1.60	0.11	95.86^c	102.30
1st-2nd	4.73+α^d	BG	2.27	0.22	79.13^e	86.35
1st-3rd	4.73+α	BG	2.94	4.16	_^{f}	-
1st-4th	4.73+α	BG	3.43	6.63	66.18^g	80.97
1st-5th	4.73+α	BG	3.88	7.98	-	-
1st-6th	4.73+α	BG	4.01	9.98	47.50^h	66.22

^{a}^{14}C mineralized and volatilized during the period of 21 weeks of the first year after application of [^{14}C]azimsulfuron. ^{b}Background. ^{c}Down to 0-40 cm soil layer. ^{d}Loss by $^{14}CO_2$ evolution not measured from the second to the sixth cultivation year. ^{e}Down to 0-50 cm soil layer. ^{f}Not measured. ^{g}Down to 0-70 cm soil layer. ^{h}Down to 0-90 cm soil layer.

and that detected in the total leachates amounts to 9.98%. About half of the ^{14}C applied remained in soil, where the ^{14}C-radioactivity was distributed mainly in the upper 0-20 cm layer of the lysimeter soil with high organic matter content. The low recovery in the second to the sixth year would be due to the loss by the $^{14}CO_2$ evolution indicated by the letter α in the table.

Conclusions

Considering the ^{14}C-radioactivity leached through the lysimeter soil (9.98%),absorbed and translocated by rice plants for six years after the [^{14}C]azimsulfuron treatment(4.01%), and remaining in soil in the 6th year(47.5%), most of the ^{14}C loss from the lysimeter would be due to mineralization to $^{14}CO_2$ in the 0-10 cm topsoil, mostly by the microbial and chemical degradation.

The parent compound azimsulfuron was not detected in the leachates percolated through the lysimeter soil in the 6th year (the data are not presented). Based on the originally treated amount four times larger than the ordinary dosage, the residual azimsulfuron reaching underground water would be very little.

In addition, the harvested rice grain contained the residue far below the maximum residue limit(MRL), indicating its safety as food.

Acknowledgments

The authors greatly acknowledge the financial support of the Korea Science and Engineering Foundation and Deutsche Forschungsgemeinschaft. They also express their thanks to DuPont for supplying [^{14}C]azimsulfuron.

References

1. Hallberg, G. R. *Amer. J. Altern. Agric.* **1988**, *2*, 3-15.
2. Führ, F.; Steffens, W.; Mittelstaedt, W.; Brumhard, B. In *Pesticide Chemistry*, Frehse, H., Ed.; VCH: Weiheim·New York·Basel·Cambridge, **1991**, 37-48.
3. Lee, J. K.; Führ, F.; Kyung, K. S. *Chemosphere* **1994**, *29*, 747-758.
4. Lee, J. K.; Führ, F.; Kyung, K. S. *J. Environ. Sci. Health* **1996**, *B31*, 179-201.
5. Burauel, P.; Führ, F. In *Environmental Pollution* **2000**, *108*, 45-52.
6. Kim, I. S.; Beaudette, L. A.; Shim, J. H.; Trevors, J. T.; Suh, Y. T. *Plant and Soil* **2002**, *239*, 321-331.
7. Lee, J. K.; Führ, F.; Kwon, J. W.; Ahn, K. C. *Chemosphere* **2002**, *49*, 173-181.
8. Mikata, K.; Schnöder, F.; Braunwarth, C.; Ohta, K.; Tashiro, S. *J. Agric. Food Chem.* **2003**, *51*, 177-182.

9. Lee, J. K.; Führ, F.; Ahn, K. C.; Kwon, J. W.; Park, J. H.; Kyung, K. S. In *Environmental Fate and Effects of Pesticides*, ACS Symposium Series 853; Coats, J.R. and Yamamoto, H., Eds; American Chemical Society, Washington, DC, USA, **2003**; pp 46-64.
10. Chun, J. C.; Ma, S. Y. *Korean J. Environ. Agric.* **1996**, *15*(4), 501-505.
11. Tomlin, C., Ed. In *The Pesticide Manual*, 12th ed.; Crop Protection Publications, British Crop Protection Council, The Royal Society of Chemistry: London, U.K., **2000**; pp 48-50.
12. Anderson, R. L.; Barrett, M. R. *J. Environ. Qual.* **1985**, *14*, 111-114.
13. Beyer, E. M. Jr.; Duffy, M. J.; Hay, J. V., Schlueter, D. D. In *Herbicides: chemistry, degradation and mode of action*, Kearney, P. C.; Kaufman, D. D. Eds.; Marcel Dekker, Inc.: New York and Basel, **1988**; 3, pp 117-189.
14. Fredrickson, D. R.; Shea, P. J. *Weed Sci.* **1986**, *34*, 328-332.
15. Barefoot, A. C.; Armbrust, K.; Fader, T.; Kato, Y.; Sato, K. *10th Symposium Pesticide Chemistry-Basic Processes*. Italy, **1996**; pp 97-104.
16. Bossi, R.; Seiden, P.; Andersen, S. M.; Jacobsen, C. S.; Streibig, J. C. *J. Agric. Food Chem.* **1999**, *47*, 4462-4468.
17. Ryang, H. S.; Moon, Y. H.; Lee, J. H.; Jang, I. S. *Proc. 12th Asian-Pacific Weed Science Society Conference*, Asian-Pacific Weed Science Society: Taipei, **1989**; 1, pp 107-112.
18. Carter, A. D. *Weed Research* **2000**, *40*, 113-122.

Chapter 6

Quick Analysis of Fipronil and Its Metabolites in Gauze and Soil Samples

Sonia Campbell and Qing X. Li

Department of Molecular Biosciences and Bioengineering, University of Hawaii at Manoa, 1955 East West Road, Honolulu, HI 96822

This study focuses on the development of a quick method for the extraction and detection of fipronil residues and its main three metabolites in Hawaiian soil (Helemano series) and cotton gauze swipe samples. Pressurized fluid extraction was used for its ease of use and automated state, its reduction in organic solvent consumption, and time saving interests. Gas chromatography-mass spectrometry (GC-MS) in selected ion monitoring mode was employed for the detection and quantification of the extracts. The extraction method was optimized for the Hawaiian soil for the simultaneous extraction of the four compounds, and was then applied to soil and cotton gauze samples collected from Maui, Hawaii, after a residential spray of fipronil.

Introduction

Fipronil as an insecticide was first introduced by Rhone Poulenc Agro in 1993. It is also used for termite and fruit fly control in Australia and throughout the Pacific Region, but not registered yet in the U.S. for that use. Fipronil acts by blocking the chloride channels in the central nervous system *(1)* and is very effective in the case of flea control on pets *(2,3)* and wild animals *(4)*. Its activity for control of insects on crops such as sorghum and alfalfa has been evaluated and found comparable to other foliar insecticides *(5)*. Fipronil is a choice candidate for pet and household treatment of pests because of its low mammalian toxicity *(6)*. However, toxicity toward non-target species is becoming a concern. *Anaphes Iole* wasps *(7)*, a biological control agent for the tarnished plant bug *Lygus Lineolaris*, and crayfish *(8)* in culture ponds located near fipronil treated rice fields have both shown to be highly sensitive to fipronil, thus, its residues should be monitored and managed adequately *(9)*.

Analytical methods for the detection of fipronil residues have been developed for such matrices as honeybees *(10)* using solid phase dispersion and gas chromatography (GC), and water and soil samples using solid phase microextraction (SPME) and GC-mass spectrometry (GC-MS) *(11)*. Pressurized fluid extraction (PFE) is an extraction technique gaining in popularity due to its ease of use and automated state, its reduction in organic solvent consumption, and time saving. It is a prime extraction method for solid or semi solid samples *(12)*. The objective of this work was to develop a quick and reliable method for the extraction and analysis of fipronil and its main three metabolites in the Hawaiian soil. The method was applied to soil and cotton gauze samples collected on the island of Maui, Hawaii, in March 2002 after a residential spray of fipronil.

Materials and Methods

The Ottawa sand (20-30 mesh size) and extraction solvents (Optima grade) were purchased from Fisher Scientific (Pittsburgh, Pennsylvania). Fipronil (($\pm$)-5-amino-1-(2,6-dichloro-α,α,α-trifluoro-p-tolyl)-4-trifluoromethylsulfinylpyrazole-3-carbonitrile) and the metabolites fipronil sulfide (A) [5-amino-1-(2,6-dichloro-α, α, α-trifluoro-p-tolyl)- 4-trifluoromethylthiopyrazole-3-carbonitrile], fipronil sulfone (B) [5-amino-1-(2,6-dichloro-α, α, α-trifluoro-p-tolyl)-4-trifluoromethylsulfonylpyrazole-3-carbonitrile], and desulfinyl fipronil (C) [5-amino-1-(2,6-dichloro-α, α, α-trifluoro-p-tolyl)- 4-trifluoromethylpyrazole-3-carbonitrile] were the US EPA standards with respective purities of 99.4%, 99.8%, 99.7%, and 98.5%.

A GC-MS system used was a series II 5890 GC interfaced with a 5989A MS from Hewlett Packard. The column was a DB5-MS (J&W Scientific, Folsom, California), 30 m in length, 0.25 mm i.d, and 0.25 μm film thickness. The GC-MS was operated in electron impact mode at 70 eV, with selected ion monitoring (SIM). The detailed operating conditions were injector temperature 250 °C, GC-MS interface 280 °C, ion source 250 °C, the quadrupole 100 °C, and column 50 °C initially which was then ramped at 30 °C/min to 200 °C and held at 200 °C for

20 min. The ions monitored were 367 for fipronil, 351 for A, 383 + 213 for B, and 388 + 333 for C. A complete GC run took 26 min.

PFE procedures for the soil were carried out in an Accelerated Solvent Extractor (ASE, Dionex Corporation, Salt Lake City, Utah). The extraction method was optimized by varying extraction temperature and number of extraction cycles. The extraction solvent was acetonitrile:methanol (1:1, v:v) and the pressure was kept constant at 1500 psi, which in previous studies pressure did not show a big influence on the extraction efficiency *(13)*. The soil used for optimization was of the Helemano series *(14)*, spiked in the laboratory by the organic solvent slurry method *(15)* at approximately 3 μg/g for each of the analytes. After extraction, the samples were filtered through sodium sulfate and 0.2 μm nylon filter and concentrated with a gentle flow of nitrogen gas before GC-MS analysis.

A portion of the spiked soil was kept in the dark at room temperature for 59 days in order to determine aging effects on extraction efficiency and the stability of the analytes in the soil.

Each gauze was placed a 25 mL Nalgene tube with 10 mL of a mixture of acetone and acetonitrile (3:7, v:v). The tubes were capped and shaken for 60 min. The solvent extract of each sample was then filtered and reduced in volume prior to GC-MS analysis.

Soil and gauze samples were collected in Kihei and at Camp Maluhia on the island of Maui, on March 21 and 22, 2003. The gauze samples were pre-wetted with acetone before the swipe; the soils were surface composite samples. The samples were were stored at 4 °C until they reached the laboratory where they were placed in a -20 °C freezer until analysis.

Results and Discussion

Figure 1 presents the results of extraction optimization studies. It is known that fipronil is thermally labile, and thus, only extraction temperatures below 120 °C were investigated. Trials were run at 25, 50, 70, and 120 °C, with 1 or 2 extraction cycles that were 5 min for each cycle. The interest was to find a suitable condition for extraction of all four compounds simultaneously by one extraction procedure. Reproducibility of the extractions was good for all conditions examined. The standard deviations for the percentage of recoveries ranged from 2.0 for metabolite C at 70 °C and 1 cycle, to 12.2 for metabolite A at the same conditions.

For recovery of fipronil alone, the best extraction condition was 50 °C and 1 cycle with an average recovery of 70% ± 3. At this condition, the recovery was 72% ± 3 for metabolite A, 63% ± 4 for B, and 67% ± 6 for C. The three metabolites showed a different optimum set of extraction parameters at 70 °C and 2 cycles, with improved recoveries of 94% ± 3 for metabolite A, 96% ± 10 for B, and 86% ± 2 for C. Fipronil recovery was then 60% ± 4.

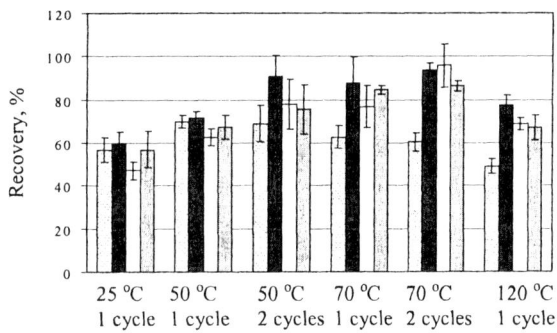

Extraction conditions

Figure 1. Optimization of ASE extraction. ☐ *Fipronil,* ■ *metabolite A,* ☐ *metabolite B,* ▨ *metabolite C. n = 3.*

A PFE condition of 70 °C and two 5-min extraction cycles appears to be a good choice, particularly when extraction time is a concern. When a limited amount of samples is available, one set of extractions can be performed with this procedure for the simultaneous analysis of fipronil and its metabolites.

Table 1 shows the comparison of the average recoveries of the four analytes fortified in the soil and incubated for 1 day and 59 days. There was no significant difference between the 1-day and the 59-day periods. This indicates that the analytes do not degrade in the soil at the incubation conditions and are recoverable from the soil. However, the extraction method remains to be tested for long-term-aged soil.

A representative GC-MS chromatogram of the four analyte standards and a mass spectrum of each are shown in Figure 2. The four analytes were well separated. The molecular ion peaks minus 1 $(M - 1)^+$ for the metabolites A, B and C are 420, 452, and 388, respectively. Fipronil did not give a molecular ion peak.

Table 1. Average recoveries of fipronil and its metabolites from fortified and aged soil

Analytes	1 day aging	2 months aging
Fipronil	60.3 % ± 4.2	75.8 % ± 10.6
Metabolite A	93.7 % ± 3.1	91.6 % ± 9.9
Metabolite B	95.7 % ± 10.0	90.5 % ± 11.5
Metabolite C	86.3 % ± 2.3	75.6 % ± 11.6

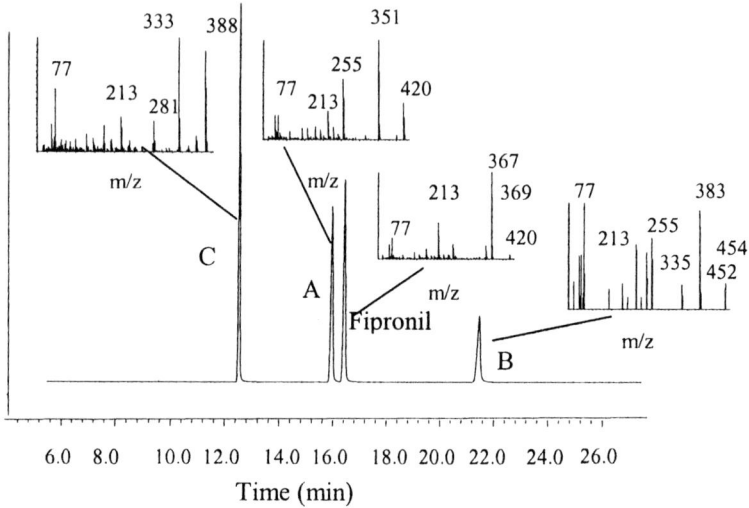

Figure 2. GC-MS chromatogram and mass spectra of fipronil and its metabolites. Fipronil 420 = $(M-O-1)^+$, Metabolite A 420 = $(M-1)^+$, Metabolite B 452 = $(M-1)^+$, Metabolite C 388 = $(M-1)^+$.

Table 2 shows the results for the field samples collected from Maui in March 2002. Five of the gauze samples showed presence of fipronil with amounts varying from 6.23 to 27.4 µg per gauze, two of those also containing the sulfone metabolite (B) in amounts of 13.7 and 14.3 µg per gauze. This metabolite is an oxidative product of fipronil.

The soil sample labeled 032202-02 was a control which was collected outside of the area of fipronil use, and as such did not show fipronil or the metabolites. It is interesting to note though that this sample contained significant amounts of the pesticide DDT and its metabolites. The soil sample collected within the area of the spray showed presence of fipronil at 16.1 µg/g and all the three metabolites from 0.54 to 1.70 µg/g.

In conculsion, this study offered a quick procedure to determine fipronil and its metabolites in fortified soil samples. It was applied successfully to native soil and cotton gauze swipe samples.

Table 2. Quantification of soil and gauze samples (results expressed in µg per gauze or in µg/g of soil)

Sample name	Sample type	Fipronil (µg or µg/g)	A (µg or µg/g)	B (µg or µg/g)	C (µg or µg/g)
032102-01	gauze	ND	ND	ND	ND
032102-02	gauze	ND	ND	ND	ND
032102-03	gauze	7.15	ND	ND	ND
032102-04	gauze	6.23	ND	13.7	ND
032102-05	gauze	7.58	ND	ND	ND
032102-06	gauze	27.4	ND	14.3	ND
032102-07	gauze	7.27	ND	ND	ND
032102-08	gauze	ND	ND	ND	ND
032202-01	soil	16.1	0.54	1.70	0.59
032202-02	soil	ND	ND	ND	ND

ND: not detected

Acknowledgements

This study was supported, in part, by a grant from the State of Hawaii Department of Agriculture - Pesticides Branch, and by a contractual agreement with the State of Hawaii Department of Health - Office of Hazard Evaluation and Emergency Response.

References

1. Bloomquist, J.R. *Annu Rev Entomol.* **1996**, 41, 163-190.
2. Payne, P.A., Dryden, M.W., Smith, V., Ridley, R.K. *Vet Parasitol.* **2001**, 120, 331-340.
3. Cardiergues, M.C., Caubet, C., Franc, M. *The Veterinary Record.* **2001**, 149, 704-706.

4. Metzger, M.E., Rust, M.K. *J Med Entomol.* **2002**, 39, 152-161.
5. Al-Deeb, M.A., Wilde, G.E., Zhu, K.Y. *J Econ Entomol.* **2001**, 94, 1353-1360.
6. Hovda, L.R., Hooser, S.B. *Small Animal Practice.* **2002**, 32, 455-467.
7. Williams, L., Price, L.D., Manrique, V. *Bio Control.* **2003**, 26, 217-223.
8. Schlenk, D., Huggett, D.B., Allgood, J., Bennett, E., Rimoldi, J., Beeler, A.B., Block, D., Holder, A.W., Hovinga, R., Bedient, P. *Arch environ Contamin Toxicol.* **2001**, 41, 325-332.
9. Chaton, P.F., Ravanel, P., Tissut, M., Meyran, J.C. *Ecotoxicol Environ Safety.* **2002**, 52, 8-12.
10. Morzycka, B. *J Chromatogr A.* **2002**, 982, 267-273.
11. Vilchez, J.L., Prieto, A., Araujo, L., Navalon, A. *J Chromatogr A.* **2001**, 919, 215-221.
12. Schantz, M., Nichols, J.J., Wise, S.A. *Anal Chem.* **1997**, 69, 4210-4219.
13. Zhu, Y., Yanagihara, K., Guo, F., Li, Q.X. *J Agric Food Chem.* **2000**, 48, 4097-4102.
14. USDASCS (U S. Department of Agriculture Soil Conservation Service). UHAES (University of Hawaii Agricultural Experiment Station). **1972**, Soil Survey of Islands of Kauai, Oahu, Maui, Molokai and Lanai, State of Hawaii; U.S. Government Printing Office: Washington, DC.
15. David, M.D., Campbell, S., Li, Q.X. *Anal. Chem.* **2000**, 72, 3665-3670.

Environmental Fate

Chapter 7

Understanding the Tropospheric Transport and Fate of Semivolatile Pest Management Chemicals

Vincent R. Hebert

Department of Entomology, Food and Environmental Quality Laboratory, Washington State University, Richland, WA 99352

Various stable tracer and elevated temperature laboratory approaches have been recently developed to assess the photochemical and oxidative fate of semi-to-low volatility agrochemicals. These specialized systems have been used to ascertain environmentally relevant reaction rates for various organochlorine, organophosphorus, and dinitroaniline agrochemicals. The use of tracer and elevated temperature approaches can provide the most useful information on tropospheric reaction rates for various semivolatile agrochemicals, subject to certain atmospheric dilution considerations and temperature constraints. When more detailed exposure information is desired for a specific agrochemical, especially in high-use agricultural areas, these specialized atmospheric fate evaluations together with structure activity relation modeling should be considered. The integration of laboratory estimates of reaction rates, atmospheric models, together with real world monitoring data will provide the best available scientific information for assessing exposure risks from airborne agrochemicals and their reaction products.

Introduction

Pesticides are applied at rates that can exceed 2 billion kilograms each year in the United States (1). Based on their physicochemical properties, it should be expected that a considerable portion of this total amount can directly enter the atmosphere. Besides the direct entry into the air at application, post-application volatilization and wind erosion also represents a second and significant source of sustained tropospheric loading from the land surface. A large data set has been gathered which details tropospheric loading resulting from drift (*2*) and post-application volatilization (*3, 4*). Majewski and Capel (*5*) have also published a detailed review of regional and national atmospheric pesticide levels along with aerosol particulate and vapor distributions.

Unfortunately, research studies that address environmentally relevant atmospheric fate processes of pesticides are relatively few in comparison to studies that measure transformations on land surfaces and in water. This scarcity of fate information is related to the difficulty in attaining relevant tropospheric photochemical and oxidative information under both environment and controlled laboratory conditions. Only a limited number of studies exist that have measured airborne pesticide reactivity under actual sunlight conditions (*6,7,8*). These studies employed photochemically stable tracer compounds of similar volatility and atmospheric mobility to compensate for physical dilution. The examined airborne sunlight-exposed pesticides in these limited studies had to react quickly to provide environmentally measurable reaction rate constants. The field examination of tropospheric reaction rates for the vast majority of agricultural pesticides is impractical since reaction rates for many of these compounds are probably too slow to yield reliable rate constant information.

Problems encountered in acquiring stable gas-phase conditions in the laboratory also contribute to the relative lack of atmospheric pesticide reaction rate and product data. Semi-volatile organics, which comprise the majority of pesticides, can sorb onto the surfaces of the laboratory reaction vessel. The "wall interference" reaction rates and products may or may not be similar to those occurring under actual atmospheric vapor-phase conditions (*9,10*). Experimental designs that can provide environmentally relevant reaction rates, characterization of gas-phase and oxidative transformation products, and maintain material balance at environmental temperatures has yet to be established.

Of the field and laboratory air studies that have been performed, sunlight-induced chemical oxidations and photochemical reaction pathways usually render pesticide residues less toxic, more polar, and more susceptible to being washed-out of the air mass (*11,12,13,14,15*). Field and laboratory atmospheric pesticide fate studies have also reported the formation of photooxidation products that can have equal or higher toxicity and/or greater environmental persistence than the parent pesticide (*16,17,18,19*). Because of the limited number of attempted atmospheric fate studies, there remains a substantial degree of uncertainty in regards to the mechanistic behavior and possible fate of many pesticide groups that can reside in the lower atmosphere.

Due to the inherent difficulties in acquiring quantifiable pesticide reactivity data, structure-activity atmospheric oxidation models are often employed by regulatory agencies for estimating reaction rates of existing and newly registered pesticides (*20,21,22*). For volatile organic air pollutants and fumigants, structure-activity relationship models can reliably provide chemical oxidation rate constant data within an acceptable (i.e., factor of 2 or less) margin of uncertainty (*22*). These bimolecular reactivity models usually rely on global averaged oxidant concentrations that are invariant with changes in regional climate and solar intensity. When extended to estimate reaction rate constants for semivolatile multifunctional pesticides, model estimation power may diminish. Because many pesticides are structurally complex, relying on structure-activity model may possibly provide erroneous rate constant information on the tropospheric fate for many pesticide groups (23).

Once in the air, pesticides can exist as vapor, liquid aerosols or be sorbed/partitioned on dust and particulates. Even though in-depth information exists for particle phase distributions of organics in the atmosphere (24, 25, 26, 27, 28), few studies have appeared in the literature that assesses the vapor/aerosol distribution of pesticides under actual tropospheric conditions (29). Further research to determine how pesticides are distributed among atmospheric phases will be needed to better gauge the overall significance of wet/dry deposition versus gas and particle-phase transformation processes occurring in air.

The remainder of this proceeding summarizes and integrates pesticide transport and fate investigations, focusing on the more recent scientific advances in understanding the behavior of pesticides in air. Although pesticides in air present a major human and ecological community exposure route (30,31), our knowledge of distributional and fate processes in this environmental medium still remains poorly understood.

Laboratory Photochemical Oxidation Pathways

Sunlight-induced transformations that lead to chemical oxidation or photolysis are likely to dominate the removal of pesticides, especially in their gaseous forms, from the lower troposphere (10,11,12,32,33). The residence time for an airborne pesticide will depend on the rate at which it will undergo direct photo-transformation and/or chemical oxidation by tropospheric reactants such as hydroxyl/peroxy radicals, NO_x radicals, or ozone. For pesticides that absorb sunlight, the loss from direct photolysis will be a function of both the extent of sunlight absorbance and the efficiency of transformation after absorbance.

Figure 1 outlines various pathways can lead to the deactivation of the excited pesticide molecule (P*) through luminescence, physical quenching, or by collisionally transferring energy to other gaseous molecules (M). This figure also illustrates electron transfer, photoionization, or direct chemical reaction processes of the excited state that can lead to dissociation and subsequent product formation.

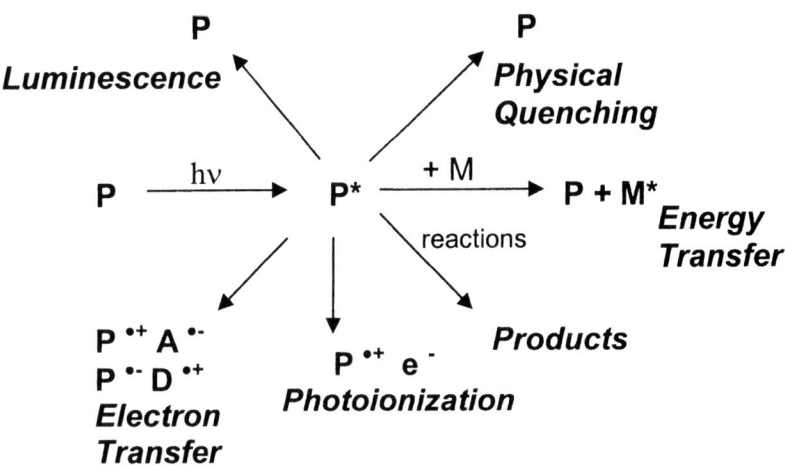

Figure 1: Photochemical Pathways

Atmospheric removal processes involving tropospheric oxidants will likely dominate for pesticides that are transparent to sunlight wavelengths. Of these oxidants, hydroxyl-radical (OH) reactions will be the primary mode for the chemical oxidation of the majority of pesticides in air (11). The rate of reaction by hydroxyl radical mediated processes can be on the order of minutes to days or years and will depend upon the reactivity of the pesticide's functional groups with the oxidant and tropospheric concentrations of OH available for reaction. Pesticides with abstractable hydrogens, or can undergo OH addition will be most susceptible to transformation. Chemical oxidation rates will decrease rapidly for electron deficient compounds, such as chlorinated aromatics or nitroaromatics (20). In many cases, heteroatoms may provide for greater reactivity. For example, Atkinson et al. (34) have reported that phosphorothio systems (P=S) typical of many organophosphorus insecticides are highly reactive towards OH. Pesticides belonging to this group can usually be expected to have tropospheric lifetimes less than four hours (32).

Tropospheric Laboratory Reactivity Evaluations

Photochemical Evaluations: Although fumigants and sterilants have vapor pressures (i.e., >1 Pa) that might allow relatively uncomplicated experimental determination of rates and products, the majority of pesticides have lower vapor pressures (<0.1 Pa) creating problems caused by surface interactions due to sorption (35). An example where wall interactions do not greatly influence reaction rates or products in confined vessels is the rapid gas-phase transformation of methyl isothiocyanate (MITC), a volatile transformation product of the fumigant metam sodium. This substance was examined in relation to a volatile, photochemically stable tracer in quartz chambers and Tedlar sampling bags (17). Reaction rates and products for other volatile fumigants have also been successfully examined in laboratory reaction chamber systems (16).

For semi-to-low volatility pesticides, the size of the reaction vessel should be as large as is practicably achievable to circumvent complications caused by wall sorption at environmental temperatures. For example, large Tedlar chambers have been successfully employed to determine oxidant reaction rates for substances with vapor pressures greater than 0.8 Pa (15, 34, 36). The use of large reaction chambers, however, is not a feasible approach for the majority of pesticides with vapor pressures lower than 0.8 Pa, especially when there is a need for determining overall material balance of reactants and transformation products.

To minimize wall sorption concerns, experimental designs that elevate air temperatures in quartz or glass reaction vessels were developed for obtaining stable gas-phase concentrations for photolysis and/or chemical oxidation determinations (37,38,39). Increasing temperature, however, can influence the activation energy for chemically mediated reactions and as a rule of thumb, elevate the vapor pressure greater than 2-fold for every $10^{\circ}C$ rise. Figure 2 illustrates an elevated reaction chamber used by Hebert et al. (13) to estimate the photochemical reaction of trifluralin in the presence of a photochemically stable gas-phase reference substance, hexachlorobenzene. The gaseous mixture was sampled by inserting a solid phase microextraction (SPME) polysiloxane coated fiber over a very short interval of time. The fiber was then thermally desorbed into a hot injection port of a gas chromatograph. The use of this technique avoided the need to consider artifact oxidation products that may occur as a result of using XAD or Tenax adsorbents and also avoided the need for considering dilution corrections. For substances, such as trifluralin, that principally undergoes direct photolysis, there were no observable activation energy effects on reaction rates seen over a limited elevated temperature range from 60 to $80^{\circ}C$ (Figure 3).

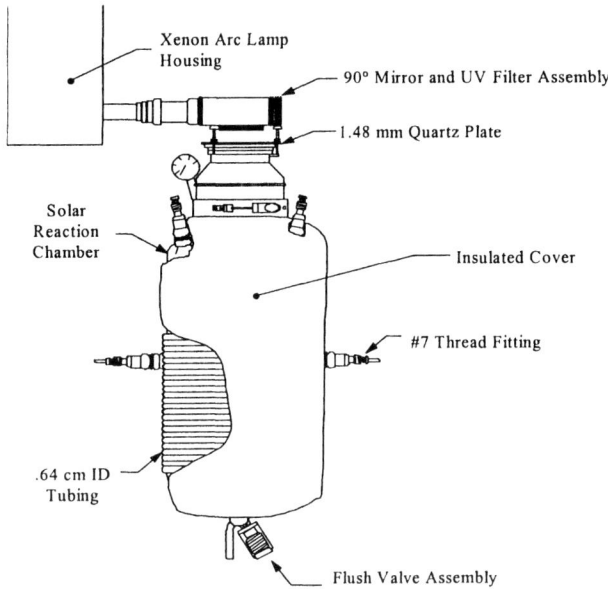

Figure 2: 60-L Elevated Temperature Reaction Chamber

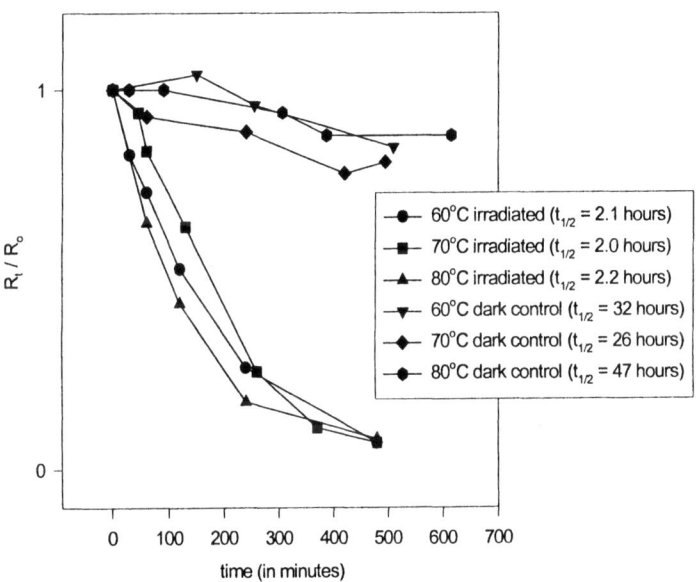

Figure 3: Trifluralin Photolysis: Elevated Temperature Evaluations at 60, 70 and 80°C: Hexachlorobenzene at Tracer

The collimated xenon light source illuminated 20% of the chamber volume. After correcting for total percent chamber illumination, estimated rates of photolysis were found to be comparable to those reported by Woodrow et al. (*8*) and Monger and Miller (*7*) who conducted stable tracer trifluralin photochemical evaluations under outdoor sunlight conditions.

Chemical Oxidation Evaluations: Direct photolysis pathways should be invariant to minor temperature changes. However, very few pesticides are expected to undergo reactions by photolysis as a primary transformation pathway. The majority of pesticides that absorb sunlight will also be susceptible to chemical oxidation. In the majority of cases, chemical oxidation will likely be a primary process leading to transformation for pesticides that either absorb or not absorb radiant energy. When evaluating pesticides that undergo chemical oxidation in elevated temperature systems, slight changes in temperature may significantly alter the activation energy either accelerating or attenuating the overall rate of reaction (*40,41*). As a result, care should be employed when extrapolating laboratory elevated rate observations to environmental temperatures.

Relative rate evaluations are often used to ascertain OH reactivity information. For most volatile organics, this method should provide an OH-radical reaction rate constant within a factor of two of the expected value (*22*). If a single oxidizing species (O) is solely responsible for the simultaneous decay of a test-substance (P) and reference compound (R) with known k_{OH}, then their relative rates of loss can be used to empirically determine the test substance's rate of reaction using the following expression:

$$\ln \frac{[P]_0}{[P]_t} = \frac{k_1}{k_2} \ln \frac{[R]_0}{[R]_t}$$

This relative rate expression is very useful since the oxidation reaction rate of the reference compound (k_2) is known and the natural log of the simultaneous loss of test and reference compound can be experimentally determined. The oxidative reaction rate for the test substance (k_1) is the lone unknown variable and can be calculated from the slope (k_1/k_2) when $\ln([R]_0/[R]_t)$ is plotted against $\ln([P]_0/[P]_t)$ (*42*).

The major advantage of the relative rate estimation method is that it is relatively easy to employ and can avoid the need for difficult experimental evaluations that must rely on estimating absolute OH concentrations over time. A disadvantage is that tropospheric OH radical concentrations can globally vary to a factor of five (*11*). The precision of calculated reaction rates may therefore be limited unless tropospheric OH concentrations can be more accurately estimated in the region of interest. Uncertainties also increase when extrapolating beyond the current database information for volatile compounds to more complex semivolatile organics. In studies performed on a series of OH reactions for dimethyl phosphoroamidates and dimethyl phosphorothioamidates, Goodman et al. (*36*) found that electronic effects between structural units may have been responsible for discrepancies up to a factor of three between measured and calculated rate constants. In their assessment, the loss in estimation power may well represent "the inherent uncertainties for more complex chemicals containing multiple substituent groups."

Starting in the 1990's, elevated temperature gas flow-through and static systems have been used to assess chemical oxidation rates for low volatility organochlorine and organophosphorus pesticides. In these systems, multiple measurements at different temperatures were performed to also evaluate the temperature effects on observed reaction rates. Anderson and Hites (*37*) elevated the temperatures within a small volume gas flow-through system with in-line MS detection to assess the gas-phase degradation of hexachlorobenzene relative to the rate of loss of cyclohexane, a reference compound with well-characterized OH reactivity. Plotting observed rate estimates for hexachlorobenzene at various elevated temperatures, these investigators observed excellent agreement to atmospheric oxidation structure-activity model predictions for this substance when extrapolated to 25 C. This system design also has the advantage of taking multiple readings over a very short time interval to increase reliability of measurement precision. Since 254 nm UV light is required for generating OH within the elevated temperature reaction flow cell, this system may be limited to evaluating thermally stable compounds that do not undergo rapid photodegradation under intense irradiation.

The use of relative rate/elevated temperature experiments have also been employed to determine tropospheric OH radical reactivity rates for the atmospherically reactive organophosphorus (OP) insecticides chlorpyrifos and diazinon employing large-volume chamber conditions as illustrated in Figure 2 (*14*). In these evaluations, the gaseous OP pesticides, photo and oxidatively stable tracer compounds, and reference compounds with well-characterized OH reaction rates were simultaneously introduced into the chamber. Hydroxyl radicals were generated by the gas-phase photolysis of methyl nitrite at wavelengths greater than 290-nm using xenon arc irradiation after the methods of Atkinson and co-workers (*15, 36, 40*)

$$CH_3ONO + h\nu \rightarrow CH_3O + NO$$
$$CH_3O + O_2 \rightarrow HCHO + HO_2$$
$$HO_2 + NO \rightarrow OH\bullet + NO_2$$

NO was added to the reaction mixture to avoid the formation of O_3 and NO_3 radicals. After dosing the vessel with the OH radical precursor, the xenon arc lamp was illuminated over the 2-hour experimental time frame. Under these conditions, sufficient methyl nitrite was present to generate OH for approximately 100 minutes.

Both of the above organophosphorus pesticides should have similar gas-phase tropospheric lifetimes based on structural activity relationship model predictions (21). OH rate measurements for the two OPs when conducted at 5° C increments between 60 and 85° C, however, showed a significant difference in reactivity. The rate of OH oxidation for diazinon was found to be ca. three to four times more rapid than for chlorpyrifos with observed tropospheric lifetimes of ca. 1 and 4 hours, respectively. The difference in observed reactivity was not due to wall sorption since both compounds behaved similarly in the gas-phase.

While OH radical reaction rate constants have been successfully generated to extend structure-activity predictive capability for thiocarbamates and chlorinated aromatics, further experimental work will be important to assess OH reaction rates for the more complex chemistries that comprise the majority of high-use urban and agricultural pesticides.

Conclusions

Many field-monitoring studies have indicated the substantial role of the troposphere as a transport medium and potential sink for semivolatile pesticides. But at the same time, it is the least studied and understood environmental compartment in regards to pesticide fate. While the fundamental principles behind tropospheric reactivity is well understood, it is becoming increasingly apparent that the ability to quantitatively measure reaction rates and product distributions from the air will continue to pose problems for researchers. To date most deterministic models try to simulate real-world conditions but may fall short in providing tropospherically relevant reaction rates since it is virtually impossible to scale down all possible interactions occurring in a near-infinite reservoir, especially for multifunctional semivolatile pesticides. More experimental data will be needed for verifying model rate predictions for the more complex chemistries that are representative of the vast majority of current and emerging pesticides. Basic laboratory and applied field research that can provide environmentally relevant reaction rate and fate information remains an urgent need in human and ecological risk assessment.

Acknowledgements

I would like to express my sincere appreciation to both Glenn Miller at the University of Nevada and James Seiber from the USDA Western Regional Agricultural Research Station, Albany CA in providing support and mentoring in this area of research. I also express my thanks to Aldos Barefoot from Dupont Agrochemical Division and Cheryl Cleveland from Dow AgroSciences for their support of pesticide research in the troposphere.

References

1. Aspelin A.L.; Grube AH.: U.S. EPA **1999**, Document #733-R-99-001
2. Lewis R.G.; Lee, R.E. Jr. **1976**, In: Air Pollution from Pesticides and Agricultural Processes. R. E. Lee Jr. (Ed), CRC Press, Cleveland, OH. pp 5-50.
3. Spencer, W.F.; Farmer, W.J.; Cliath, M.M. **1973**, Residue Rev. *49,* 1-47.
4. Taylor A.W.; Spencer, W.F. **1990**, In: Pesticides in the Soil Environment: Processes, Impacts, and Modeling. SSSA Book Series #2, Soil Science Society of America, Madison, WI., pp 213-269
5. Majewski, M.S.; Capel, P.D. **1995**, Pesticides in the Atmosphere: Distribution, Trends, and Governing Factors. (R.J. Gilliol ed) *Vol 1: Pesticides in the Hydrologic System.* Ann Arbor Press, MI. 214 pp
6. Hebert V.R.; Miller G.C.: Chemosphere **1998**, *36*, 2057-2066.
7. Mongar K.;Miller G.C.:Chemosphere **1988**, *17*, 2183-2188.
8. Woodrow J.E.; Crosby D.G.; Mast T, Moilanen K.W.; Seiber J.N.: J. Agri. Food Chem. **1978**, 26, 1312.
9. Edelstein D.M.; Spatz D.S.: Unresolved Issues In Pesticide Fate Data Guidance. **1994**, Paper presented at the Eighth IUPAC Cong. of Pest. Chem., Wash. D.C.
10. Miller G.C., Hebert V.R.: Fate of Pesticides in the Environment, J.N. Seiber (ed). **1987**, Univ. of CA, Div. of Agric. and Nat. Res. Pub: 3320.
11. Atkinson R., Guicherit R., Hites R., Palm W., Seiber J.N., de Voogt P.: **1999**, Water, Air and Soil Pollut.*115*, 219-243.
12. Crosby D.G., Moilanen K.W.: Chemosphere, **1977**, *6*, 167-172.
13. Hebert V.R., Hoonhout C., Miller G.C.: J. Agric. Food Chem. **2000a**, *48*, 1916-1921.
14. Hebert V.R., Hoonhout C., Miller G.C. (2000b). J. Agric. Food Chem. **2000b**, *48*, 1922-1928.
15. Kwok E., Atkinson R., Arey J.:Environ. Sci. Technol. **1992**, *26*, 1798.
16. Carter W.P., Luo D., Malkina I.L.: Atmos. Environ. **1997**, *31*, 1425-1439.

17. Geddes J.D., Miller G.C., Taylor G.E.: Environ. Sci. Technol. **1995**, *29*: 2590-2594.

18. Seiber J.N., Wilson B.W., McChesney M.M.: Environ. Sci. Technol. **1993**, *27*, 2236-2243.

19. Woodrow J.E.; Seiber J.N., Crosby D.G., Moilanen K.W., Soderquist C.J., Mourer C.: Arch. Environ. Contam. Toxicol. **1977**, *6*, 175-191.

20. Atkinson R.: **1986**, Environ. Toxicol. Chem. *7*, 435-442.

21. Meylan W., Howard D.: Estimation Accuracy of the Atmospheric Oxidation Program. **1996**, Syracuse Research Corporation, Syracuse, NY.

22. OECD: OECD Monographs **1992**, *61*, Organis. Econ. Coop. And Develop., Paris.

23. Guicherit R., Bakker D.J., De Voogt P., Van Der Berg F., Van Dijk H.F.G., Van Pul W.A.H.: Fate of Pesticides in the Atmosphere: Implications for Environmental Risk Assessment. (H. van Duk, W. Van Pul and P. De Voogt, eds) **1998**, Klumwer Academic Pub. London, England.

24. Bidleman, T.F. **1988**, Environ. Sci. Technol. *22*, 361-367.

25. Forman, W.T.; Bidelman, T.F. **1987,** Environ. Sci. Technol. *21*, 869-875.

26. Pankow, J.F. **1987**, Atmos. Environ. *21*, 2275-2283.

27. Pankow, J.F. 1994. Atmos. Environ. 28, 185-188.

28. Yamasaki, H.; Kuwata, K.; Miyamoto, H. **1982**, Environ. Sci. Technol. *16*, 189-194.

29. Saret N., Miribel M., Wortham, H. **2001,** http://ies.jrc.cec.eu.int./Units/cc/events/torino2001/torinocd.

30. Kurtz, D. A. (Ed) **1990**, Long Range Transport of Pesticides. Lewis Publishers, Chelsea, MI.

31. Wania F.; MacKay, D. **1993**, .Ambio *22*, 10-18.

32. Winer R., Atkinson R.: *In* Long Range Transport of Pesticides, Kurtz, D. A. (Ed). **1990**, Lewis Publishers, Chelsea, MI.

33. Woodrow J.E., Crosby D.G., Seiber J.N.: Residue Reviews, **1983**, *85*, 111-125.

34. Atkinson R., Aschmann S.M., Arey J., McElroy P.A.,Winer A.M.: Environ. Sci Technol. **1989**, *23*, 243-244.

35. Lemaire J., Campbell I., Hulpe H., Guth J.A., Merz W., Phelp J, von Waldron C.: Europ. Chem. Ind. Ecol. Toxicol. Center, **1982**, Tech. Rep. No. 3, Brussels, Belguim, 1-61.

36. Goodman M.A., Aschmann S.M., Atkinson R., Winer A.M.: Environ. Sci. Technol, **1988**,*22*, 578-583.
37. Anderson P.N., Hites R.A.: Environ. Sci. Technol., **1996**, *30*:1756-1763.
38. Brubaker W.W.,Hites R.A.: Environ. Sci. Technol. **1998**, *32*: 766-769.
39. Mill T., Mabey W.R., Bomberger D.C., Chou T.W., Hendry D.G., Smith J.H.: U.S. Environmental Protection Agency, Athens GA, **1981**, EPA Contract No. 68-03-2227.
40. Atkinson R.: J. Phys. Chem. Ref. Data. **1989**, Monograph 1, 1-246.
41. Atkinson R (1994) J. Phys. Chem. Ref. Data., **1994**, Monograph 2, 1-216.
42. Finlayson-Pitts B.J., Pitts Jr. J.N.: Atmospheric Chemistry: Fundamental and Experimental Techniques, **1986**, Wiley, New York

Chapter 8

Ecology of Pesticide-Degrading Bacteria: Degradation of Organophosphorus and Carbamate Insecticides

Masahito Hayatsu, Kanako Tago, Mitsuru Fukui, and Emi Sekiya

Faculty of Agriculture, Shizuoka University, Ohya, Shizuoka 422–8529, Japan

Microorganisms play an important role in the pesticide degradation in the environment. Microorganisms that are capable of degrading pesticides detoxify pesticides and influence the fate of pesticides in the soil environment. The population size, activity and diversity of the bacteria that degrade organophosphorus (fenitrothion) and carbamate (carbaryl) pesticides were investigated using soils treated with the pesticides. There are several types of fenitrothion and carbaryl degradation by microorganisms, including complete degradation by a single strain, synergistic utilization by two or more strains, and cometabolism.

Introduction

Microorganisms are thought to play a significant role in the breaking down and detoxification pesticides in the environment. Methylcarbamate and organophosphate insecticides are currently being used widely and extensively in agriculture and public health programs in Japan. Extensive use of insecticides results in pollution of soil and ground water systems. Microorganisms able to degrade organophosphate and methylcarbamate insecticides may provide a practical solution to the insecticides pollution. On the other hand, repeated application of several pesticides such as carbofuran (*1*) to soil has been shown to induce rapid microbial degradation of the pesticides and thus to reduce the efficacy of the pesticides against the target pest. This phenomenon, known as enhanced degradation or accelerated degradation of pesticide (*2*), is caused by microorganisms that have adapted to environmental changes such as the repeated application of pesticide to soil. Some genetic events such as horizontal gene transfer and mutation are thought to be involved in the microbial adaptation to pesticide degradation. Many microorganisms capable of degrading organophosphate and methylcarbamate insecticides have been isolated, but knowledge about the molecular genetics and ecology of these bacteria is limited in Japan.

In the present study, we investigated the molecular genetics and ecological aspects of bacteria that degrade organophosphorus and methylcarbamate insecticides.

Isolation of Pesticide-Degrading Bacteria

Experiments were performed using microcosms (500 ml plastic beaker) containing 300 g of the A horizon soil. The soil was taken from corn fields in Kumamoto, Shizuoka and Tochigi Prefectures in Japan. The soil type was ando soil, and the soils had previous history of treatment with the pesticide used in the present study. Pesticide-degrading bacteria were isolated from soil samples that had been treated with fenitrothion (O,O-dimethyl-O-4-nitro-m-tolyl phosphorothioate) or carbaryl (1-naphthyl-N-methylcarbamate) at intervals of two weeks. The number of fenitrothion degrading bacteria was counted using a selective agar medium for detection of fenitrothion-degrading ability at intervals of two weeks. Ten fenitrothion-degrading colonies on the agar plate were selected at random and purified for identification. The number of carbaryl-degrading bacteria could not be counted because we had no suitable selective medium for detection of carbaryl-degrading ability. Therefore, carbaryl-degrading bacteria were isolated by enrichment culture using carbaryl-treated soil as described previously (*3,4*). Ten colonies on the agar medium were randomly selected. The 16S rDNA sequences of these isolated bacteria were determined to analyze diversity of the bacteria. The sequences of the 16SrDNA of the isolated strains were compared with other published 16SrDNA sequences using the BLAST search option of DNA Data Base of Japan (DDBJ), and phylogenetic analysis was performed using CLUSTAL W.

Diversity of Fenitrothion-Degrading Bacteria

The fenitrothion-degrading bacteria were isolated from the three soil samples regardless of the fenitrothion treatment. The results of phylogenetic analysis showed that these strains belonged to the genus *Sinorhizobium*. Comparison of the the partial 16SrDNA sequences of the isolated strains showed that there were four types of strain with low sequenes divergence of 16S rDNA (less than 2%). The highest degree of similarity was found in the 16SrDNA genes of *Sinorhizobium meliloti* and *Sinorhizobium fredii*. The isolated strains cometabolically hydrolyzed fenitrothion to 3-methyl-4-nitrophenol and did not utilize the pesticide and its metabolites. *Rhizobium* bacteria, including *Sinorhizobium* genera, are known to be nitrogen-fixing bacteria, but results of a hybridization experiment showed that the isolates had no *nifH* and *nodA* genes, which are essential for nirogen fixiation and nodule formation. The addition of fenitrothion to soil did not affect the population sizes (less than 100 cfu / g of soil) of these strains in the soil. On the other hand, the ability of the strains to hydrolyze fenitrothion was unstable and was lost irreversibly. The ability of bacteria to degrade some pesticides is known to offten be controlled by plasmids. Thus, the strains were examined for their plasmid contents by the alkaline-SDS plasmid extraction method and agarose gel electrophoresis. The strains were found to harbor some high-molecular-weight plasmids.

The number of fenitrothion-degrading bacteria increased from below the detection limit to 100,000 or more per gram of soil after several additions of fenitrothion. Comparison of the partial 16SrDNA sequences of the isolated colonies revealed that there were two types of strains, strain SFEE100 in soil obtained from Shizuoka and strain KFEE100 in soil obtained from Kumamoto. The strains were identified as strains of the genus *Burkholderia* based on the 16SrDNA sequences (about 1400 bp). Strain SFEE100 and strain KEFF100 showed the highest degree of similarity to *Burkholderia caribiensis* (99%) and *Burkholderia glathei* (97%), respectively. Both strains were found be able to use fenitrothion as a carbon and energy source.

Burkholderia sp. NF100, which was previously reported to be capable of utilizing fenitrothion (5), was found to possess a fenitrothion-hydrolyzing enzyme gene (*fed*) and two oxygenases genes (*mhqA* and *mhqB*) that encoded oxygenases degrading the intermediate metabolites of fenitrothion. NF100 firstly hydrolyzed an organophosphate bond of fenitrothion to 3-methyl-4-nitrophenol, which was further metabolized via methylhydroquinone. The ability to degrade fenitrothion was found to be encodes on two plasmids, pNF1 and pNF2. Plasmid NF2 encodes fenitrothion-hydrolyzing ability, and pNF1 encodes the methylhydroquinone metabolizing pathway (Figure 1).

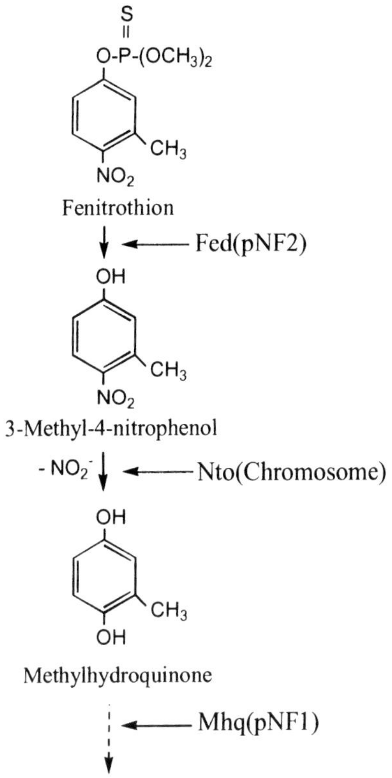

```
              S
              ‖
         O-P-(OCH₃)₂
            ⌬ CH₃
            NO₂
         Fenitrothion
              ↓ ◄──── Fed(pNF2)
            OH
            ⌬ CH₃
            NO₂
    3-Methyl-4-nitrophenol
       -NO₂⁻ ↓ ◄──── Nto(Chromosome)
            OH
            ⌬ CH₃
            OH
       Methylhydroquinone
              ↓ ◄──── Mhq(pNF1)
```

Figure 1. Metabolic pathway of fenitrothion by Burkholderia sp. NF100

Fed, fenitrothion hydrolase encoded on pNF2
Nto, nitro group- removing oxidative pathway
MhqA, monooxyganase
MhqB, dioxygenase

The ability to remove the nitrite from nitrophenol was encoded on the chromosome. A combination of enzymes encoded on the chromosome and two plasmids leads to total degradation of fenitrothion.The *fed* has three open reading frames (ORF 1, 2, 3). ORF 1 seemed to contain a catalytic site of the enzyme reaction, and the other ORFs seemed to participate in localization of the enzyme to the cell membrane. MhqA is a member of the family of flavoprotein type monooxygenases which have FAD-binding and ADP-binding sequences.

MhqB is an extradiol-type dioxygenase that catalyzes the aromatic ring fission of methylhydroquinone. The highest degree of identity found was with catechol-2,3-dioxygenase (57%) from *Alcaligenes eutrophus* (6). The results of a hybridization experiment showed that strains SFEE100 and KFEE100 had genes similar to *mhqA* and *mhqB* and did not have *fed*. The two strains had plasmids of different sizes from those of NF100. These results indicated that fenitrothion-degrading enzyme system is diverse, even in the *Burkholderia* genus.

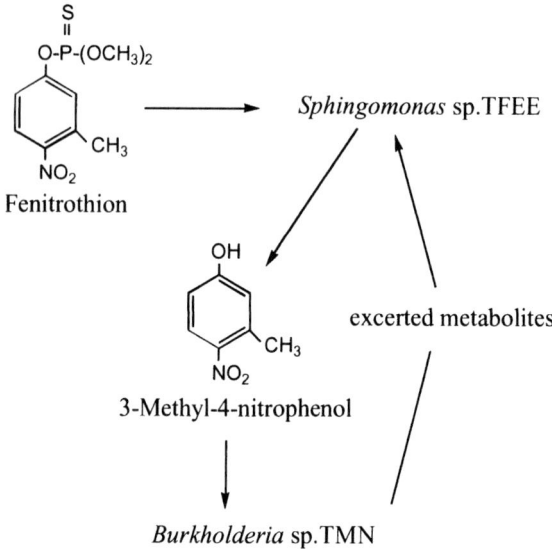

Figure 2. *Proposed synergistic fenitrothion utilizing relationship between Sphingomonas sp. TFEE and Burkholderia sp. TMN*

Synergistic fenitrothion-utilizing bacteria, strains TFEE and TMN, were isolated from soil obtained from Tochigi. TFEE hydrolyzed fenitrothion to 3-methyl-4-nitrophenol, which was further degraded and utilized by TMN as a sole source of carbon and energy (Figure 2). Strains TFEE and TMN were identified as strains of the genera *Sphingomonas* and *Burkholderia*, respectively, based on the 16SrDNA sequences (about 1400 bp). *Sphingomonas* sp. TFEE and *Burkholderia* sp. TMN showed the highest degree of similarity to *Sphingomonas agrestis* (99%) and *Burkholderia glathei* (99%), respecvively. These two bacterial strains seemed to synergistically degrade and utilize fenitrothion and its metabolites in the soil because both bacterial population sizes were increased by the addition of fenitrothion.

Diversity of Carbaryl-Degrading Bacteria

A carbaryl-degrading mixed culture was enriched from the soil sample. Synergistic carbaryl-utilizing bacteria, strains NC100 and NC200, were isolated from the culture. Strains NC100 and NC200 were identified as strains of the genera *Mesorhizobium* and *Pseudomonas,* respectively, based on the 16SrDNA sequences (about 1400 bp). *Mesorhizobium* sp. NC100 and *Pseudomonas* sp. NC200 showed the highest degree of similarity to *Mesorhizobium huakuii* (96%) and *Pseudomonas putida* (99%), respecvively. NC100 hydrolyzed carbaryl to 1-naphthol and methylamine. NC100 utilized methylamine and NC200 utilized 1-naphthol as a sole source of carbon and energy. NC100 and NC200 synergistically degraded and utilized carbaryl in the culture medium supplemented with carbaryl.

Sphingomonas sp. SB100 was isolated from soil to which fenobucarb *(o-sec-*butylphenyl methylcarbamate) had been added instead of carbaryl. Carbaryl hydrolases (CehSB and CehNC) were purified to homogeneity from strains SB100 and NC100. The properties of CehSB and CehNC were compared with those of the carbaryl hydrolase (CehAC) purified from *Rhizobium* sp. AC100 previously reported (*3*). Each of these native enzyme had a molecular mass of 167 kDa and was composed of two identical subunits with molecular masses of 84 kDa. The properties of CehSB, except substrate specificity, were similar to those of CehNC and CehAC. Purified CehNC and CehAC showed higher levels of activity toward carbaryl than fenobucarb. However, CehSB showed a higher level of activity toward fenobucarb(Table 1), though the amino acid sequence deduced from the DNA sequence of *cehSB* was different from the sequence of *cehAC* in only two places.

The involvement of naturally occurring plasmids in the degradation of synthetic organic compounds has been extensively documented (*7*). Catabolic plasmids are thought to play an important role in the evolution of pesticide-degrading ability in microorganisms. Therefore, the plasmid contents of these strains were examined. These strains had several large plasmids, and the cells that had lost carbaryl hydrolyzing activity also had lost the plasmid or had a smaller plasmid. These results suggest that the carbaryl hydrolase gene is encoded on the plasmid.

Table I. Substrate Specificity of Purified Carbaryl Hydrolase*

Pesticide		CehAC % (U/mg of protein)	CehSB
carbaryl	OCONHCH₃ (naphthyl)	100 (26)	100 (24)
fenobucarb	OCONHCH₃, CH(CH₃)CH₂CH₃ (phenyl)	11 (2.8)	141 (34)
propanil	NHCOC₂H₅ (3,4-dichlorophenyl)	0 (0)	0 (0)

*The rates of hydrolysis are expressed as % of the rate observed with carbaryl

Rhizobium sp.AC100 was found to have a carbaryl hydrolase gene (*cehAC*) encoded on a transposon-like structure (Tn*cehAC*) on the plasmid AC1 (*8*). The DNA fragment cloned from plasmid AC1 encoded six open reading frames containing *cehAC*. The deduced amino acid sequence encoded by two ORFs located upstream of *cehAC* gene showed significant homology to IstA and IstB of IS*1600* from *Alcaligenes eutrophus* NH9 and of IS*1326* from *Pseudomonas aeruginosa*, which belong to the IS*21* family (*9, 10*). The nucleotide sequence of the region containing IstA- and IstB-like sequences showed 67.7% homology with IS*1600* and 60.5% homology with IS*1326*. The gene encoding *cehNC* and *cehSB* was found to have a DNA sequence and a transposon-like structure similar to those of Tn*cehAC* based on hybridyzation and nucleotide sequences analysis. Thus, the Tn*ceh AC* type gene seems to be distributed widely in the soil ecosystem.

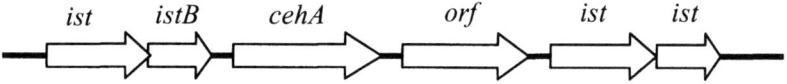

Figure 3 Transposon- like DNA structure coding carbaryl hydrolase gene

Arthrobacter sp. RC100 hydrolyzed carbaryl to 1-naphthol, which was used as a sole carbon and energy source (*4*). Strain RC100 had 110-, 120- and 130-kbp plasmids designated pRC1, pRC2 and pRC3, respectively. Plasmid RC1 encoded carbaryl-hydrolyzing ability, and pRC2 encoded the metabolizing pathway of 1-naphthol to gentisic acid (Figure 3). The gentisic acid metabolic

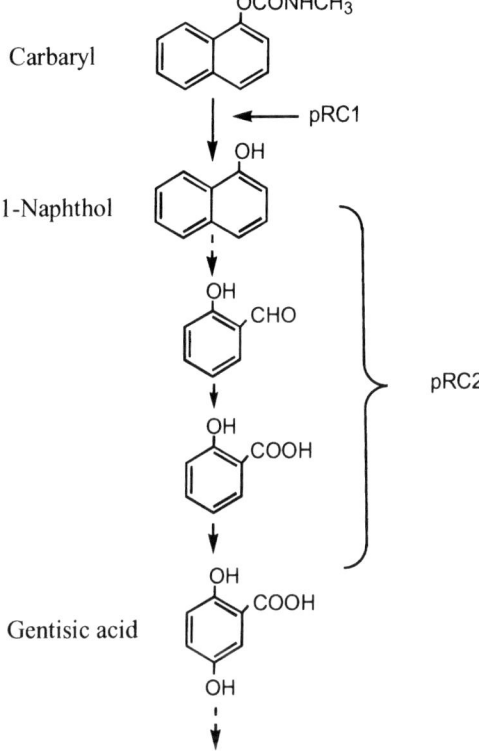

Figure 3. Pathway of carbaryl degaradation by Arthrobacter sp.RC100.

pathway was encoded on the chromosome.The carbaryl hydrolase gene (*cahA*) was cloned from pRC1. The deduced amino acid sequence of *cahA* showed significant similarity to sequences of amidase genes. Pyrazinamidase (37.0%) from *Mycobacterium smegmatis* (*11*) and enantiomer-selective amidase (35.9%) from *Rhodococcus* sp. (*12*) showed higher homology with the entire region of CahA. These enzymes, belonging to an amidase family, conserve the amidase signature sequence (AS sequence). The AS sequence was also found in the amino acid sequence of CahA. CahA purified from strain RC100 had a molecular mass of 100 kDa and was composed of two identical subunits with molecular masses of 51 kDa (*13*). The enzyme hydrolyzed four *N*-methylcarbamate insecticides (carbaryl, xylylcarb, metolcarb and XMC) and also hydrolyzed 1-naphthylacetamide and exhibited no activity toward 1-naphthylacetate. These results indicate that CahA belongs to the amidase family.

Conclusions

Analysis of the diversity of pesticides-degrading bacteria indicated that there are several types of pesticide degradation by microorganisms, including complete degradation by a single strain, synergistic utilization by two or more strains, and cometabolism. Mobile genetic elements such as plasmids and transposons were shown to encode enzymes responsible for the fenitrothion and carbaryl degradation. The DNA probe specific for the genes cloned in this study can be used to monitor the spread and persistence of the insecticide-degrading bacteria in various soil environments. The isolated bacteria, plasmids and purified hydrolases in this study may help us to understand how the metabolic pathway and enzymatic reaction evolved within the soil microbial community.

References

1. Felsot, A.S.; Mddox, J.V.; Bruce, W.: *Bull. Environ. Contam. Toxicol.* **1981**, *26*, 781 - 788.
2. Racke,D.D.; Coats, J.R.: *ACS Symposium Series.* **1990.**
3. Hayatsu, M.; Nagata, T. : *Appl. Environ. Microbiol.* **1993**, *59*, 2121 - 2125.
4. Hayatsu, M.; Hirano, M.; Nagata, T. : *Appl.Environ. Microbiol.* **1999**, *65*, 1015 - 1019.
5. Hayatsu, M.; Hirano, M.; Nagata, T. : *Appl. Environ. Microbiol.* **2000**, *66*, 1737-1740.
6. Kabisch,M.;Fotnagel,P. : *Nucl.Acid.Res.* **1990**, *18*, 5543
7. Sayler, G. S., S.; Hooper, W.; Layton, A. C.; Henry King. J. M.:*Microb.Ecol.* **1990**. *19*, 1-20.

8. Hashimoto, M.; Fukui, M.; Hayano, K.; Hayatsu, M.: *Appl. Environ. Microbiol.* **2002**, *66*, 1220 - 1227.
9. Brown, , H.; Stokes, W.; Hall, R. M.: *J. Bacteriol.***1996**, *178*, 4429-37.
10. Ogawa, N.; Miyashita, K.: *Appl. Environ. Microbiol.* **1999**, 65,724-31.
11. Boshoff, H. I. M.; Mizrahi, V.: *J. Bacteriol.* **1998**, *180*, 5809-5814.
12. Mayaux, J. F.; Cerbelaud, E. ; Soubrier, F. ; Yeh, P.; Blanche, F. ; Petre, D.: *J. Bacteriol.* **1991**, *173*, 6694-6704.
13. Hayatsu, M.; Mizutani, A.; Sato, K.; Hayano, K.: *FEMS Microbiol. Lett.* **2001**, *201*, 99-103.

Chapter 9

Polymerization of the Herbicide Bentazon and Its Metabolite with Humic Monomers by Oxidoreductive Catalysts

Jong-Soo Kim and Jang-Eok Kim*

Division of Applied Biology and Chemistry, Kyungpook National University, Daegu 702–701, South Korea
*Corresponding author: telephone: 82–53–950–5720, fax: 82–53–953–7233, e-mail: jekim@knu.ac.kr

Summary and Backgrounds

The formation of covalent bonds between pesticides and their metabolites, which structurally resemble phenolic and aniline compounds and soil organic matters, is mediated by biotic or abiotic catalysts and results in stabilization against bioavailability. The herbicide bentazon in the soil environment has shown that it can be degraded by hydroxylation at the 6 or 8 positions on the phenyl ring and its metabolites can be incorporated into soil organic matters.

To elucidate the binding mechanism of the bentazon and its metabolite 8-hydroxybentazon to humic substances, bentazon and 8-hydroxybentazon were incubated with catechol in the presence of laccase an oxidoreductive catalyst. A product with a molecular weight of 348 identified as a dimer was composed of one catechol and one bentazon molecule. A second reaction product with a molecular weight of 586 appeared to be a trimer, consisting of one catechol molecule and two bentazon molecules. Data of ^{1}H-NMR and ^{13}C-NMR spectroscopy indicate that the herbicide bentazon can be incorporated into soil organic matter by nucleophilic addition to quinone-like substrate during humification process. In case of the 8-hydroxybentazon, the three intermediates were confirmed in polymerization reaction mixture of 8-hydroxybentazon by HPLC and LC-MS

One of them was found as dimeric species (MW: 510) of 8-hydroxybentazon, which was stabilized at neutral pH. In acidic condition, the main products were considered as two trimeric isomers (MW: 764). Oxidative copolymerization of 8-hydroxybentazon and catechol resulted in the formation of tetrameric oligomer during initial stage of the reaction. The structure of the tetrameric oligomer (MW: 728) was thought to be cross-coupled with dimeric species of 8-hydroxybentazon and catechol-derived dimmer.

Experimental Section
Reaction of bentazon and catechol
Transformation reactions were conducted at 25 ℃ for 4 h in 100 mL of 0.1 M citrate-phosphate buffer containing 4 unit/mL laccase with 1 mM bentazon (including 65,000 dpm/mL of [^{14}C]bentazon) and 1 mM catechol. The enzymatic reaction was stopped by adjusting the reaction mixture to pH 2.0 with acetic acid; subsequently, the reaction mixtures were centrifuged at 12,000g for 10 min to remove the precipitates. The supernatants were filtered through 0.45 µm nylon membrane, and the residual bentazon was quantified by HPLC. Boiled enzyme used in the control sample. After incubation for 4 h, the reaction mixture was extracted three times with an equal volume of methyl chloride. The extract was dehydrated by anhydrous sodium sulfate and evaporated to 1-2 mL by rotary evaporation at 35 ℃ and then dried under a stream of N_2. The disappearance of bentazon and the formation of reaction products were monitored by a HPLC system equipped with a Model 6000A and a Model 501 solvent delivery system using a 4.0 mm × 25 cm reverse phase column (Nucleosil C_{18} 5 µm, Macherey Nagel) with UV detection at 254 nm. The mobile phase set at a flow rate of 0.8 mL/min was a mixture of an aqueous component [A] (1% acetic acid in Milli-Q water) and [B] methanol. The reaction products were isolated by a gradient elution in which the initial mobile phase composition of 70/30 (A/B) was changed to 35/65 (A/B) over 18 min. This ratio was then maintained for 5 min. Next, the solvent was taken from 35/65 (A/B) to 0/100 (A/B) in 7 min.

Molecular weights of the reaction products obtained following the oxidation of the bentazon and catechol were determined by electron ionization (70 eV) mass analysis. A Kratos MS 9/50 double-focusing mass spectrometer was used for the analysis, and sample was introduced by direct insertion probe with the source temperature between 250 and 350 ℃. ^{1}H- and ^{13}C-NMR spectroscopy were used to elucidate the structure of the reaction products. 1-D and 2-D experiments were carried out using high-resolution NMR spectrometers operating in the quadrature mode. All of the 1-D experiments were performed on a Bruker AM-500 spectrometer except for the ^{13}C analysis of the bentazon standard, for which a Bruker WM-360 spectrometer was used. The ^{13}C-NMR measurements were made with continuous WALTZ decoupling of the protons.

Reaction of 8-hydroxybentazon and catechol

Enzymatic reaction was conducted at 25 ℃ for 3, 6, 15, 45 min in 80 mL of citrate-phosphate buffer containing 0.5 unit/mL of laccase with 0.1 mM 8-hydroxybentazon. Enzymatic reaction was stopped by adjusting the reaction mixture to pH 0.4 with hydrochloric acid. The reaction mixture was extracted two times with 40 mL of ethyl acetate. The extract was dehydrated by anhydrous sodium sulfate and evaporated to 2 mL by rotary evaporator and then dried under a steam of N_2. The disappearance of substrate and formation of reaction products were quantified with a Shimadzu LC-10A high performance liquid chromatography system using a 4.0 mm × 15 cm reverse phase column (Inertsil phenyl-3, 5μm, MethChem) with UV detection at 254 nm. The reaction products were monitored by gradient elution in which the at flow rate of 1 mL/min was mixture of 1% acetic acid in deionized water (A) and methanol (B). The initial mobile phase composition of 25/75 (A/B) was then brought to 50/50 for 12 min and held for 5 min. For 25 min, the mobile phase composition was brought from 50/50 to 65/35 and this final composition was held 10 min. Molecular weight of reaction products following the oxidation of the polymerization and copolymerization were confirmed by LC-MS. Atmospheric pressure ionization (API) mass spectrometry was performed in the negative mode using an Agilent 1100. The collision energy was in the range of 50–200 V.

Results and Discussion

The time course for the reaction of bentazon with catechol by laccase at different pH values is shown in Fig. 1. Bentazon was completely transformed in 30 min of incubation at pH 4.0, but at pH 6.0 the transformation of bentazon was 17% after 4 hours, and no additional transformation occurred.

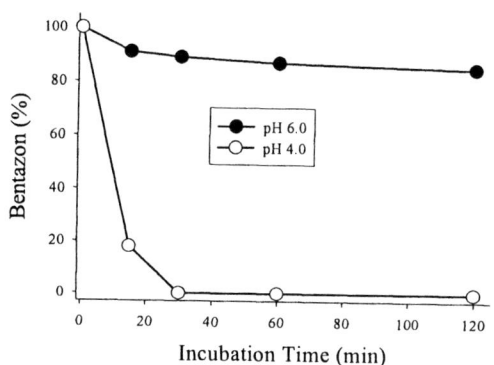

Fig. 1. Time course for transformation of bentazon at pH 4.0 and pH 6.0 by laccase in the presence of catechol.
(Reproduced from ref. 1.)

To isolate products, a reaction mixture containing ^{14}C-bentazon was extracted by methylene chloride. Through the use of a radioscanner for detection of components separated by TLC, a strongly radioactive band of products from the reaction of bentazon, catechol, and the laccase was detected at Rf 0.83. This band was subsequently analyzed by HPLC and was found to contain three components (Fig. 2): unreacted bentazon (retention time, 18 min), product A (retention time, 22 min) and product B (retention time, 30 min).

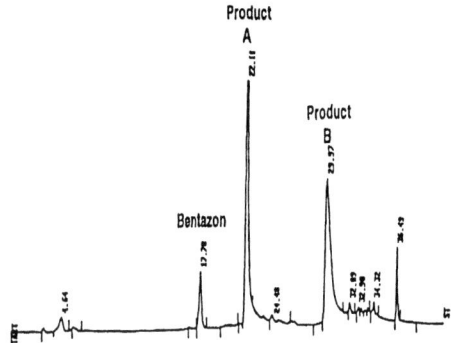

Fig. 2. HPLC analysis of the reaction products of bentazon and catechol after incubation by laccase.
(Reproduced from ref. 1.)

Product A with a molecular weight of 348 was found to be dimer and was produced by the binding of the protonated nitrogen of one bentazon molecule to a carbon of an *o*-quinone molecule generated from oxidized catechol. Similarly, product B was identified as a trimer of a single catechol molecule bound to two bentazon molecules para to the hydroxyl groups of catechol. ^{1}H-NMR spectra of this product also showed the aromatic ring of bentazon was not involved in the coupling reaction while ^{13}C-NMR showed the disappearance of the two protonated catechol carbon signals as a result of substitution at these two sites. The reaction of bentazon with catechol in the presence of a laccase is not a radical reaction, but proceeds via nucleophilic addition to a quinone structure (Fig. 5).

In the presence of catechol, laccase-mediated transformation rate of 8-hydroxybentazon was decreased in comparison to the transformation observed in 8-hydroxybentazon as unique substrate (Fig. 3).

a)

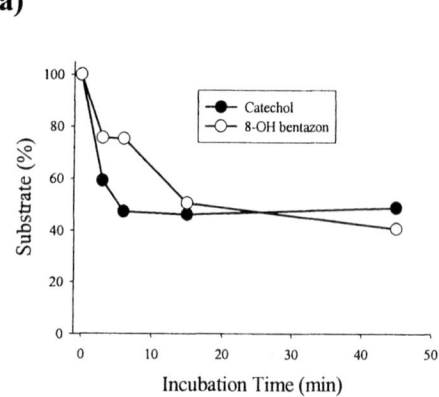

b)

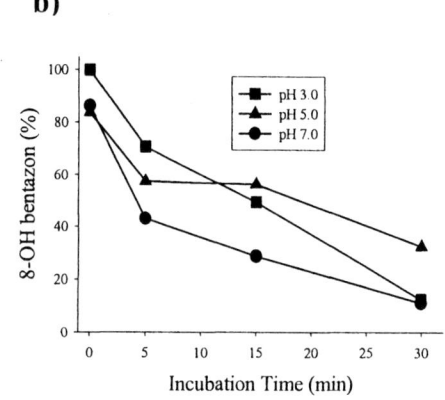

Fig. 3. Time course for the transformation of (a) 8-hydroxybentazon with catechol and (b) 8-hydroxybentazon alone by laccase.

Enzymatic transformation of 8-hydroxybentazon in aqueous solution at pH 7.0 and product formation was investigated by HPLC. Copolymerization of 8-hydroxybentazon and catechol resulted in the formation of tetrameric oligomer (Product C) during initial stage of the reaction (Fig. 4a). The structure of the tetrameric oligomer (MW: 728) was thought to be cross coupled with dimeric species of 8-hydroxybentazon and catechol-derived dimer. Fig. 4b shows that at least three coupling products were formed in the early stage of polymerization of 8-hydroxybentazon. Further polymerization is possible. 8-Hydroxybentazon-derived products D and E were monitored after 3 min of incubation with laccase and their maximum yield was investigated at 6 and 15 min of incubation, respectively. Additional main product (F) was clearly observed after 15 min of incubation time, but all the products (D, E and F) were decreased after 45 min of incubation time. Polymerization products of 8-hydroxybentazon by lacccase at pH 7.0 was assessed by LC-mass spectrometric analysis. The molecular weights of the products D, E and F were 510 (dimer), 764 (trimer) and 764 (trimer), respectively. Identical molecular weight of the products E and F suggests that product F is an isomeric form of the product E.

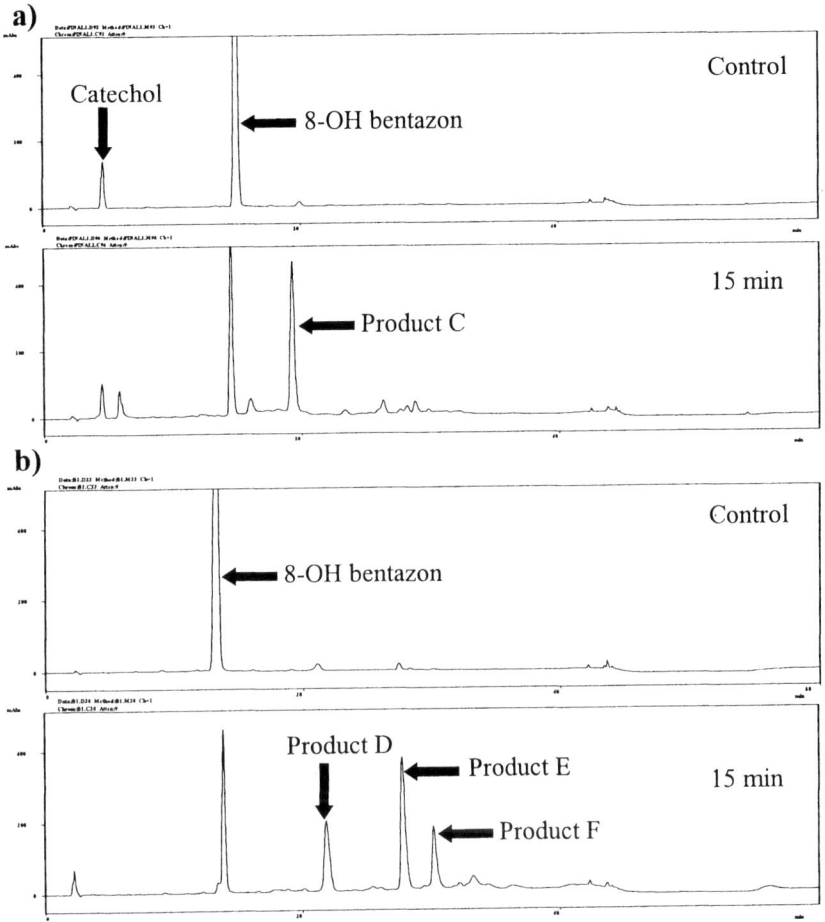

Fig. 4. HPLC chromatograms of reaction products obtained with a 15 min incubation of (a) 8-hydroxybentazon with catechol and (b) 8-hydroxybentazon alone by laccase at pH 7.0.

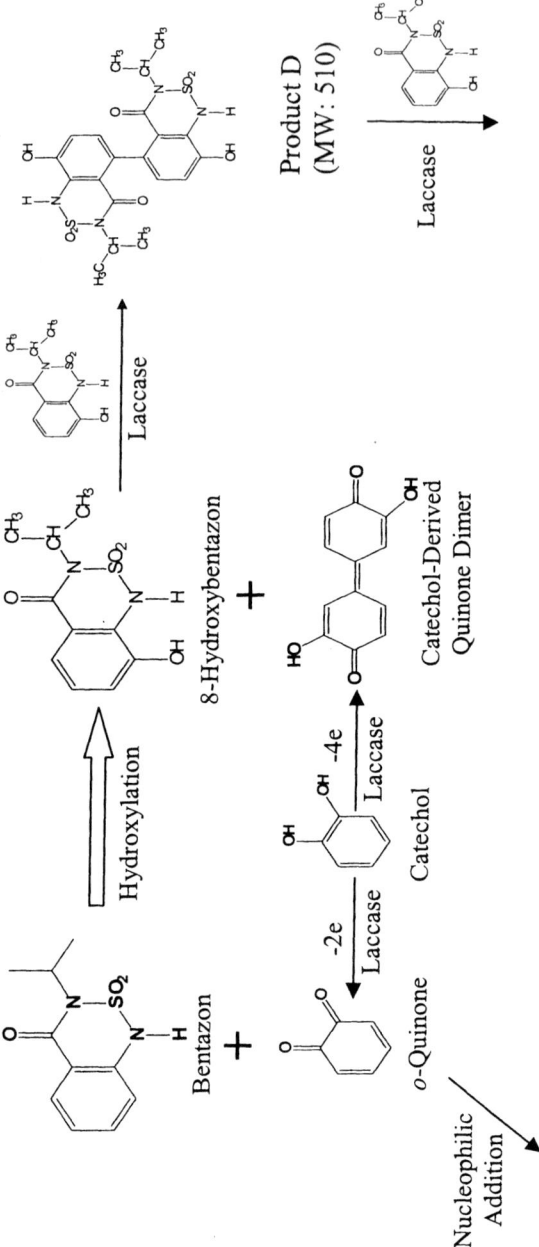

99

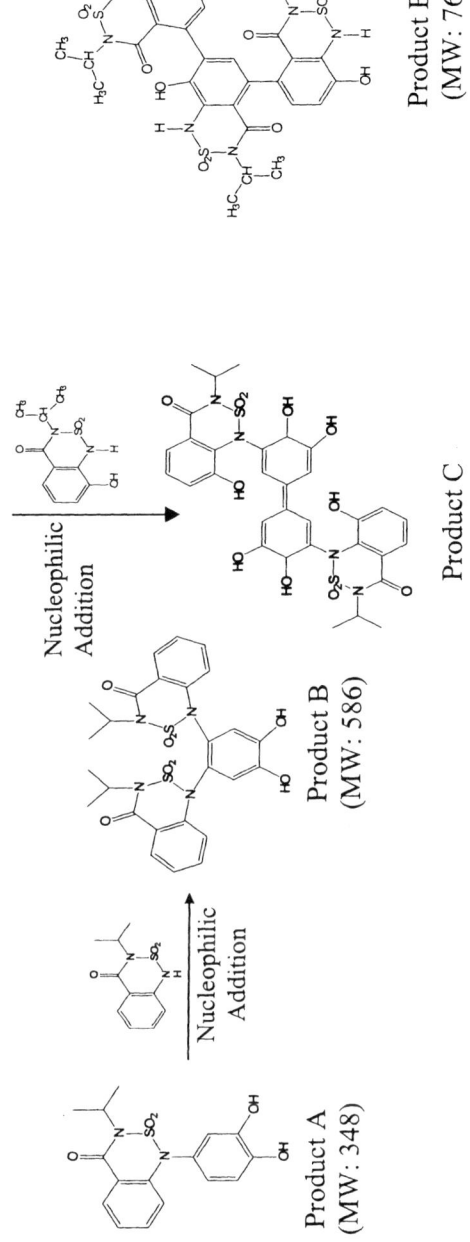

Fig. 5. Possible transformation products of bentazon and 8-hydroxybentazon with catechol by laccase

On the basis of the observed products and literatures on the oxidation of phenolic compounds by laccase, the reaction scheme for 8-hydroxybentzon oxidation by laccase is proposed as shown in Fig.5. Initially the phenol moiety of 8-hydroxybentazon is oxidized by laccase and formed phenoxy radicals. Further reactions of these phenoxy radicals are cross coupled with each other and formed dimeric products. Ketone moiety of 8-hydroxybentazon-derived dimers are converted to enol in acidic condition. The coupling reactions most likely occur via C-C coupling at para positions because of their electronic (resonance and inductive) and satiric effects. Proposed reaction pathway leading to the formation of co-polymerization product appears to be initiated by the coupling of two catechol radicals, and subsequent oxidation of that dimmer, resulting in catechol-derived benzoquinone dimer. It appears that a tetrameric copolymerization product is formed by the non-enzymatic addition of two 8-hydroxybentazon to a catechol-derived benzoquinone dimer (Fig. 5).

Reference

1. Kim, J.-E.; Fernandes, E.; Bollag, J.-M. Environ. Sci. Technol. 1997, 31, 2392-2398.

2. Kim, J.-S.; Park, J.-W.; Lee, S.-E.; Kim, J.-E. J. Agric. Food Chem. 2002, 50, 3507-3511.

3. Huber, R.; Otto, S. Rev. Environ. Contam. Toxicol. 1994, 137, 111-134

4. Simmons, K. E.; Minard, R. D.; Bollag, J.-M. Environ. Sci. Technol. 1989, 23, 115-121.

Chapter 10

Predicting Pesticide Volatilization from Bare Soils

Scott R. Yates, Sharon K. Papiernik, and William F. Spencer

U.S. Department of Agriculture, Agricultural Research Service,
George E. Brown Jr. Salinity Laboratory, 450 West Big Springs
Road, Riverside, CA 92507-4617

Understanding the processes and mechanisms that affect pesticide volatilization from fields is important in developing methods to control emissions. Changes in ambient air temperature and atmospheric stability can strongly affect volatilization. A field study was conducted to measure the volatilization rate. A numerical model was developed to simulate pesticide fate, transport and volatilization. Three volatilization boundary conditions were used to assess the accuracy in predicting the volatilization rates. First, a stagnant boundary layer and isothermal conditions are assumed. Second, a temperature-dependent stagnant boundary layer is considered. A third boundary condition that couples soil and atmospheric processes was found to provide an accurate and realistic simulation of the instantaneous volatilization rates compared to the other boundary conditions. For certain information, such as cumulative emissions, all the boundary conditions yielded similar results and suggest that simpler methods may be useful for this information.

The occurrence of pesticides in the atmosphere or in water supplies is an important national issue (*1-3*). Numerous monitoring studies have demonstrated that agricultural use of pesticides can contribute to both atmospheric and water contamination. Pesticide movement in the soil zone is affected by many interrelated factors such as the pesticide application methods, soil and environmental conditions, and water management practices. It is expected that with an understanding of pesticide fate and transport, water management and pesticide application practices could be developed that reduce the movement of pesticides outside of the root zone.

Volatilization is an important route of dissipation for pesticides with large vapor pressures or, similarly, large Henry's Law constants (*4-6*). Volatilization has also been shown to be important for pesticides with low to moderate volatility (*7-9*). Volatilization reduces the pesticide available to control pests and reduces the potential of ground water contamination, but increases contamination of the atmosphere. This poses an increased risk to persons living near treated fields, since many pesticides are considered to be carcinogenic (*10*). To protect public health, there is need for more information on the important processes and mechanisms that affect pesticide fate and transport under typical field conditions.

The purpose of this paper was to study the influence of the surface boundary condition on the pattern of pesticide emissions into the atmosphere and to assess the ability to simulate accurate and realistic emission rates. A numerical model was developed to simulate triallate movement in soil, and volatilization into the atmosphere. Three volatilization boundary conditions, with increasing complexity, have been explored: (1) volatilization under isothermal conditions, (2) volatilization in response to solar-driven temperature changes at the soil surface, and (3) volatilization coupled to atmospheric processes.

Methods

The field site was located at the University of California's Moreno Valley Field Station. The soil is a Greenfield sandy loam and is classified as a coarse-loamy-mixed-thermic-typic-Hyploxeralf. The field site was treated with a herbicide, triallate (Fargo 4E; S-(2,3,3-trichloroallyl) diisopropyl thiocarbamate; CAS: 2303-17-5), by spraying the bare-soil surface with 11.5 kg/ha using 0.1514 m^3 (i.e., 40 gallons) of water. Table 1 gives several properties of triallate. Atmospheric and soil measurements of triallate concentration were collected for a six day period. Atmospheric triallate concentrations were obtained at 10, 30, 51, 80, 120 and 160 cm above the soil surface and represent averages over a two or four hour sampling interval. The four hour sampling intervals occurred during the night-time hours of 2000-2400 and 0000-4000. Meteorological measurements of incoming radiation, net radiation, air temperature, wind speed, wind direction, relative humidity, were obtained for each 10 min interval during this period.

Table 1. Properties of Triallate (*11*)

Molecular Weight	305	g mol^{-1}
Boiling Point	117	°C
Half life	82	d
Henry's Law Constant, K_H	0.00045	
Organic Carbon Coefficient, K_{oc}	2400	mL g^{-1}
Saturated Vapor Pressure	1.1x10^{-4}	mm Hg
Solubility	4	mg L^{-1}

Triallate concentration in air was obtained using a series of two polyurethane foam (PUF) plugs held in a glass tube (*12*). A vacuum system was used to draw triallate-laden air into the foam at a prescribed flow rate, 15 L min^{-1}. After each sampling interval, the PUF was stored in a freezer until transport to the laboratory for analysis of the triallate concentration. Soil samples were taken using a coring device 2.5 cm in diameter. Twice each day, a total of 31 samples were collected randomly within the field to a depth of 1 cm.

The triallate in the PUF was removed using soxhlet extraction for 2 h at 60 °C with a azeotropic mixture of hexane and acetone (250 mL total solution). The extracts were concentrated to about 10 mL in a rotary evaporator and analyzed on a gas chromatograph using an electron capture detector. A DB-608 (30m - 0.53mm ID) column with helium as carrier gas was used. The injector and detector temperatures, respectively, were 230 °C and 280 °C. Extracting triallate from the soil samples followed a similar procedure except the extraction time was 8 h. A complete description of the sampling methodology and laboratory analysis is given by Cliath et al. (*13*) and Spencer et al. (*14*).

Model Description

A numerical model was developed to simulate MeBr movement in soil and volatilization into the atmosphere. The model simultaneously solves partial differential equations for nonlinear transport of water, heat, and solute in a variably saturated porous medium. Henry's Law is used to express partitioning between the liquid and gas phases and both liquid and vapor diffusion are included in the simulation. Soil degradation is simulated using a first-order decay reaction and the rate coefficients may differ in each of the three phases (i.e., liquid, solid, or gaseous).

The simulation of saturated-unsaturated water movement in soils is accomplished using the Richard's equation. The simulation of heat movement in soil is obtained using the approach of Sophocleous (15) and included diurnal variations in the surface temperature. The governing equation for the fate and transport of triallate is

$$\frac{\partial \theta C}{\partial t} + \frac{\partial \rho_b S}{\partial t} + \frac{\partial \eta G}{\partial t} = \frac{\partial}{\partial z}\left[D_\ell \frac{\partial C}{\partial z} + D_G \frac{\partial G}{\partial z} - qC\right] \\ - \mu_\ell \theta C - \mu_s \rho_b S - \mu_G \eta G \qquad (1)$$

where C, S, G are solute concentrations for the liquid, solid, and gaseous phases, respectively (g cm^{-3} or g g^{-1}); q is the volumetric flux density (cm s^{-1}); D_ℓ and D_G are the liquid and vapor phase diffusion coefficients (cm^2 s^{-1}), respectively, μ is a first-order degradation coefficient (s^{-1}) and θ, ρ_b, η, respectively, are the water content (cm^3 cm^{-3}), bulk density (g cm^{-3}) and the air content (cm^3 cm^{-3}). The subscripts: ℓ, s and G indicate liquid, solid, and gaseous phases, respectively. The soil-atmosphere boundary condition is described using

$$-D_E \frac{\partial C}{\partial z}\bigg|_{z=0} + qC\bigg|_{z=0} = -h\left(G - G_{air}\right)\bigg|_{z=0} \qquad (2)$$

where G_{air} is the concentration in the atmosphere, $D_E = D_\ell + K_H D_G$ is the effective dispersion coefficient and h is the mass transfer coefficient (cm s^{-1}). The mass transfer coefficient is often related to the binary diffusion coefficient (16) as

$$h = \frac{D_G^{air}}{b} \qquad (3)$$

where b (cm) is the a stagnant boundary layer thickness controlling vapor transport at the soil surface. In Eq. 3, the boundary layer thickness, b, embodies all the processes that affect the transport of chemical across the soil-atmosphere interface.

An alternative formulation for the mass-transfer coefficient that includes atmospheric resistance terms is (8).

$$h = \frac{u^*}{7.3 R_e^{1/4} S_c^{1/2} + \left(\frac{U_r}{u^*} - 5\right)\Phi_m} \qquad (4)$$

where R_e and S_c, respectively, are the roughness Reynolds and Schmidt numbers, u^* is the friction velocity (cm s^{-1}), U_r is the wind speed at the measurement height (cm s^{-1}) and Φ_m is an atmospheric stability correction. The denominator of Equation 4, consists of two atmospheric resistance terms, one representing diffusive resistance near the surface and the other aerodynamic resistance from the diffusive layer to the measurement height. To use Eq. 4, many meteorological measurements are needed including gradients of wind speed and temperature.

Three simulations were conducted. Case 1 assumes both D_g^{air} and b in Eq. 3 were constant with D_g^{air} obtained using the average experimental temperature of 9.8 °C. Case 2 assumes b is constant, but D_g^{air} varies according to the ambient temperature as given by the Fuller et al. (*17*) correlation. Case 3 uses Eq. 4 and measured micrometeorological data to evaluate the mass transfer coefficient, h.

Results

Shown in Fig. 1 is a time series of the volatilization rate at the field site during the experimental period. The volatilization rate has a cyclic behavior with peaks occurring predominately during the midday with low values at night. The integer values on the time axis indicate midnight and the vertical dotted lines are at noon. The daily maximum volatilization rate ranged from a high of 59.6 g ha^{-1} h^{-1} on day 1 to a low of 24.2 g ha^{-1} h^{-1} on day 3. The minimum daily volatilization rates were generally less than 5 g ha^{-1} h^{-1}. The low values are primarily due to the effect of the atmospheric stability term (Φ_m in Eq. 4) which tends to suppress the flux during periods of stable conditions ($\Phi_m > 1$, for stable conditions); generally at night. The maximum daily volatilization rates slowly decrease over the course of the experiment, but at a much slower rate than for more volatile compounds such as methyl bromide (*18*). The slow decrease in the volatilization rate is due to a reduction in the concentration gradients at the soil surface as material volatilizes, degrades and diffuses away from the surface.

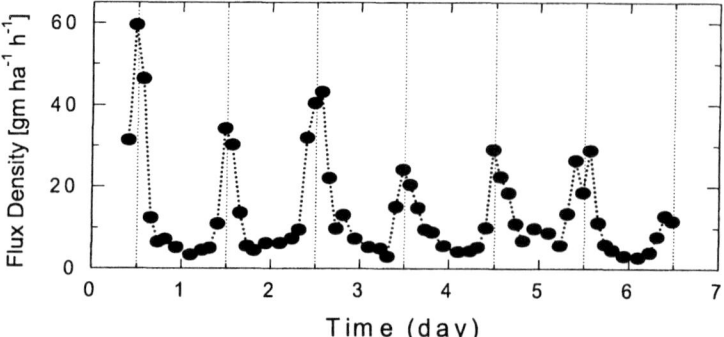

Fig. 1. Volatilization of triallate as a function of time after application.

The relationship between volatilization and atmospheric stability, Φ_m, is clearly shown in Fig. 2, where high volatilization rates generally follow low values of Φ_m, indicating unstable conditions. Low atmospheric concentration occurs during unstable conditions, where buoyancy forces dominate. For stable conditions, mixing in the atmosphere is suppressed causing higher concentrations in the lower atmosphere that reduce gradients at the soil-atmosphere interface. This yields lower volatilization rates compared to unstable conditions.

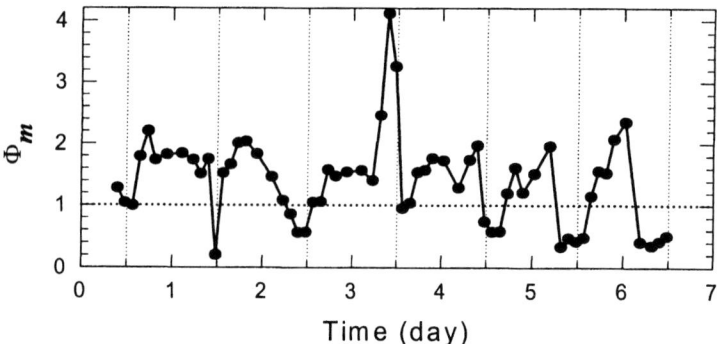

Fig. 2. Stability of the atmosphere as a function of time. Stable atmospheric conditions exist for $\Phi_m > 1$, unstable conditions for $\Phi_m < 1$.

Shown in Fig. 3 is a comparison of simulations for each of the three boundary conditions (i.e., Cases 1-3). The dotted line is the predicted rate of volatilization for Case 1 with a constant boundary layer thickness of 0.5 cm and isothermal conditions. A nearly constant volatilization rate occurs because the triallate was applied to the soil in a very thin layer and the concentration exceeded the solubility limit. To simulate this, the mass in excess of the solubility limit was partitioned to an immobile phase, that was unavailable for transport. At each time step, if the liquid phase concentration was less than the solubility limit, a portion of the immobile phase was added to the liquid phase to raise the concentration to the solubility limit. This was repeated until the immobile phase was consumed.

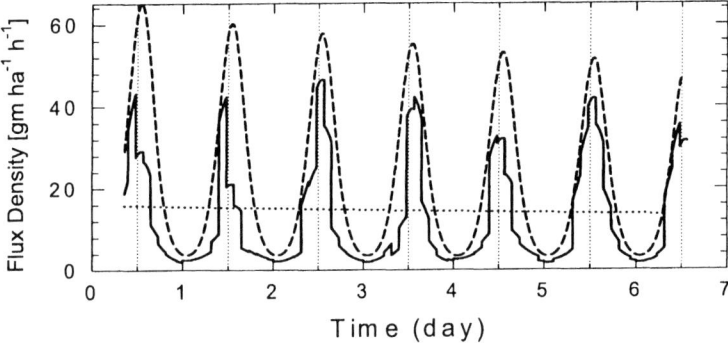

Fig 3. Comparison of predicted volatilization rates from three simulations. The dotted line is Case 1 and assumes the mass transfer coefficient, h, is constant. The dashed line is Case 2 and assumes that h depends on ambient temperature. The solid line is Case 3 and uses Eq. 4 to express h.

Comparing Fig. 1 to the dotted line in Fig. 3 shows a significant disagreement between simulated and measured volatilization rate. This suggests that it is not appropriate to assume isothermal conditions when predicting volatilization rates.

The dashed line in Fig. 3 signifies the predicted volatilization rate for a constant boundary layer thickness and temperature dependent D_g^{air} (i.e., Case 2). Since the air temperature during the experiment varied smoothly, the predicted volatilization rate has a similar behavior. Comparing Case 2 to Fig. 1 shows a general agreement between simulated and measured volatilization rates. This also shows the importance of ambient temperature variations in the simulation.

The solid line in Fig. 3 represents predictions for Case 3. Clearly, the additional micrometeorological information produces a volatilization curve that has commonly-observed erratic variations. This is caused by relatively rapid changes in the wind speed and the stability parameter, Φ_m, and suggests that the effect of ambient temperature and meteorological conditions are important.

Shown in Fig. 4 are measured and predicted cumulative volatilization for the three simulations. Case 1 and Case 3 produce cumulative emission curves that are nearly identical to the measured values. Case 2 produces a curve that deviates from the measurements and has a greater slope. In this instance, it appears to make little difference which approach is used to characterize the surface boundary condition. Even a constant mass transfer coefficient will produce accurate cumulative volatilization provided that the value of h is determined for average experimental conditions.

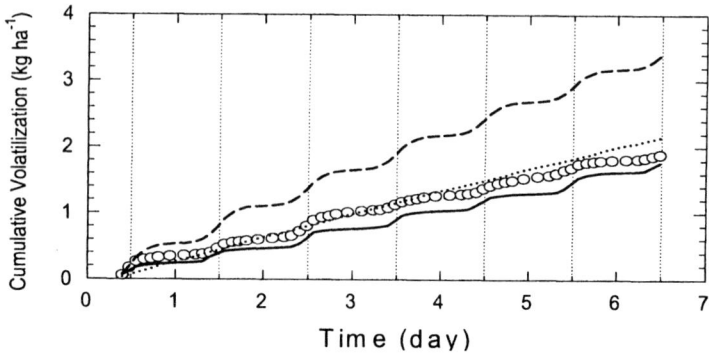

Fig 4. Comparison of cumulative volatilization from three simulations. The dotted line are results from Case 1, the dashed line from Case 2 and the solid line from Case 3. The open circles are measured values.

Figures 3 and 4 can be used to compare the three methods for predicting the volatilization rate, but they do not provide quantitative information that can be used to select the best case, and thereby determine the best choice for characterizing the boundary condition. Shown in Table 2 are three measures of the accuracy of a simulation predicting the measured volatilization rate.

Table 2. Three measures of the accuracy of predicting the volatilization rate.

Method	Area, g ha^{-1}	Area, %	SSQ (g ha^{-1} h^{-1})^2	RMS g ha^{-1} h^{-1}	Regression Intercept	Slope	r^2
Measured	1808	100					
Case 1	2096	115.9	8941	94.6	-57.85	4.944	0.079
Case 2	3221	178.2	16841	129.8	0.252	0.531	0.663
Case 3	1642	90.8	5487	74.1	5.048	0.663	0.545

SSQ: sum of squares error; RMS: root-mean square error.

One method for evaluating the accuracy of a simulation is by comparing the total volatilization, obtained by integrating the volatilization rate with time. This is shown in Fig. 4 as a function of time and is reported in Table 2 as areal mass (i.e., g ha^{-1}) and percent relative to the measured volatilization rate. The deviation for Case 1 was approximately 16% and for Case 2 was nearly 80%. Case 3 was the most accurate and was within 10% of the measured values.

Another method for determining accuracy is by comparing the sum of square error (or, equivalently, the root mean square error). For this approach, lower values indicate more accurate predictions. Case 3 had the smallest sum of square error and was nearly 1/3 the magnitude of Case 2.

Regression offers a third method for comparing simulation accuracy. A simulation that fits the measured values perfectly would have a slope, intercept and r^2, respectively, of 1, 0, and 1. Case 3 had a slope that was nearest 1, but Case 2 had an intercept and r^2, respectively, that was closest to 0 and 1. Comparing all three indicators and visual inspection of Figs. 1, 3 and 4 suggest that Case 3 is the most accurate simulation.

Including micrometeorological information into simulations of the volatilization rate should provide more accurate and realistic emissions. This requires considerably more information about atmospheric process, and may be justified when the rate of volatilization is of interest. For the cumulative or total emissions, however, a simulation using micrometeorological data may not be justified because simple boundary conditions were found to be accurate.

Acknowledgments

We acknowledge the assistance of Mark M. Cliath, RoseAnn Lemert, and Christian Taylor in obtaining and analyzing the experimental data.

References

1. United States Department of Agriculture (USDA). 1999. Reducing pesticide risk, Proceedings of the Reducing Pesticide Risk Workshop. USDA, Agricultural Research Service. Washington, DC **1999**, 62 pp.

2. United States Environmental Protection Agency (USEPA). 1987. Agricultural chemicals in ground water: Proposed pesticide strategy. USEPA Office of Pesticides and Toxic Substances. Washington DC **1987**, 150 pp.

3. Kaufmann, C.; Matheson, N. Protecting ground water from agricultural chemicals: Alternative farming strategies for northwest producers. Alternative Energy Resources Organization. Helena, MT. **1990**, 28 pp.

4. Yagi, K.; Williams, J.; Wang, N.Y.; Cicerone, R.J. *Science* **1995**, *267*, 1979-1981.

5. Majewski, M.S.; McChesney, M.M.; Woodrow, J.E.; Pruger, J.H.; Seiber, J.N. *J. Environ. Qual.* **1995**, *24*, 742-752.

6. Yates, S.R., Gan, J., Ernst, F.F., Wang, D. and Yates, M.V. In: *Fumigants: Environmental Fates, Exposure and Analysis*, Am. Chem. Soc. Symp. No. 652, Seiber, J.N. et al., Eds. Washington, DC **1997**, pp. 116-134.

7. Majewski, M.S.; Glotfelty, D.E.; Seiber, J.N. *Atmos. Environ.* **1989**, *23*, 929-938.

8. Baker, J.M.; Koskinen, W.C.; Dowdy, RH. *J. Environ. Qual.* **1996.**, *25*, 169-177.

9. Tabernero, M.T.; Alvarez-Benedi, J.; Atienza, J.; Herguedas. A. *Pest Manag. Sci.* **2000**, *56*, 175-180.

10. Doull, J. 1989. American Chemical Society. Washington, DC. 246 pp.

11. Wauchope, R.D.; Buttler, T.M.; Hornsby, A.G.; Augustijn-Beckers, P.W.M.; Burt, J.P. *Rev. Environ. Contamin. Toxic.* **1992**, *123*, 1-155.

12. Turner, B.C.; Glotfelty, D.E. *Anal. Chem.* **1977**, *49*, 7-10.

13. Cliath, M.M.; Spencer, W.F.; Farmer, W.J.; Shoup, T.D.; Grover, R. *J. Agric. Food Chem.* **1980**, *28*, 610-613.

14. Spencer, W.F.; Cliath, M.M.; Jury, W.A.; Zhang, L. *J. Environ. Qual.* **1988**, *17*, 504-509.

15. Sophocleous, M. *Water Resour. Res.* **1979**, *15*, 1195-1206.

16. Jury, W.A.; Spencer, W.F.; Farmer, W.J. *J. Environ. Qual.* **1983**, *12*, 558-564.

17. Fuller, E.N.; Schettler, P.D.; and Giddings, J.C. *Ind. Eng. Chem.* **1966**, *58*, 19-27.

18. Yates, S.R.; Wang, D.; Papiernik, S.K.; Gan, J. *Environmetrics* **2002**, *13*, 569–578.

Environmental Risk Assessment

Chapter 11

Concentrations of Herbicides Used in Rice Paddy Fields in River Water and Impact on Algal Production

S. Ishihara, T. Horio, Y. Kobara, S. Endo, K. Ohtsu, M. Ishizaka, Y. Ishii, and M. Ueji

National Institute for Agro-Environmental Sciences, Tsukuba, Ibaraki 305–8604, Japan

Plankton and periphyton algae communities are important for maintaining the proper function of aquatic ecosystems. The purpose of this study is to investigate exposure characteristics of rice herbicides in Japanese river and the effects of these herbicides to fresh water micro algae. Distribution and variation of 16 rice herbicides in the Sakura River and the Lake Kasumigaura were monitored during rice growing periods for two years (2001-2002). The highest concentrations were observed at midstream of Sakura R. and these levels were from 0.12 to 8.8 ppb in the middle of May. They were diluted about 20 to 50 times at the center of L. Kasumigaura. The sensitivities of the 4 unicellular algal species to 14 rice herbicides were compared, by using 72-h EC_{50} (50% growth inhibition concentration at 72-h after treatment) values. The order of relative sensitivity to these herbicides were *Selenastrum capricornutum* > *Merismopedia tenuissima* ≧ *Achnanthes minutissima* ≧ *Chlorella vulgaris*. Further, to find the sensitivity range of native species, the effects of 6 rice herbicides (Bensulfuronmethyl, Imazosulfuron, Simetryn, Dimethametryn, Pretilachlor and Cafenstrole) to the freshwater algal species isolated from the natural environment such as rice paddy and streams were investigated. As a result, some *Senedesmus* strains indicated low sensitivity to the pretilachlor and cafenstrole. In addition, some pennate diatoms and *Senedesmus* strains indicated low sensitivity to the triazine herbicide.

Plant communities are important for maintaining the proper function of freshwater, estuarine and marine ecosystems. Algae associated with the plankton and periphyton form the base of most food chains, produce oxygen, and play important role in cycling of nutrients (1).

In Japan, more than half of agricultural land is paddy fields. The rice herbicides are easy to run off especially because those are applied directly to the surface water of the paddy field. Thus, some rice herbicides are detected frequently in river water of Japan at the low ppb levels for some months after the rice-planting period (2-6). Alterations of a phytoplankton community as a result of toxic stress may affect the structure and functioning of the whole ecosystem (7). However, there have been few reports on the influence of the rice herbicides to phytoplankton (8-11). Therefore it is important to assess the effects of these herbicides on primary production in riverine, lake and marine ecosystems.

Since 2000, MAFF (Ministry of Agriculture, Forestry and Fisheries) Japan has been required that registrants submit phytotoxicity data (Algal, Growth Inhibition Test) for CPPs (Crop Protection Products) registration. This test results show only the adverse effect on unicellular green algae, and cannot estimate the adverse effect on a phytoplankton community by CPPs. One of the key issues that need to be resolved in toxicity testing is the great variability in sensitivity among species (12-13). Because of geographical condition, Japanese rivers are steep and its flow is very fast. Residence time of the floating speases at certain point in a river seems much shorter than the life span of planktonic algae. Therefore, the periphytonic algae, mainly diatoms are important organisms in aquatic ecosystem in Japan. Also, it is important to select appropriate species in order to evaluate the proper function of native environment. However, the data on phytotoxicity testing are limited, which makes realistic and scientifically sound risk characterizations difficult.

The objectives of this study are to investigate exposure characteristics of rice herbicides in Japanese river, to compare the relative sensitivity of various algal taxa to herbicides, and to conduct more realistic and scientifically sound risk characterizations.

Materials and Methods

Herbicide concentrations in the river

The location of monitoring stations is shown in Figure 1. Lake Kasumigaura is the second largest lake in Japan, and located on the 60 km northeast of Tokyo. Sakura River is one of the main rivers which flow into L. Kasumigaura and many paddy fields are situated in the river basin. Herbicides concentrations were monitored at 7 locations in 2001. St. 1 is located at mountainside of Mt. Tsukuba and one of the headstream of Saka R, which is a tributary of sakura R. There are neither paddy fields nor houses above this point. St. 2 is located at down stream of Saka R., which frow trough extensive paddy field area. St. 3 is located at the midstream of Sakura R. and many paddy fields are also situated around the riverside. St. 4 is located on the downstream of Sakura R. This station is located at the center of a Tsuchiura City. St. 5 is located at the river mouth of the Sakura R. St. 6 is located in Tsuchiurairi Bay. St. 7 is located in the center of the L. Kasumigaura.

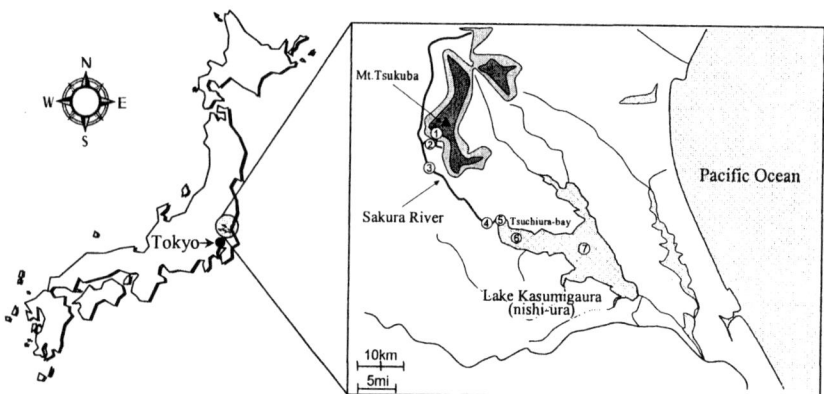

Figure 1. The location of monitoring station

Water sampling period was from March 20 to September 19 in 2001 and St. 2, St. 3 and St. 4 were investigated from April 22 to August 19 in 2002 again. Twelve rice herbcides (cafenstrole, dimepiperate, dimethametryn esprocarb, mefenacet, molinate, pentoxazone, pretilachlor, pyributicarb, pyriminobac-methyl, simetryn and thiobencarb) were analyzed by gas chromatography (GC/FTD: SIMAZU GC-17A), and 4 rice herbcides (bensulfuronmethyl, daimuron, imazosulfuron and pyrazosulfuronethyl) were analysed by Liquid Chromatograph/Mass Spectrometer/Mass Spectrometer (LC-ESI-MS/MS:

API3000LC/MS/MS System (Applied Biosystems/MDS SCIEX)). For GC analysis, each of water sample was filtered through a glass filter (Whatman GF/F 0.7μm) and its pH was ajusted to be 6.5. One litter of filtrate was passed through solid-phase extraction cartrige (Waters Sep-Pak tC18) and eluted with dichloromethane. The eluate was concentrated in vacuo and dissolved in 2ml acetone. For LC-ESI-MS/MS analysis, each of water sample was filtered through a glass filter (Whatman GF/F 0.7μm) and kept frozen until the herbicides analysis. Equal amount of internal standard substance contain acetonitrile was added to the sample and centrifuged for 10minutes at 20,600×g and then 5μl of clear supernatant liquid was injected to LC-ESI-MS/MS.

Toxicity tests on standard species

Algae

Tested 4 algae were axenic unicellular freshwater algae, *Selenastrum capricornutum* Printz ATCC 22662 (Clorophyceae), *Chlorella vulgaris* Beijerinck NIES-227 (Clorophyceae), *Achnanthes minutissima* Kuetzing NIES-71 (Bacillariophyceae) and *Merismopedia tenuissima* Lemmermann NIES-230 (Cyanophyceae). *S. capricornutum* was obtained from the American Type Culture Collection, USA. The other three strains were obtained from the Microbial Culture Collection in the NIES (National Institute for Environmental Studies), Japan. *S. capricornutum* and *C. vulgaris*, have been recommended as standard strains for the OECD algal toxicity test (14). Each strain was maintained on Medium Csi (pH7.5) (15) 1.5% agar slants.

Herbicides

The 14 herbicides tested are listed in Table 1, with their chemical properties, amounts of shipment (2000), typical aplication rates (g/10a) and the registration year in Japan. All herbicides used were analytical grade.

Algal assay

The algal inoculum was taken from preculture in the exponential growth phase. Preculture and assay were conducted at 23 ± 2 ℃ under continuous illumination of approximately 3,000lux using a shaking incubator at 100rpm (SIBATA, RS-200). The test algae were cultivated in 300ml sterile foam-plugged Erlenmeyer flasks, each containing 100ml of Medium C (pH7.5) for *S.*

capricornutum C. vulgaris and *M. tenuissima*, 100ml of Medium Csi (pH7.5) for *A. minutissim*. The initial algal cell concentration in the test culture was approximately 1×10^4 cells/ml for *S. capricornutum* and *C. vulgaris*, 3×10^5 cells/ml for *M. tenuissima*, 5×10^3 cells/ml for *A. minutissim*. The concentration range in which effects were likely to occur was determined from range-finding tests. For the definitive test (EC_{50} value finding tests), five concentrations arranged in a geometric series in a ratio of 1.6 to 2.5 depending on test conditions. However, definitive testing was not performed if the highest herbicide concentration tested (water saturation concentration or 1,000mg/L) results in less than a 50% reduction in growth. DMSO (dimethyl sulfoxide) was used as the vehicle to dissolve herbicides at a concentration of 0.1%(maximum). Algal assays were performed by three replicates at each concentration. There were also three replicates of the controls that contained 0.1% DMSO without test herbicides. Every test species cell counts was performed after 0,24,48 and 72 h using a PAS-FCM (Particle Analysing System-Flow Cytometer, partec GmbH). For the EC_{50} value finding tests, herbicides concentrations were measured at the beginning and the end of the assay using a GC/FTD or HPLC/UV.

Table I. Testing herbicides and their mode of action, chemical family, common name, chemical properties, amount of shipment (2000), typically applied quantities (g/10a) and registration year in Japan

No.	Common name	Mode of action	Chemical family	Water solubility (mg/l)	Kow logP	Common registrated application rates (g/10a)	Shipment in 2000 (a.i., t or kl)	Registration year in Japan
1	Thiobencarb	Inhibition of lipid synthesis - not ACCase inhibition -	Tiocarbamates	30	3.4	150	470	1969
2	Esprocarb	"	"	4.9	4.6	150	310	1988
3	Molinate	"	"	900	2.9	240	260	1971
4	Dimepiperate	"	"	20	4.0	300	75	1986
5	Pretilachlor	Inhibition of Very Long Chain Fatty Acids (Inhibition of cell division)	Chloroacetamides	50	4.1	40	280	1984
6	Mefenacet	"	Oxyacetamides	4	3.2	110	470	1986
7	Cafenstrole	"	Others	2.5	3.2	70	110	1996
8	Dimethametryn	Inhibition of photosynthesis at photosystem II	Triazines	50	3.8	6	22	1975
9	Simetryn	"	Triazines	400	2.6	45	77	1969
10	Bentazone	"	Benzothiadiazinones	570	-0.5	330	230	1975
11	Quinoclamine (ACN)	"	Quinones	22	1.5	270	86	1968
12	Bensulfuronmethyl	Inhibition of acetolactate synthase	Sulfonylureas	120	0.62	17	63	1987
13	Imazosulfuron	"	Sulfonylureas	310	0.05	9	14	1993
14	Daimuron	Unknown	Ureas	1.2	2.7	150	550	1974

Toxicity test on native species

Algae

Unialgal clonal strains (pennate diatoms and green alga, *Senedesmus* sp.) were isolated from natural population samples collected at the paddy field, the drainage canal (near the St. 2, see Figure 1) and the river (St. 1, 3 and 4) in August or October, 2002. Each of strains was maintained on Medium Csi (pH7.5) 1.5% agar slants. Total number of isolated strains were 42(pennate diatoms) and 20 (*Senedesmus* sp.)

Herbicides

The 6 rice herbicides (Bensulfuronmethyl, Imazosulfuron, Simetryn, Dimethametryn, Pretilachlor and Cafenstrole) were used for the algal assay.

Algal assay

The algal inoculum was taken from preculture in the exponential growth phase. Preculture and the assay were conducted at $23 \pm 2°C$ under continuous illumination of approximately 3,000lux. The test algae were cultivated in 96-well microplate, each containing 0.2ml of Medium Csi (pH7.5). The initial algal cell concentration in the test culture was approximately 5×10^3 cells/ml for *Senedesmus* sp. and 3×10^3 cells/ml for pennate diatoms. For the EC_{50} value finding tests, ten concentrations arranged in a geometric series in a ratio of 2.0. Cell density was measured after 96 h using a spectrophotometer (BIO-RAD, Model 550). The absorbance values of the sample was measured at 680nm and 96 h EC_{50} value was calculated.

Results and Discussion

Herbicides concentrations

At the St. 1, any herbicides were not detected through the period. At the St. 2, fourteen kind of herbicides were detected. The herbicides included cafenstrole, dimepiperate, dimethametryn, esprocarb, mefenacet, pentoxazone, pretilachlor, pyributicarb, pyriminobac-methyl, simetryn, bensulfuronmethyl, daimuron, imazosulfuron and pyrazosulfuronethyl. The herbicides detection period of this station was about one month. Summed concentrations of detected herbicides in water samples of Sakura R. and L. Kasumigaura is shown in Figure 2. Total of 16 herbicides (The above-mentioned herbicide plus molinate and thiobencarb) were detected at Sakura R. The highest concentrations were observed at midstream of Sakura R. (St. 3) and these levels ranged from 0.12 ppb for pyriminobac-methyl on May 8, 2002 to 8.8 ppb for daimuron on May 14, 2002. ppb in the middle of May. The concentrations of detected herbicides were approximately the same level in the midstream (St. 3) and downstream (St. 4) of Sakura R. They were diluted about 2 times at the river mouth, about 10 times at Tsuchiurairi Bay, and about 20-50 times at the center of L. Kasumigaura. For most rice herbicides detected in Sakura R. and L. Kasumigaura, the maximum detected concentration levels were below low ppb lebel. Recently, the concentration level of individual herbicide detected in surface water tend to decrease in Japan (3,4,16). Especially during 90's, this tendency become obvious. This phenomenon is probably the results of reduction of rice cultivation areas and increase in number of active ingredient in the market. Also, it was probably affected by the development/application of low-dose, high-potency herbicides such as sulfonylurea herbicides has advanced.

Herbicide susceptibility in four unicellular freshwater algae

Effects of 14 kinds of rice herbicides to 4 kinds of algae were investigated, and 39 EC_{50} values were determined (12; *S. capricornutum*, 10; *A. minutissim*, 9; *M. tenuissima*, 8; *C. vulgaris*). *S. capricornutum* is highly sensitive to tested herbicides and it is the most popular test algal strain for toxicity tests. Also, *S. capricornutum* have been recommended as standard strains for the OECD algal toxicity test. On the other hand, *C. vulgaris* is insensitive to tested herbicides. Although the trend of *C. vulgaris* EC_{50} is very different from the case of *S. capricornutum*. *C. vulgaris* have still been recommended as standard strains for the OECD algal toxicity test. *A. minutissim* is one of periphytonic pinnate diatoms which commonly observed in Japanese rivers. Similar to *C. vulgaris*, it has low sensitivity on most of herbicides as compared with *S. capricornutum*. Especially, sensitivity to sulfonylurea herbicides is very low. *M. tenuissima* is a kind of blue-green algae and is the same order of algae with Genus microcystis which causes blue-green algae waterbloom. Usually blue-green algal speces is difficult to do the subculturing on solid medium however this *M. tenuissima* has an advantage of performing subculturing easily on solid medium and it is highly applicable for toxicity tests. Similar to algae explained above such as *C. vulgaris* and *A. minutissim*, *M. tenuissim* has also low sensitivity on most of herbicides as compared with *S. capricornutum*. Though, it has very high sensitivity to

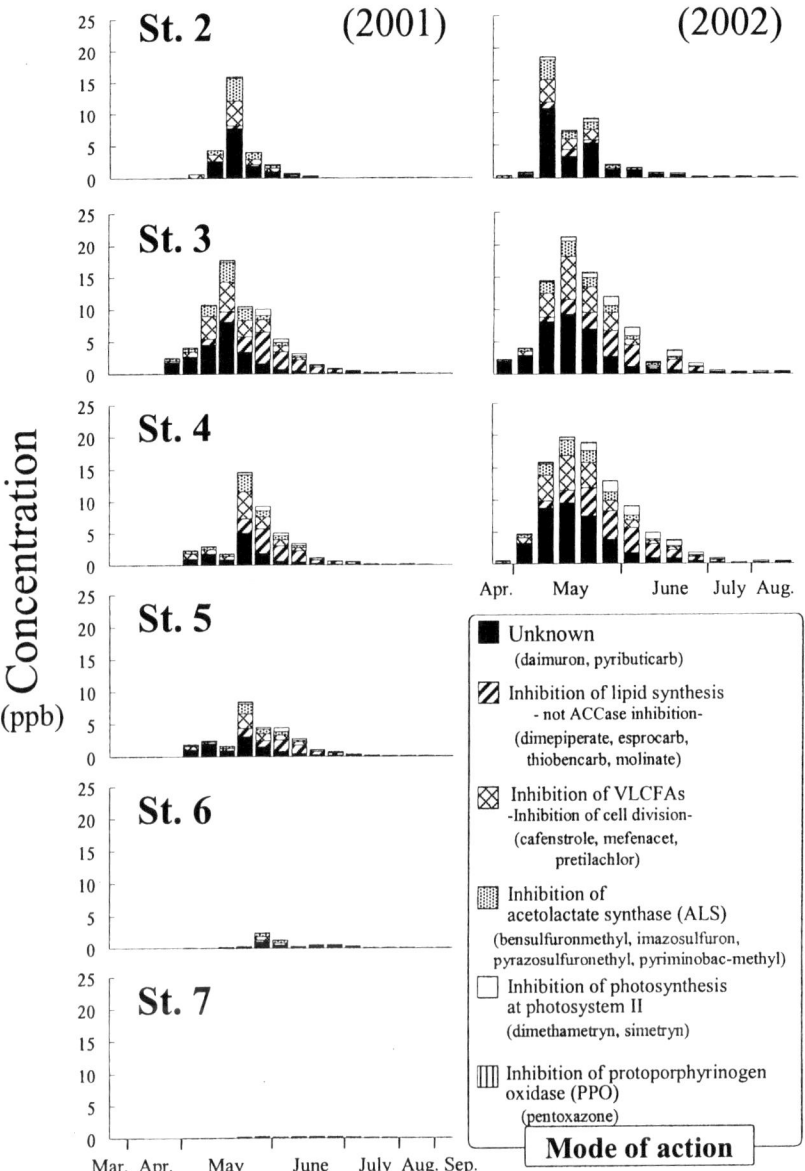

Figure 2. Herbicide concentrations in water samples of Saka R, Sakura R. (2001, 2002) and L. Kasumigaura (2001)

sulfonylurea herbicides. The order of relative sensitivity was *S. capricornutum* > *M. tenuissima* ≧ *A. minutissim* ≧ *C. vulgaris* (Figure 3).

Triazine herbicides and quinoclamine, having the mode of action in inhibition of PS II (Photosynthesis at photosystem II) had low variability on sensitivities in different algal taxa. On the other hand, Amide herbicide such as Pretilachlor and Cafenstrole as well as sulfonylurea herbicides of bensulfuronmethyl and imazosulfuron had great variability on sensitivities in different algal taxa. These herbicide have in other the mode of action in inhibition of cell division or in inhibition of acetolactate synthase rather than in inhibition of PS II. Carbamate herbicide showed relatively low toxicity on algae. Daimuron and bentazone exhibit low toxicity on all the tested species (Figure 3).

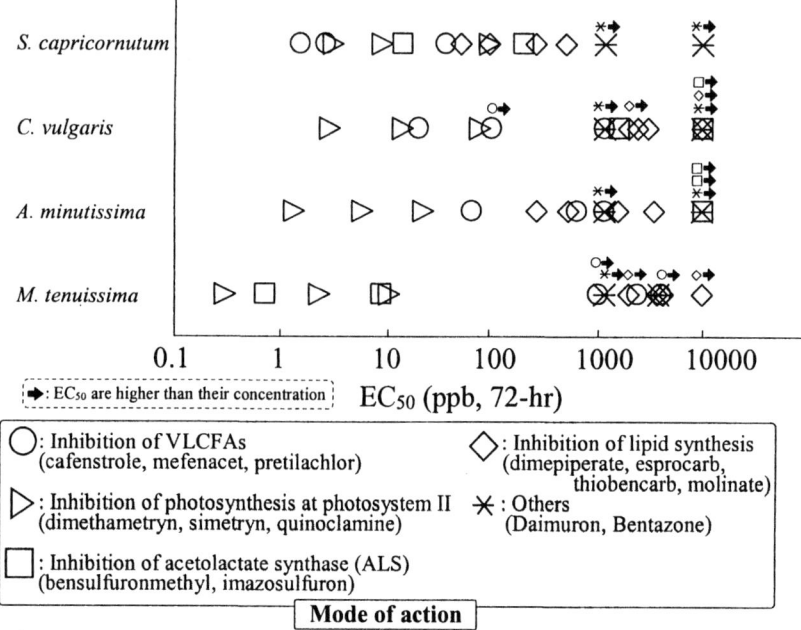

Figure 3. *EC$_{50}$ values of four unicellular freshwater algae at fourteen rice herbicides*

Herbicide susceptibility in freshwater pennate diatoms and green alga, Senedesmus sp.

Figure 4 shows the EC$_{50}$ values of 6 herbicides for individual pennate diatoms and *Senedesmus* sp. isolated from natural population strains.

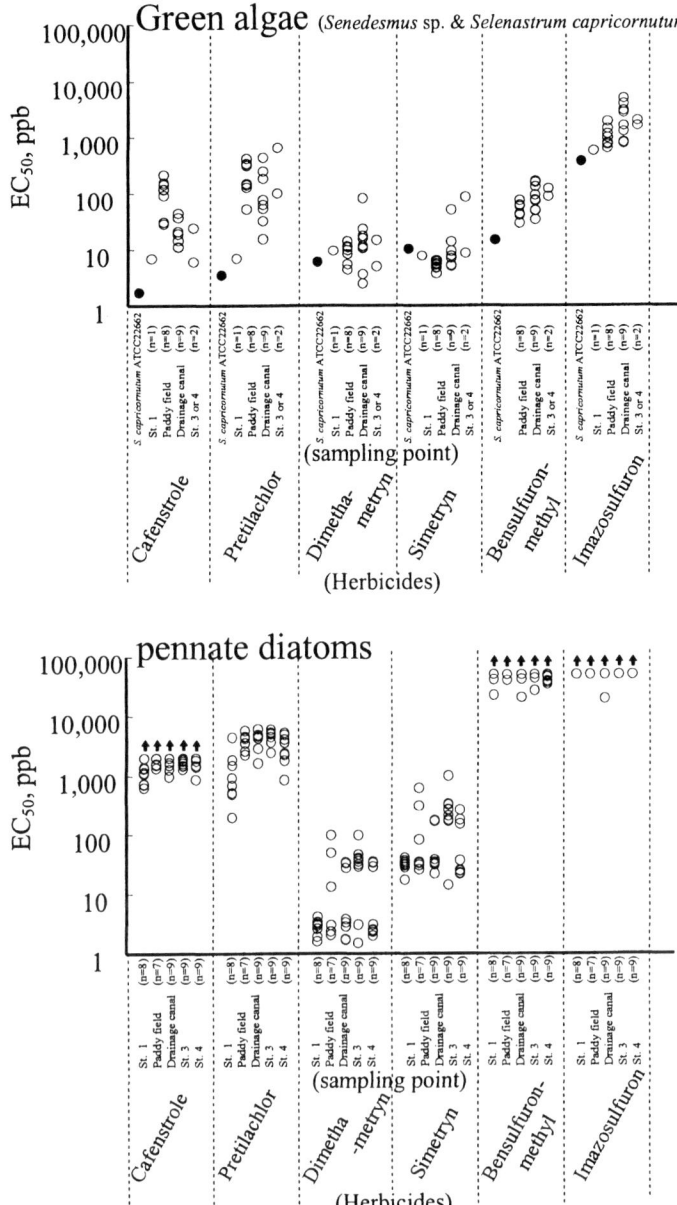

Figure 4. EC50 values of six herbicides for pennate diatoms and Senedesmus sp. isolated from natural environment.

Senedesmus sp. is one of the dominant gleen algal species in paddy field, and pennate diatoms are the dominant species in Japanese river. Some pennate diatoms and *Senedesmus* strains isolated from the paddy fields, drainage canal and river showed low susceptibility to the triazine herbicide. Especially, the susceptibility of the pennate diatoms was dropped to about 1/10 times compared with the high susceptibility strains. The susceptibility of *Selenastrum capricornutum* to the Pretilachlor, Cafenstrole and sulfonylurea herbicides (Bensulfuronmethyl and Imazosulfuron) was higher than that of other strains of *Senedesmus* sp. On the Triazine herbicides (Simetryn and Dimethametryn), some *Senedesmus* strains were more susceptible than *Selenastrum capricornutum*. Some *Senedesmus* strains isolated from the paddy fields showed low susceptibility to the Cafenstrole. It might be a cause that the herbicide which contain the Cafenstrole were applied in the paddy field in the previous year (2001). The toxicity of the sulfonylurea herbicides to the pennate diatoms were low.

These data indicates that these six rice herbicides are unlikely to cause acute adverse impacts on primary productivity of the Sakura River. However, further investigations are necessary in order to clarify the fact that appearance of low sensitive strains has triggered by the use of herbicide and existence of low sensitive strains can be an ecologically acceptable phenomenon or not.

Conclusion

For most rice herbicides detected in Sakura R. and L. Kasumigaura, the maximum detected concentration levels were below low ppb lebel. Recently, the concentration level of individual herbicide detected in surface water tend to decrease in Japan. the probable reasons are the reduction of rice cultivation areas and increase in number of active ingredient in the market.

Effects of 14 kinds of rice herbicides to 4 kinds of algae were investigated. The order of relative sensitivity was *S. capricornutum* > *M. tenuissima* $\geq$ *A. minutissim* $\geq$ *C. vulgaris*. Six rice herbicides to the freshwater algal native species. As a result, some strains indicated low sensitivity to rice herbicides. However, these herbicides are unlikely to cause acute adverse impacts on primary productivity of the Sakura R. It is difficult to evaluate the realistic and scientifically sound risk characterizations on the primary production in aquatic system of herbicides using only a result of the toxicity test to unicellular green alga. Therefore, we should take into account the species/strain differences in the susceptibility when the aquatic plant risk was assessed. Furthermore, to establish phytotoxicity assessment scheme (including level of progression for phytotoxicity testing), we should decide test species/methods of aquatic plants (including algae) at an early date.

Acknowledgements

We would like to express our sincere thanks to a number of colleagues at Tokyo University of Agriculture and Technology, National Agricultural Research Organization, Japan and Agricultural Chemicals Inspection Station, Japan for supporting this research.

References

1. Gary M. R.; Fundamentals of aquatic toxicology Second edition. Taylor & Francis. London, UK (1995)
2. Shiraishi H., Pura F., Otsuki A. and Iwakuma T.; The science of the Total Environment. 1988, 72, 29-42.
3. Maru S.; Special bulletin of the chiba prefectural agricultural experiment station. 18, Chiba-shi, Japan, 1991.
4. Nakamura K.; Bulletin of the saitama agricultural experiment station. 46, Kumagaya, Japan, 1992.
5. Nohara S. and Iwakuma T.; Chemosphere. 1996, 33, 1417-1424.
6. Okamoto Y., Fisher R.L., Armbrust K.L. and Peter C.J.; Journal of pesticide science. 1998, 23, 235-240.
7. Mosser. J.L., N.S. Fisher and C.F. Wurser.; Science. 1972, 176, 533-535.
8. Hatakeyama S., Fukushima S., Kasai F. and Shiraishi H.; The Japanese Journal of Limnology. 1992, 53, 327-340.
9. Kasai F., Takamura N. and Hatakeyama S.; Environmental Pollution. 1993, 79, 77-83.
10. Hatakeyama S., Fukushima S., Kasai F. and Shiraishi H.; Ecotoxicology. 1994, 3, 143-156.
11. Nishimoto H.; Research Bulletin of the Aichi ken Agricultural Research Center. 1998, 30, 71-77.
12. Roshon R, McCann JH, Thompson DG, Stephenson GR.; Canadian journal of forest research. 1999, 29, 1158-1169.
13. Susan A. F.; Impacts of low-dose, high-potency herbicides on nontarget and unintended plant species. SETAC PRESS. Pensacola, FL. (2001)
14. OECD; OECD Guidelines for Testing of Chemicals, Alga, growth inhibition test. Organization for Economic Co-operation and Development, Paris. (1984)
15. GEF; GEF List of Strains '97, Microalgae and Protozoa. Global Environmental Forum, Tokyo, Japan. (1997)
16. NIES, Japan; Report of special research from the National Institute for Environmental Studies, Japan. SR-29-'99. (1999)

Chapter 12

Ecotoxicological Risk Assessment of Atrazine in Amphibians

Keith R. Solomon[1], J. A. Carr[2], L. H. du Preez[3], J. P. Giesy[4],
T. S. Gross[5], R. J. Kendall[6], E. E. Smith[6], and
G. J. Van Der Kraak[7]

[1]Centre for Toxicology, University of Guelph, Guelph, Ontario
N1G 2W1, Canada
[2]Department of Biological Sciences, Texas Tech University,
Lubbock, TX 79409–3131
[3]School of Environmental Sciences and Development, North-
West University, Potchefstroom Campus, Potchefstroom, South
Africa
[4]Department of Zoology, Michigan State University, East
Lansing, MI 48824–1115
[5]Florida Caribbean Science Center, U.S. Geological Survey,
Biological Resources Division, and University of Florida,
Gainesville, FL 32611
[6]The Institute of Environmental and Human Health and
Department of Environmental Toxicology, Texas Tech
University, Lubbock, TX 79409–1163
[7]Department of Zoology, University of Guelph, Guelph, Ontario
N1G 2W1, Canada

Although the ecological risks of atrazine in surface waters have been extensively reviewed with respect to its potential effects on a number of components of ecological systems, there is a lack of specific data on amphibians, especially for endpoints related to development and reproduction. To assess whether atrazine causes adverse effects in frogs through endocrine-mediated mechanisms, several hypotheses were tested in laboratory and field studies, using guidelines for the identification of causative agents of disease and eco-epidemiology derived from Koch's postulates and the Hill criteria. These criteria were: Temporality; Strength of

Association; Consistency; and Biological Plausibility. Data from the literature and from studies conducted for the purpose of this assessment were used to test the following hypotheses: Atrazine causes adverse effects in amphibians through; 1) estrogen-mediated mechanisms, 2) androgen-mediated mechanisms, 3) thyroid-mediated mechanisms, 4) adverse effects on gonadal development in amphibians, or 5) adverse effects at the population level in exposed amphibians. In a lines-of-evidence approach to address the causal link between atrazine and effects in amphibians, no temporal correlation exists between the occurrence of gonadal effects and the introduction and use of atrazine. The strength of association is not strong. While some concentration-responses have been observed in some studies for some endpoints, these have not been observed under different conditions. The incidence rate of gonadal anomalies and other effects on populations was not related to atrazine exposures. Controlled studies in different laboratories produce inconsistent results and observations are inconsistent among laboratories as well as between laboratory and field studies. The biological plausibility of either of the proposed mechanisms of endocrine disruption is not supported by the laboratory or field observations. The data showed no evidence of effects through thyroid hormone mediated mechanism and little evidence for a mechanism mediated through androgens or estrogens. Overall, there was no compelling evidence to suggest that atrazine causes adverse effects in amphibians that are mediated by endocrine or developmental mechanisms.

Introduction

The potential for ecological risks from atrazine in surface waters has been extensively reviewed with respect to its potential effects on organisms in a number of ecological systems (*1,2*). Fish full life-cycle studies and mesocosm studies have not suggested effects on reproduction and development at environmentally relevant concentrations. However, there is a lack of specific data on some non-target aquatic and semiaquatic species such as reptiles and amphibians, especially for the newly identified sublethal endpoints related to development and reproduction. A robust and tested framework within which to conduct risk assessments for endocrine and developmental endpoints is also absent (*3*). Specifically, standardized tests have not been developed and validated for studies with amphibians or for endpoints other than survival, growth and early development. In addition, there is no guidance available on how to interpret non-linear dose-response relationships or for relating subtle effects at molecular or physiological levels of organization to ecologically-relevant assessment endpoints. With this in mind, we developed several assays for endocrine and developmental disruptors in fish, amphibians, and reptiles, both to refine general assays using relevant methods of exposure and to investigate the possible endocrine and reproductive effects of atrazine. The primary focus of this chapter is on characterizing and assessing endocrine-mediated effects of atrazine on amphibians. Much of this work has been presented and deliberated at a United States Environmental Protection Agency (USEPA) Science Advisory Panel (SAP) meeting in June 2003 (*4,5*) and this chapter is a summary of key issues related to our findings in the context of other work in the literature.

Laboratory studies using traditional endpoints indicate that atrazine is not highly toxic to frogs and that it does not cause mortality during larval development and metamorphosis at environmentally relevant exposures (*2,6-8*). Atrazine only causes reproductive effects in birds at large exposure concentrations (*9*). Acute and chronic exposure studies in mammals do not indicate great sensitivity for the traditional assay endpoints. The exception to this is the increased incidence of mammary tumors in female Sprague Dawley (SD) rats, an effect that was strain-, sex-, and species-specific and judged to not be relevant to humans (*10*) or other species. Many of the factors controlling amphibian reproduction and sexual development are modulated by the endocrine system (Figure 1).

Amphibians exhibit a variety of reproductive adaptations that include some forms of hermaphroditism in frogs and toads. Hermaphroditism has been reported since the 1860s and the development of this condition varies among species and the magnitude of expression varies among locations (*11*). Sex reversal has been reported in some species in response to temperature changes, other factors, or spontaneously (*12*). Other environmental factors, such as disease or parasite loads that may influence this are unknown. Sexual differentiation is a highly labile process in some species of amphibians which illustrates that cross-species extrapolations are not simple nor is it easy to establish cause and effect relationships between various endogenous and exogenous signals that control sexual differentiation.

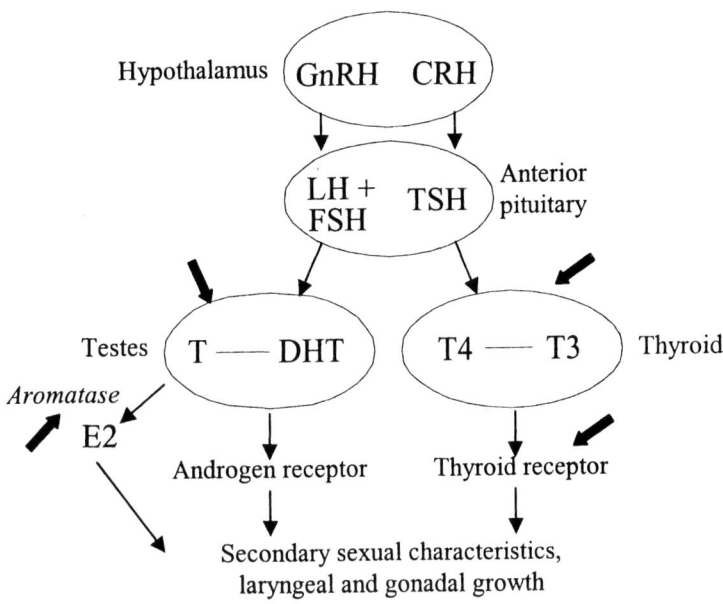

Figure 1. Hypothetical mechanism of induction of secondary sexual characteristics in Xenopus laevis showing possible targets for modulation by xenobiotics (heavy arrows). CRH, corticotropin-releasing hormone; DHT, dihydrotestosterone; E2, estradiol; FSH, follicle-stimulating hormone; GnRH, gonadotropin-releasing hormone; LH, luteinizing hormone; T, testosterone; T3, triiodothyronine; T4, thyroxine; TSH, thyroid stimulating hormone.

To assess whether atrazine causes adverse effects in frogs through endocrine-mediated mechanisms, several hypotheses were proposed to address specific mechanisms of action for endocrine modulating substances. In testing these hypotheses, guidelines based on those employed in the identification of causative agents of disease and eco-epidemiology (13,14) were used. These criteria were: Temporality; Strength of Association; Consistency; Biological Plausibility; and Recovery (15).

Assessment of Possible Mechanisms

With respect to potential effects of atrazine on endocrine function in amphibians, the question of possible mechanism can be reduced to several hypotheses which address specific mechanisms of action for endocrine modulating substances. Also, test endpoints can be defined which include both changes in hormone concentrations or in hormone-mediated processes or structural or functional endpoints that are under the control of steroid hormones such as are illustrated in Figure 2. These are discussed in the following sections:

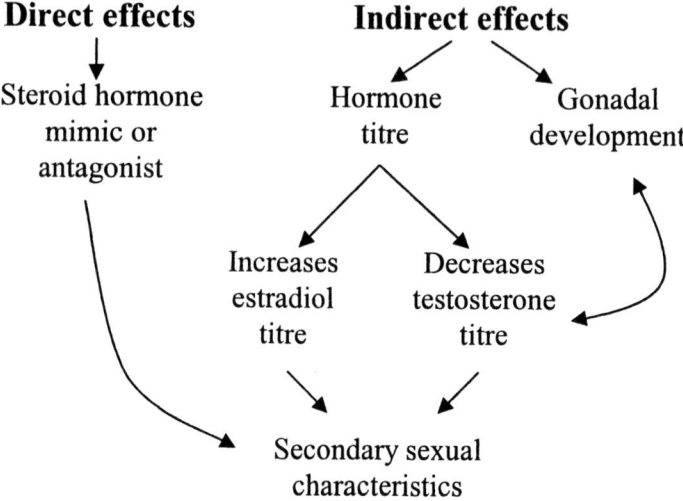

Figure 2. Graphical illustration of the possible pathways through which atrazine may affect the expression of secondary sexual characteristics and gonadal development

Estrogen and Anti-estrogen Mediated Mechanisms

In vitro studies conducted with mammalian estrogen receptor (ER) have indicated that atrazine does not bind to the ER (*16*). Based on effects on estrogen-mediated determination of sex ratios, atrazine does not mimic the effects of estradiol on sex ratio in frogs, such as *Xenopus laevis* (*17,18*). Atrazine exposure did not increase plasma estradiol titres in *X. laevis* frogs, exposed in the laboratory (*19*) or collected from corn-growing areas in South Africa (*18*), or in *Bufo marinus* collected in sugar cane-growing sites in Florida where atrazine was present in the environment (*20*). Atrazine has been reported to up-regulate aromatase activity *in vitro* in a human adrenocarcinoma (H295R) cell line and fish hepatocytes at relatively high concentrations (*21,22*). Although atrazine has been postulated to affect hormone titres through induction of aromatase in frogs (23), it did not significantly increase the activity of gonadal or brain aromatase in laboratory and field studies on exposed frogs (18,19). The lack of response of both estradiol titres and aromatase is consistent with the known role of aromatase in the biosynthesis of estradiol. For the hypothesis that atrazine acts through estrogen or anti-estrogen mediated mechanisms, there was little evidence for a strength of association, for consistency, and biological plausibility. Overall, there was little support that atrazine affects (anti)estrogen-mediated processes in amphibians.

Androgen and Anti-androgen Mediated Mechanisms

Atrazine has been reported to bind to the androgen receptor in mammals in vitro, but with very low affinity (24,25). No studies on binding of atrazine to the androgen receptor or androgen receptor-mediated gene activation have been conducted in amphibians, but on the basis of low-affinity binding in mammals, effects via this mechanism would not be expected. Although one study reported that exposure to atrazine decreased plasma testosterone concentrations (23), studies using laboratory and field exposures did not observe this response in *X. laevis* (18,19) or *B. marinus* (20).

Laryngeal development in frogs is a sexually dimorphic process, and the formation of a larynx capable of male calling behavior is androgen-dependent (26). Under normal conditions, male frogs have greater larynx muscle size than do females and this has been used as a measure of possible androgen-mediated effects of atrazine. In the single study reporting that atrazine exposure in the laboratory caused a decrease in larynx dilator muscle size in male *X. laevis* frogs, there was no consistent concentration-response relationship (23). Two other laboratory studies in the same species did not observe this effect (17,19) and no effects on larynx weight were observed in the same species from corn-growing sites in South Africa (27). For the hypothesis that atrazine acts through androgen or anti-androgen mediated mechanisms, there was no conclusive evidence for a strength of association, for consistency, and biological plausibility. Overall, there was little support that atrazine affects (anti)androgen-mediated processes.

Thyroid Mediated Mechanisms

Binding to or interference with the thyroid hormone receptor has not been directly tested in amphibians. No effects on metamorphosis were reported in laboratory studies in *X. laevis* (17,19,23) *Rana clamitans*, or *R. pipiens* (6,28).

Changes in plasma corticosterone and thyroxine concentrations were reported in salamander larvae (*Ambystoma tigrinum*) exposed to atrazine at concentrations of 75 and 250 :g/L (*29*). The authors did not suggest that this was an atrazine-specific mechanism but rather that it was a compensatory response to stress from an environmental contaminant. In addition, since the development of the laryngeal dilator muscle is dependent on both androgen and thyroid hormone, the fact that effects were not observed in male *X. laevis* in all but one of the studies conducted with atrazine supports the conclusion that atrazine did not disrupt thyroid hormone status in these frogs (*17,19,27*). Overall, there is no evidence that atrazine affects thyroid-mediated processes in amphibians.

Mechanisms Mediated Through Gonadal Development

Several studies have examined the gonads of developing male and female frog larvae exposed to atrazine, both under laboratory and field conditions. Most of the effects that have been attributed to exposure to atrazine have been reported to have occurred in males and have ranged from morphological anomalies in the gonads to the presence of ovarian tissue in the testes. Two laboratory studies reported gross gonadal anomalies in two species of frogs (*X. laevis* and *R. pipiens*) at concentrations as low as 0.1 :g/L (*23,30*). However, other laboratory studies have not shown significant increases in these responses in *X. laevis* or *R. clamitans* exposed up to 25 :g/L atrazine (*19,28*) or have only observed these at greater concentrations (25 :g/L in *X. laevis*) and at lower frequencies (*17*). Where histology was conducted in these studies, testicular oocytes have been observed, but in small numbers (per testis) and at a relatively low frequency of individuals.

A series of laboratory studies on the effects of atrazine on gonadal and kidney development in male and female *X. laevis* tadpoles were reported in a thesis and two published papers (*31-33*). In males, reductions in testicular size were reported in one experiment. Unfortunately, the data in the published papers and in the thesis on which they are based are inconsistent and some experiments showed responses while others (only published in the thesis) did not. These data cannot be interpreted.

A field and a microcosm study on *X. laevis*, reported no relationship between exposure to atrazine and other agricultural chemicals used in corn production and gonadal anomalies (*27*) and the incidence and frequency of testicular oocytes (*34*). Studies on *X. laevis* exposed to time-weighted mean concentrations of atrazine of 0, 1, 12 and 30 μg/L in semi-field microcosms from shortly after hatch until after metamorphosis showed a low frequency of metamorphs (stage 66) with abnormalities but no concentration response to atrazine. Frequency of incidence of gross gonadal anomalies (visible without histology) ranged from 8% of males in the controls to 7% of males exposed to 30 μg/L atrazine. Rigorous histological examination of the testes in the microcosm-exposed metamorphs revealed the presence of one or more testicular oocytes per frog at frequencies ranging from 57% in the unexposed control metamorphs to 40% in those exposed to 30 μg/L. The mean number of oocytes per testis ranged from 9.5 in the controls to 12 in metamorphs exposed to atrazine at 30 μg/L. The total number of oocytes per testis ranged from 1 to 60 but no concentration response was observed for either frequency of occurrence or number of oocytes. Although the number of frogs with testicular oocytes was similar to those in metamorphs after 10 months of additional exposure, the mean number of testicular oocytes per testis was smaller (>3 per testis) in control and exposed frogs examined 10 months after metamophosis.

In a field study of *R. pipiens* carried out in the US Midwest, the greatest frequencies of gonadal anomalies occurred in sites with the lowest concentrations of atrazine (*30*). The authors suggested that this could be due to an inverted "U" concentration response, resistance developed in some populations, or that it was caused by other agents. In fact, the measured atrazine concentrations likely had no relation to exposures during development as the samples for residue analysis were taken at the same time as the metamorphosed frogs. Thus, exposures during development were not known. The most plausible conclusion is that there was no relationship with atrazine. The reported frequency of anomalies was different between *R. pipiens* and *X. laevis*, but there was no indication of a concentration-response nor any effects on the frequency of other cell types in testes of field collected *X. laevis* from corn-growing sites in South Africa (*27*). The historical incidence of intersex in the cricket frog (*Acris crepitans*) did not show a statistically significant relationship to concurrent atrazine exposure concentrations (*35*) or a temporal relationship with the introduction of atrazine (*36*).

In a preliminary study, adult male *B. marinus* from sugar cane-growing areas where atrazine was used showed female-specific skin coloration patterns and the presence of oocytes in the testes, possibly as a result of development of Bidder's organ. Bidder's organ is a rudimentary ovary which is known to be responsive to a wide variety of endocrine factors and, in some genera of amphibians, such as *Bufo*, is retained in association with the testes or ovaries in adults. Exposure measurements were incomplete and give the large number of pesticides used in sugar production, causality cannot be assessed from this study.

There was no evidence for temporality, little evidence for strength of association, some evidence of consistency, and little evidence of biological plausibility, particularly as gross gonadal anomalies and testicular oocytes have been reported from unexposed controls and from historical observations. Overall, there is little evidence that atrazine affects gonadal development in amphibians, however, there is uncertainty with respect to background incidence of some of the responses observed in the gonads of male frogs. Also, it is unclear how atrazine can cause effects on gonads above a threshold of 0.1 :g/L without any apparent concentration-response relationship, such as has been suggested.

Mechanisms Mediated Through Effects at the Level of the Population

Although few studies have addressed this question relative to causality, there is no evidence of effects at the population level that have been linked to atrazine exposures. Apparently robust populations of *R. pipiens* were observed across a range of atrazine exposures in the US Midwest (*30*). Preliminary observations in *B. marinus* from South Florida showed an abundance of frogs in sugarcane-growing areas where atrazine was present (*20*). No differences in sex and age classes in *X. laevis* from reference and corn-growing sites in S. Africa were observed (*37*). There was no evidence for temporality, for strength of association, for consistency, and for biological plausibility. Overall, there is no evidence that atrazine affects frogs at the population level. Based on the frequency and magnitude of gonadal anomalies in frogs that have been reported to have occurred in laboratory studies and similar observations in apparently robust field populations, responses in these endpoints would not be expected to translate into changes in the assessment endpoint of populations. Because most frogs are 'r-selected', they produce many eggs, few of which survive to become reproductive adults. Survival of California red-legged frog (*Rana aurora draytonii*) tadpoles to metamorphosis under field conditions has been reported as low as 5% (*38*). Thus, even if the subtle effects described to have occurred in the gonads of the frog species studied were attributable to atrazine, they may have little relevance at the population level. Without consistent, robust, and mechanistically understood responses, it is not possible to determine if they might be adverse responses in other amphibian species or not.

Conclusions

Using the guidelines suggested for assessing endocrine disruptors by the International Program on Chemical Safety (*15*) in a lines-of-evidence approach to address the causal link between atrazine and effects in amphibians, the following general conclusions are possible:

With respect to temporality, no correlation exists between occurrence of gonadal effects and the introduction and use of atrazine. Intersex, other gonadal anomalies, and testicular oocytes were observed before the introduction and use of atrazine as herbicide. There is little evidence to support the strength of association. While some concentration-responses have been observed in some studies for some endpoints, these have not been found under slightly different conditions. The incidence rate of gonadal anomalies and other effects on populations is not related to atrazine exposures and the possible influence of a number of plausible confounders has not been specifically addressed. The observations reported are not consistent. Controlled studies in different laboratories produce very different results and the outcomes for many of the endpoints and observations are inconsistent between laboratory and field studies. Similarly, the biological plausibility of the proposed mechanisms is not supported by the laboratory or field observations. There is no evidence of effects through an (anti)thyroid hormone mediated mechanism and little evidence for a mechanism mediated through (anti)androgens or (anti)estrogens. The criterion of recovery cannot be specifically addressed since no consistent and robust responses have been identified from which to measure recovery.

Application of Koch's postulates (*13*) as modified for assessment of substances instead of infections leads to similar conclusions. That is, 1) That exposure to atrazine is always associated with effects in amphibians is not satisfied as there are no consistent or robust effects observed in the field. 2) That the substance must be isolated from the environment of the affected organism is not satisfied because of the background incidence of many of these effects and their occurrence prior to the introduction and use of atrazine. 3) That the substance causes the responses experimentally is not fully satisfied by the inconsistent laboratory and field observations in controlled exposure situations.

Understanding of mechanisms of action of endocrine disruptors reduces uncertainties and may offer useful biomarkers for further study. These biomarkers can be useful in risk assessment and for confirming responses under field conditions. Thusfar, laboratory responses are inconsistent and, where some of these potential biomarkers have been measured in the field, no trends have been apparent. The lines of evidence thus does not support a causal link between atrazine exposures and possible endocrine-mediated effects in amphibians. The lack of consistent biomarkers makes it impossible to conduct a risk assessment as no hazard has been identified.

A number of uncertainties have been identified. These have been discussed in more detail and expanded upon by the SAP (5). These relate to laboratory assays and method studies on the influence of husbandry and source of animals on responses to atrazine as well as positive and negative controls and the effects of seasonal and circadian cycling on hormonal and other responses in frogs. Several uncertainties exist with respect to knowledge of basic biology and stages of development of amphibians. Efforts should be made to standardize tests so that gonadal anomalies such as gross morphology and the presence of testicular oocytes are assessed at similar stages of gonadal development. None of the laboratory studies performed to date have controlled for stage of gonadal development at sampling. Samples were collected at the same point in larval development (at complete metamorphosis), but different species and different populations of frogs vary in the timing of gonadal development relative to somatic development. Some of the anomalies observed may be "normal" and their type and incidence may change with stage of gonadal development and the time of observation.

A significant uncertainty exists in our basic understanding of reproductive and endocrine responses in frogs under natural conditions that are free of anthropogenic contaminants as well as the relevance of any effects caused by known endocrine-active substances on reproductive success of frogs. Uncertainty also exists with regard to temporal variability of exposure of amphibians to atrazine under field conditions and, although the bioconcentration of atrazine is known to be small, uptake and depuration kinetics as well as tissue and organ distribution of atrazine in larval and adult frogs has not been studied.

If and when consistent and robust responses to atrazine are identified in frogs, the mechanisms by which these effects are mediated should be elucidated. This will better allow the application of biomarkers to assess the significance of responses in the field situations and increase the precision and accuracy of extrapolation of effects to other amphibians and assessment of their relevance at the level of the population.

Acknowledgements

The authors were members of the Atrazine Endocrine Ecological Risk Assessment Panel which was convened by Ecorisk Inc. We wish to acknowledge Ecorisk Inc. and Syngenta Crop Protection, Inc. for their support of this research

References

1. Solomon, K. R.; Baker, D. B.; Richards, P.; Dixon, K. R.; Klaine, S. J.; La Point, T. W.; Kendall, R. J.; Giddings, J. M.; Giesy, J. P.; Hall, L. W. J.; Weisskopf, C.; Williams, M. *Environ. Toxicol. Chem.* **1996**, *15*, 31-76.
2. Giddings, J. M.; Anderson, T. A.; Hall, L. W. J.; Kendall, R. J.; Richards, R. P.; Solomon, K. R.; Williams, W. M. "Aquatic Ecological Risk Assessment of Atrazine - A Tiered Probabilistic Approach: A Report of An Expert Panel," Novartis Crop Protection, Inc., **2000**.
3. Kendall, R. J.; Dickerson, R.; Giesy, J. P.; Suk, W., Eds. *Principles and Processes for Evaluating Endocrine Disruption in Wildlife*; SETAC Press: Pensacola, FL, **1998**.
4. USEPA; U.S. Environmental Protection Agency: Washington, DC, USA, **2003**; p 95.
5. USEPA; U.S. Environmental Protection Agency: Washington, DC, USA, **2003**; p 34.
6. Allran, J. W.; Karasov, W. H. *Environ. Toxicol. Chem.* **2000**, *19*, 2850-2855.
7. Allran, J. W.; Karasov, W. H. *Environ. Toxicol. Chem.* **2001**, *20*, 769-775.
8. Diana, S. G.; Resetarits, W. J., Jr.; Schaeffer, D. J.; Beckman, K. B.; Beasley, V. R. *Environ. Toxicol. Chem.* **2000**, *19*, 2961-2967.
9. USEPA; U.S. Environmental Protection Agency, Office of Pesticide Programs, Environmental Fate and Effects Division, U.S. EPA, Washington, D.C., **2001**; Vol. 2003.
10. USEPA; U.S. Environmental Protection Agency: Washington, DC, USA, **2000**; p 7.
11. Atz, J. W. In *Intersexuality in Vertebrates Including Man*; Armstrong, C. N., Marshall, A. J., Eds.; Academic Press: New York, NY, USA, **1964**; pp 145-232.
12. Witschi, E. *J. Exp. Zool.* **1929**, *543*, 157-122.
13. Koch, R. In *Source book of medical history*; Clark, D. H., Ed.; Dover Publications, Inc.: New York, NY, USA, **1942**; p 392–406.
14. Hill, A. B. *Proc. R. Soc. Med.* **1965**, *58*, 295-300.
15. IPCS "Global Assessment of the State-of-the-Science of Endocrine Disruptors," International Programme on Chemical Safety of the World Health Organization, **2002**.
16. USEPA; U.S. Environmental Protection Agency: Washington, DC, USA, **2001**; p 117.
17. Carr, J. A.; Gentles, A.; Smith, E. E.; Goleman, W. L.; Urquidi, L. J.; Thuett, K.; Kendall, R. J.; Giesy, J. P.; Gross, T. S.; Solomon, K. R.; Van Der Kraak, G. J. *Environ. Toxicol. Chem.* **2003**, *22*, 396-405.

18. Hecker, M.; Giesy, J. P.; Jones, P. D.; Jooste, A. M.; Carr, J. A.; Solomon, K. R.; Smith, E. E.; Van der Kraak, G. J.; Kendall, R. J.; Du Preez, L. H. *Environ. Toxicol. Chem.* **2004**, *23*, In press.
19. Coady, K. K.; Murphy, M. B.; Villeneuve, D. L.; Hecker, M.; Jones, P. D.; Carr, J. A.; Solomon, K. R.; Smith, E. E.; Van Der Kraak, G. J.; Kendall, R. J.; Giesy, J. P. *Environ. Toxicol. Chem.* **2004**, *23*, In Press.
20. Sepulveda, M. S.; Gross, T. S. "Characterization of atrazine exposures and potential effects in Florida ecosystems dominated by sugarcane agriculture: A reconnaissance survey of amphibians in South Florida for the assessment of potential atrazine effects," Ecotoxicology Laboratory, University of Florida, **2003**.
21. Sanderson, J. T.; Seinen, W.; Giesy, J. P.; van den Berg, M. *Toxicol. Sci.* **2000**, *54*, 121-127.
22. Sanderson, J. T.; Letcher, R. J.; Heneweer, M.; Giesy, J. P.; van den Berg, M. *Environ. Hlth. Perspect.* **2001**, *109*, 1027-1031.
23. Hayes, T. B.; Collins, A.; Mendoza, M.; Noriega, N.; Stuart, A. A.; Vonk, A. *Proc. Nat. Acad. Sci. U.S.* **2002**, *99*, 5476-5480.
24. Kniewald, J.; Osredecki, V.; Gojmarac, T.; Zechner, V.; Kniewald, Z. *J. Appl. Toxicol.* **1995**, *15*, 215-218.
25. Danzo, B. J. *Environ. Hlth. Perspect.* **1997**, *105*, 294-301.
26. Tobias, M. L.; Marin, M. L.; Kelley, D., B *J. Neurosci.* **1993**, *13*, 324-333.
27. Smith, E. E.; du Preez, L. H.; Gentles, B. A.; Solomon, K. R.; Tandler, B.; Carr, J. A.; Van Der Kraak, G. J.; Kendall, R. J.; Giesy, J. P.; Gross, T. S. *J. Herpetol.* **2004**, Submitted.
28. Coady, K. K.; Murphy, M. B.; Villeneuve, D. L.; Hecker, M.; Jones, P. D.; Carr, J. A.; Solomon, K. R.; Smith, E. E.; Van Der Kraak, G. J.; Kendall, R. J.; Giesy, J. P. *J. Toxicol. Environ. Hlth. A.* **2004**, *67*, In Press.
29. Larson, D. L.; McDonald, S.; Fivizzani, A. J.; Newton, W. E.; Hamilton, S. *J. Physiol. Zool.* **1998**, *71*, 671-679.
30. Hayes, T. B.; Haston, K.; Tsui, M.; Hoang, A.; Haeffele, C.; Vonk, A. *Environ. Hlth. Perspect.* **2003**, *111*, 568-575.
31. Tavera-Mendoza, L. E. In *Department of Biology*; Concordia University: Montreal, PQ, Canada, **2001**; p 73.
32. Tavera-Mendoza, L.; Ruby, S.; Brousseau, P.; Fourier, M.; Cyr, D.; Marcogliese, D. *Environ. Toxicol. Chem.* **2002**, *21*, 1264-1267.
33. Tavera-Mendoza, L.; Ruby, S.; Brousseau, P.; Fourier, M.; Cyr, D.; Marcogliese, D. *Environ. Toxicol. Chem.* **2002**, *21*, 527-531.
34. Du Preez, L. H.; Jooste, A. M.; Solomon, K. R. "Exposure of *Xenopus laevis* larvae to different concentrations of atrazine in semi-natural microcosms.," School of Environmental Sciences and Development, Potchefstroom University for CHE, **2003**.

35. Reeder, A. L.; Foley, G. L.; Nichols, D. K.; Hansen, L. G.; Wikoff, B.; Faeh, S.; Eisold, J.; Wheeler, M. B.; Warner, R.; Murphy, J. E.; Beasley, V. R. *Environ. Hlth. Perspect.* **1998**, *106*, 261-266.
36. Beasley, V. R.; Reeder, A. L.; Pessier, A. P.; Post, M. A.; Beckmen, K. B.; Kunkle, K. E.; Fick, S. T.; Pikas, B. B. In *SETAC Annual Meeting*; SETAC, Pensacola, FL, USA: Baltimore, MD, USA, **2001**; pp Abstract 353, p 376.
37. Du Preez, L. H.; Solomon, K. R.; Carr, J. A.; Giesy, J. P.; Gross, T. S.; Kendall, R. J.; Smith, E. E.; Van Der Kraak, G. J.; Weldon, C. *J. Herpetol.* **2004**, *Submitted*.
38. Lawler, S. P.; Dritz, D.; Strange, T.; Holyoak, M. *Conserv. Biol.* **1999**, *13*, 613-622.

Chapter 13

Dietary Risk Assessment of the Organophosphate Insecticide/Acaricide Methamidophos

Derek W. Gammon, Wesley C. Carr, Jr., and Keith F. Pfeifer

Medical Toxicology Branch, Department of Pesticide Regulation, Cal/EPA, Sacramento, CA 95812

A toxicological profile of methamidophos was integrated into two computer models, (TAS® and DEEM®) to calculate dietary margins-of-safety (MOSs) for cotton, potato and tomato. No Observed Effect Level(s) (NOELs) of 0.3 mg/kg/day (rat, acute) and 0.02 mg/kg/day (dog, estimated chronic), for brain AChE inhibition, were used. Crop residue data were employed to estimate exposure/MOSs for 20 population sub-groups. Deterministic (point estimate) and probabilistic (Monte Carlo) simulations compared acute exposure (95th percentile) using two consumption databases. Chronic exposures/MOSs were calculated using annualized means. Acute and chronic MOSs were >100 for all population sub-groups, which is generally considered adequate. For residues at tolerance, acute MOSs were above 100 for cotton and potato, but only 34 to 94 (DEEM®/TAS®) for tomato.

Note: The opinions expressed are the authors' and do not necessarily reflect the policies of Cal/EPA.

Methamidophos was developed in the mid-1960s, controlling insects and mites in a range of vegetable crops. It was also used on non-food crops such as alfalfa, clover, Bermuda grass, cut/outdoor flowers and in greenhouses. It has high acute toxicity (USEPA Category 1), with LD_{50} values ~15 mg/kg (rat, oral) and ~100 mg/kg (rabbit, dermal). It acts by inhibiting acetylcholinesterase (AChE), in insects and mammals. In 1996, following concerns about occupational safety, all (13) registrations except cotton, potato and tomato were voluntarily canceled by Bayer, the principal registrant. In California, methamidophos usage fell from 500,000 (1995) to 47,000 lbs. (2001). In 2002, an IRED (Interim Reregistration Eligibility Document) *(1)* and a Risk Management Decision Document *(2)* were issued for methamidophos by USEPA. The main conclusions were: 1/ dietary exposure to methamidophos did not require mitigation, 2/ cotton uses were to be discontinued by 2007, with three mitigation issues: a) monitoring data were to be collected to address possible surface water contamination (as calculated using PRZM-EXAMS and GENEEC models); b) engineering controls were needed to reduce worker exposure; c) ecological risks to birds and mammals were excessive but would be mitigated by removing cotton uses. Further, this IRED was "interim" pending the evaluation of acephate, an organophosphate (OP) that is toxic only after conversion to methamidophos, and other insecticides, under a "cumulative" risk assessment, as required by the Food Quality Protection Act (FQPA) *(3)*. Because such a complete database is available for methamidophos, USEPA has chosen it as the "Index Chemical" for the relative potency factor (RPF) approach that it proposes to adopt in its cumulative OP risk assessment *(4)*. Paradoxically, under the terms of the IRED, tolerances for cottonseed were increased from 0.1 ppm to 0.2 ppm, for tomato from 1 ppm to 2 ppm and potato, to remain at 0.1 ppm.

Risk assessment for a chemical is a function of toxicity and exposure. As such, toxicity is determined in a series of studies and from these, no observed effect levels (NOELs) are chosen to represent acute and chronic toxic effects. The key studies are described in detail. Dietary exposure, in turn, is a function of two parameters, *i.e.* the pesticide residue in the treated crop and the amount of treated food consumed by a person. Computer models have been developed to estimate dietary exposure to pesticides and two of these are compared, TAS® *(5)* and DEEM® *(6)*. Both programs divide the US population into groups, based principally on age, gender and ethnicity. They make use of USDA surveys known as CSFII – Continuing Surveys of Food Intake by Individuals. The two surveys used here cover the years 1989-1992 and 1994-1996, 1998. The later CSFII was conducted around the time that FQPA was passed (1996) and is biased towards collecting more data on infants and children. The programs model dietary exposure in two ways. First, a deterministic (point estimate) model was used, which assumes that the crop contains a specific amount of pesticide contamination. This value is based on crop residue studies and is generally chosen to be at the 95th percentile of the available residue data.

Secondly, a probabilistic (Monte Carlo) simulation was used. This considers a complete set of crop residue data and samples data points randomly. This approach is generally considered to be more "realistic" for assessing dietary exposure to a pesticide. Tolerance assessments used TAS® and DEEM® in order to establish the safety of tolerances (MRLs or maximum residue limits) for cotton, potato and tomato and also to compare the output of the two programs.

Environmental Fate

Methamidophos has low persistence in soil ($t_{1/2}$ values of $\leq$ 4 days). It is highly hydrophilic ($\log K_{ow}$ = -0.66) *(7)*, is weakly adsorbed by soil with a high potential to leach but, because of its rapid breakdown, it is unlikely to leach in practice. In California, there have been no detections of methamidophos in groundwater monitoring studies. It has a high vapor pressure, but because of its high water solubility, it has a low Henry's Law constant, thus giving it a low tendency to volatilize under field conditions. The results of crop residue studies are summarized in Table I.

Table I. Summary of Methamidophos Crop Residue Data.

Raw Agricultural Commodity (RAC)	Tolerance, ppm	Residue acute, ppm[a]	Residue chronic, ppm	%-Crop treated (%CT)
Cottonseed, meal[b]	0.1 (N)[e,f]	0.044 (n=32)	0.042(N)	15%
Cottonseed, oil[b]	0.1 (N)[f]	0.01 (n=4)	0.005	15%
Potato[c]	0.1 (N)	0.0091 (n=1401)	0.0019	30%
Tomato[d]	1[g]	0.082 (n=849)	0.013	20% (processed) 85% (fresh)

a/ Pre-harvest Interval: 7 days, tomato, 14 days potato and 50 days, cottonseed; (n=number of composite samples analyzed)
b/ from Bayer, the primary registrants; LOQ = 0.01 ppm
c/ from USDA - PDP program, 1994-5; Limit of detection (LOD) = 0.003 ppm
d/ from USDA - PDP program, 1996-7; Limit of detection (LOD) = 0.001 ppm
e/ negligible residue
f/ tolerance increased to 0.2 ppm in *(1)*.
g/ tolerance increased to 2 ppm in *(1)*.

Toxicology Profile

Chronic Toxicity and Oncogenicity

Summary: Methamidophos given in the diet for periods of one or two years, resulted in body weight loss in the rat and mouse, but not the dog. There was an increase in relative brain weight at the highest dose tested (HDT) in these rodents, in both sexes, but not in the dog. There were few clinical signs and no evidence of oncogenicity. In all three species, inhibition of AChE in plasma (which also includes butyryl cholinesterase), red blood cell (RBC) and brain was reported. The LOEL for inhibition of brain AChE was 2 ppm, equivalent to 0.1, 0.3 or 0.06 mg/kg/day in the rat, mouse and dog, respectively. An estimated NOEL of 0.02 mg/kg/day, after 11-18% inhibition of brain AChE at 0.06 mg/kg/d in a 1-yr. dog study, was the critical one selected for chronic risk characterization. The use of an uncertainty factor of 3 (rather than 10) in calculating an estimated NOEL from a LOEL was supported by Benchmark Dose (BMD) calculations.

In the critical study, methamidophos was fed to beagle dogs (6/sex/level) at 0, 2, 8, or 32 ppm (0, 0.06, 0.24 or 0.90 mg/kg/day, M and 0, 0.06, 0.22 or 0.88 mg/kg/day, F) for 1-yr *(8)*. Mean body weight was not significantly affected by treatment. There was increased lacrimation at 8 and 32 ppm, significant ($p<0.05$) only with combined sexes (Table II). The signs of salivation, diarrhea and vomiting were also increased in incidence (n.s.). There were no significant changes in absolute or relative organ weights, in either sex, or abnormal histopathology. Plasma, erythrocyte and brain AChE were inhibited at all doses. The LOEL for inhibition of brain AChE in both sexes was thus 2 ppm, (or 0.06 mg/kg/day). The chronic NOEL was estimated by dividing the LOEL by 3 *i.e.* 0.02 mg/kg/day. Support for the use of an uncertainty factor of 3 rather than 10, which is often used in estimating the NOEL from the LOEL, was provided by the calculation of Benchmark Doses (BMDs). The mean and S.D. for brain AChE inhibition were analyzed using BMD software developed for USEPA (2002) by the National Centre for Environmental Assessment. Four programs were run (Hill, Polynomial, Power and Linear models) and the BMD and BMDL (95% lower confidence limit of the BMD) were determined by each one, based on 10% enzyme inhibition. For male dogs, BMDLs were 0.025, 0.033, 0.11 and 0.11 mg/kg/day, for the four models, respectively. From the AIC (Akaike Information Criteria) values and an inspection of the dose/response curves, the Hill model provides the best analysis of the data, thus giving a BMDL of 0.025 mg/kg/day. For female dogs, the BMDLs were 0.035, 0.037, 0.11 and 0.11, respectively. The lowest AIC value of -8.0 was again associated with the Hill plot. It is thus concluded that using 0.02 mg/kg/day as an estimated NOEL from this study should be relatively close to the "true" NOEL.

Table II. Mean Inhibition of AChE by Methamidophos at (1-mon. and) 1-yr. and Clinical Signs in the Dog [a,b/]

AChE Assay		0	2	8.	32 ppm
Blood plasma	M	---	(5.6%) 20%	(26%**) 35%***	(44%***) 56%***
	F	---	(2.8%) 12%	(13%) 7%	(31%**) 39%***
RBC	M	---	(7.9%) 10%	(56%***) 62%***	(73%***) 81%***
	F	---	(16%) 11%	(47%**) 56%***	(74%***) 83%***
Brain	M	---	18%***	55%***	71%***
	F	---	11%*	45%***	66%***
Clinical sign					
Lacrimation	M	1/6	3/6	3/6	4/6
	F	1/6	1/6	4/6	3/6
	M+F	2/12	4/12	7/12*	7/12*
Salivation	M	0/6	0/6	1/6	3/6
	F	0/6	0/6	0/6	0/6
	M+F	0/12	0/12	1/12	3/12
Diarrhea	M	1/6	1/6	2/6	2/6
	F	0/6	0/6	0/6	3/6
	M+F	1/12	1/12	2/12	5/12
Vomiting	M	1/6	2/6	2/6	4/6
	F	1/6	0/6	1/6	1/6
	M+F	2/12	2/12	3/12	5/12

a/   data from (8).
b/   n = 6 dogs/sex/dose/time point
*   p<0.05 (Student's t test, AChE), Fisher's exact, clinical signs
**   p<0.01 (Student's t test, AChE)
***   p<0.001 (Student's t test, AChE).

Genotoxicity

Genotoxic effects were examined in both mammalian and microbial cells. The assays included: *Salmonella* and CHO/HGPRT (Chinese hamster ovary/hypoxathine-guanine phosphoribosyl transferase) gene mutation (± liver microsome S9), *in vitro; in vivo* and *in vitro* chromosome aberration assays, in mouse and CHO cells, and mouse micronucleus and dominant lethal assays, *in vivo;* in DNA repair and UDS (unscheduled DNA synthesis) in *E. coli* and in rat primary hepatocytes. Most of these tests were negative for genotoxicity, with only a few positive results e.g. gene mutation was reported in the CHO/HGPRT assay (+S9), in vitro, but only at the HDT, and there was only one trial. In an acceptable study with replicate assays, methamidophos did not give mutations in this assay. No chromosome aberrations were detected in a CHO cytogenetic assay (±S9), in vitro. It appears, from a consideration of all of the data (1) that methamidophos has a low potential to be genotoxic to humans.

Reproductive Toxicity

In a standard reproductive toxicity study, SD rats dosed at 0, 1, 10 and 30 ppm for two-generations, with two-litters/generation. Reduced body weight in adults and pups was observed, with a NOEL for both of 1 ppm. No reproductive toxicity was observed. At 10 ppm and 30 ppm, there were significant reductions in plasma, RBC and brain AChE for both adults and pups of both sexes in both generations. At 1 ppm, there was no significant inhibition of plasma AChE, in either adults or pups and for RBC AChE, significant inhibition ($p<0.05$) was only evident in adult males, of 21% (F0) and 18% (F1). An estimated NOEL was therefore considered to be 0.03 mg/kg/day for adult RBCAChE inhibition with a pup NOEL, 0.1 mg/kg/d. For brain AChE, significant inhibition ($p<0.05$) was observed only in female adults, of 7% (F0) and 5% (F1); for pups, mean brain AChE activity ranged from 2% stimulation to 3% inhibition, relative to concurrent controls. It was concluded that these low levels of inhibition were probably not toxicologically relevant, thus 1 ppm (0.1 mg.kg/day) was the NOEL for adults and pups for brain AChE inhibition. There was no indication from this experiment (1) that pups were more sensitive than adults to the toxic effects of methamidophos.

Developmental Toxicity

Oral gavage studies were conducted in the rat (n=4) and the rabbit (n=3). There were no fetal malformations in either species. In the rat, reduced maternal and fetal body weight had LOEL/NOEL values of 5.49 and 1.41 mg/kg/day. Maternal clinical signs had the same LOEL/NOEL values. Inhibition of maternal plasma, RBC and brain AChE had NOEL values of <0.05, 0.05 and 0.15 mg/kg/day, respectively. In the rabbit, reduced maternal body weight and clinical signs had LOEL/NOEL values of 2.47/0.65 mg/kg/day. Reduced fetal body weight had LOEL/NOEL values of 7.73/4.90 mg/kg/d. Reduced plasma and RBC AChE in dam had LOEL/NOEL of 0.46/0.2 mg/kg/d. There was thus no indication from these experiments (*1*) that possible adverse effects occurred in the fetus at doses that were not maternally toxic.

Developmental Neurotoxicity

Following FQPA, OPs were required in 1999 to have data from a new type of study, the developmental neurotoxicity test, in order to maintain their registrations in the USA. This study was part of an effort made by USEPA to ensure that neurotoxic pesticides would not present a greater threat to infants/children than to adults. It most closely resembles the multi-generation reproductive toxicity test using the rat. Some of the issues and challenges associated with the planning and execution of this type of study were recently discussed at a Society of Toxicology workshop *(9)*. In the absence of data obtained from such a study, then it can be assumed that an immature human is up to 10-fold more susceptible to the toxic effects of an OP.

Mated female Wistar rats were fed diets containing methamidophos, from Day 0 of gestation to Day 21 of lactation, at 0, 1, 10 or 30 ppm (n=30/dose level) *(10)*. No effects were observed on any parameters associated with developmental or reproductive toxicity or on FOB (Functional Observational Battery) tests in dams (Table III). There was no reduction in body weight in dams. A dose-related delay was observed in preputial separation, 4.6% (p<0.01) at 30 ppm, and a fall in body weight in pups on days 11, 17 and 21 (p<0.01) at 30 ppm, of 8.5 to 12%. Effects on the FOB test in pups were observed: reduced motor activity (M and F), at 10 and 30 ppm, on PND 13 (n.s.). Organ weights were unaffected and there were no histopathological abnormalities. Significant inhibition of plasma and RBC AChE was recorded at 10 and 30 ppm in pups and dams, but not at 1 ppm. The NOEL for pup brain AChE inhibition was 1 ppm, equivalent to 0.1 mg/kg/day, based on 5% to 53% (p<0.05) inhibition at the LOEL of 10 ppm. For dams, 1 ppm was the LOEL, since inhibition of brain AChE was 8% (p<0.05) at this dose. The evidence indicates that developmental neurotoxicity only occurs at doses causing maternal toxicity. However, because of its high hydrophilicity, it would be anticipated that methamidophos would not be present in dams' milk at high concentrations. For example, acephate administered to lactating goats was eliminated almost entirely in urine, with only *ca.* 1% appearing in milk as acephate or methamidophos *(11)*. Moreover, it is also somewhat surprising to see such a high degree of brain AChE inhibition for a compound that would not *a priori* be anticipated to penetrate the blood-brain-barrier.

Table III. Rat Developmental and Maternal Neurotoxicity of Methamidophos (10)

	0	1	10	30 ppm
Mean body wt, g (M+F) [a]				
Dams, Day 20	308±6.2	323±3.5	308±4.8	300±6.0
Pups, Day 0	5.7	5.7	5.7	5.6
Pups, PND 4	8.9	8.9	8.7	8.2
Pups, PND 11'	21.5	21.7	20.8	18.9** (12%)
Pups, PND 17	32.8	33.6	32.3	30.0** (8.5%)
Pups, PND 21	42.8	43.2	42.1	38.6** (9.8%)
Pups, PND 28, M	75.3	76.9	73.0	67.8* (10%)
Pups, PND 28, F	74.9	73.3	71.4* (5%)	67.9* (9%)
Pups, PND 42, M	169	170	167	157* (7%)
Pups, PND 42, F	140	137	134* (4%)	130* (7%)
Pups, PND 56, M	256	253	250	236* (8%)
Pups, PND 56, F	176	172	168* (5%)	164* (7%)
Pups, PND 70, M	312	306	306	288* (7%)
Pups, PND 70, F	197	194	189* (4%)	188* (4%)
Preputial separation, d	45.9	46.1 (0.0%)	46.7 (1.7%)	48.0**(4.6%)
FOB, motor activity, % [b]				
PND 13 (M,F)	N/A	+29,-8	-25,-33	-45,-27
PND 17 (M,F)	N/A	-23,+12	-15,+29	-22,+8
PND 21 (M,F)	N/A	+10,-12	+3,-10	+7,-10
PND 60 (M,F)	N/A	+11,+4	-4,+9	+6,+9
ChE % I (%↓ vs. control)	(n=10)	(n=10)	(n=10)	(n=10)
Dams, Day 21: Plasma	N/A	5	50*	77*
RBC	N/A	12	64*	84*
Brain	N/A	8*	63*	83*
Pups, PND 4 (M+F)	(n=21)	(n=18)	(n=19)	(n=18)
Plasma	N/A	+3	5	12*
RBC	N/A	3	20*	40*
Brain	N/A	0	5	14*
Pups, PND 21 (M/F)	(n=10/10)	(n=10/10)	(n=10/10)	(n=10/8)
Plasma	N/A	+2 / +10	22* / 8	34* / 40*
RBC	N/A	+6 / 4	12 / 16	37* / 53*
Brain	N/A	3 / 2	37* / 53*	34* / 43*

a/ Rats were dosed until 21 days and then given an untreated diet.
b/ Percent motor activity relative to controls.
*/** p<0.05 or p<0.01 (Dunnett's test)

Neurotoxicity

The acute neurotoxicity of methamidophos was studied in oral gavage studies in the rat (n=2) and hen (n=4) plus one dermal toxicity study in the rat. Subchronic neurotoxicity studies were also conducted, lasting from 3 to 13 weeks, in the rat (n=4) and the hen (n=2). The inhibition of AChE (plasma, RBC, brain) was measured, along with NTE (neuropathy target esterase) inhibition in some of them, also. A critical, acute NOEL of 0.3 mg/kg/day, for inhibition of rat AChE (plasma, RBC and brain) and FOB (functional observational battery) effects, after a single dosing with 0.6 mg/kg/day, was used for estimating acute dietary risks. Inhibition at the LOEL was 15 (p<0.05) – 27% (p<0.05) and at the NOEL, was 0 (n.s.) – 24% (n.s.), considering both sexes (Table IV).

Dietary Exposure & Risk Characterization

Acute dietary exposure is summarized in Table V. Using TAS EX-4,® at the 95th percentile of exposure, considering the consumption of tomato, potato and cotton, combined, the estimations ranged from 0.238 to 0.674 µg/kg-d (point estimate), or 0.047 to 0.170 µg/kg-d (Monte Carlo). Using DEEM,® the corresponding estimates were 0.088 to 0.630 µg/kg-d (point estimate), or 0.027 to 0.127 µg/kg-d (Monte Carlo). The sub-populations with the lowest and highest dietary exposure were seniors (55+ yrs.) and nursing infants, respectively, using TAS EX-4® and seniors (55+ yrs.) and children (1 – 6 yrs.) using DEEM.® These population categories were the same, for each software type, whether the point estimate or Monte Carlo method was used.

The corresponding MOS values ranged from 445 – 1262 (point estimate) and 1770 – 6320 (Monte Carlo) using TAS EX-1® and from 464 – 1260 (point estimate) and 2320 – 6230 (Monte Carlo) using DEEM.® A MOS of ≥ 100 is generally considered adequate to protect people from the toxic effects of a pesticide, when the toxicity data used for risk assessment are from animal studies. However, under FQPA, cumulative (other OPs) and aggregate (other sources of exposure) also need to be taken into account, whenever conducting an acute or chronic risk assessment for an OP pesticide.

Selected MOS values are also given in Table V using the earlier (1989-1992) CSFII database. These values (in parenthesis) indicate a decreased MOS for nursing infants, by approximately 50%, and an apparent increase in MOS for non-nursing infants, by a similar amount. There is little difference in MOS values for children (1 - 6 yrs.) between the two CSFII surveys, and for other population sub-groups.

Table IV. Acute, Oral Neurotoxicity of Methamidophos in the Rat (12, 13)

Observation	Dosage (mg/kg/d)			Dosage (mg/kg/d)			
	0	0.3	0.6[a]	0	0.9	3.3	9.0[b]
AChE (%↓)[a] (M)							
Plasma	0	6%	27%*	0	39%*	81%*	91%*
RBC	0	5%	21%*	0	32%*	73%*	92%*
Brain	0	0%	15%*	0	33%*	70%*	82%*
AChE (%↓)[a] (F)							
Plasma	0	24%	25%	0	24%*	67%*	89%*
RBC	0	8%	26%*	0	33%*	68%*	86%*
Brain	0	6%	26%*	0	29%*	72%*	84%*
FOB effects (M)							
landing footsplay[b]							
Day -1 (mm)	81	69	84	--	--	--	--
Day 0	75	63	92* (23%↑)	83	89	92	--[c]
Day 7	77	68	84	85	80	92	92
Day 14	78	72	86	84	83	88	90
FOB effects (F)							
landing footsplay[b]							
Day -1 (mm)	59	61	67	--	--	--	--
Day 0	60	64	61	82	92	87	--[c]
Day 7	62	63	65	84	88	86	90
Day 14	63	62	63	78	77	79	85
Clin. signs (M)							
muscle fascicul.	--[d]	--[d]	--[d]	0/24	0/24	7/24**	12/24***
urine stained fur				0/24	1/24 [e]	9/24***	12/24***
oral stain				0/24	1/24 [e]	3/24	8/24**
nasal stain				0/24	1/24 [e]	10/24***	6/24*
Clin. signs (F)							
muscle fascicul.	--[d]	--[d]	--[d]	0/24	0/24	0/24	7/24**
urine stained fur				0/24	0/24	5/24*	12/24***
oral stain				0/24	0/24	0/24	7/24**
nasal stain				0/24	0/24	2/24	6/24*

a/ mean % inhibition vs. control, measured 2 h after dosing (n=6/sex/dose).
b/ measured 2 h after dosing (n=12/sex/dose).
c/ rats were showing cholinergic signs, unable to complete the (FOB) test
d/ none of rats in Ref. 12 exhibited any clinical signs (n=18/sex/dose).
e/ the same rat showed all of these clinical signs.
* significantly different from control at p<0.05 (Dunnett's test, raw AChE data).
* p<0.05; ** p<0.01; *** p<0.001 (Fisher's exact test for clinical signs and FOB).

Table V. MOSs for Potential Acute Dietary Exposure to Methamidophos[a]

Population subgroup	MOS for Acute Exposure [b,c]			
	$TAS^{® \ d}$	$TAS^{® \ e}$	$DEEM^{® \ d}$	$DEEM^{® \ e}$
US Pop. all seasons	908	4430	928	4500
Western Region	888	4380	844 (962)	4220 (4820)
Pacific Region	921	4470	--------	-------
Hispanics	698	3670	762	3600
Non-Hispanic Whites	926	4430	955	4620
Non-Hispanic Blacks	1020	5070	934	4930
Non-Hispanic Other	850	4110	850 (864)	4210 (4380)
All Infants	471	3180	522	3430
Infants (nursing)	**445**[f]	**1770**[f]	818 (432)	5070 (1930)
Infants (non-nursing)	622	3530	496 (723)	3250 (4440)
Children (1-6 yrs)	485	2450	**464**[f] (523)	**2320**[f] (2670)
Children (7-12 yrs)	660	3040	734	3480
Females, 13-19 yrs, NN, NP	853	4390	1000	4770
Females, 13+ yrs, P, NN	1133	5660	1030	4910
Females (13+ yrs, N)	1022	5040	976	5380
Females, 20+ yrs, NP, NN	1172	5760	1210	5830
Females (13-50 yrs.)	1085	5270	1150 (1160)	5430 (5750)
Males (13-19 yrs)	968	4220	871	3970
Males (20+ yrs.)	1046	5090	1070	5120
Seniors (55+ yrs)	**1262**[f]	**6320**[f]	**1260**[f]	**6230**[f]

Note: NN= non-nursing, NP = non-pregnant, P = pregnant, and N = nursing.
a/ Residues on the raw agricultural commodities (RACs) tomato, potato, and cottonseed at the 95th percentile of exposure.
b/ Calculations of exposure are based on % user-days, *i.e.*, consider only consumers of RACs in a dietary survey. A *per capita* user-day considers both consumers and non-consumers of the RACs on any particular day.
c/ MOS = NOEL / Acute Dietary intake (Pesticide residue x commodity consumption) NOEL of 0.3 mg/kg/day based on brain and plasma AChE inhibition and FOB effects at 0.6 mg/kg/day in a rat neurotoxicity study (*12*).
d/ TAS EX-4® (1989-1992 CSFII survey) or DEEM® (1994-1996, 1998 CSFII) point estimate of dietary exposure. For DEEM® 1989-1992, CSFII, in parenthesis.
e/ TAS EX-4® (1989-1992 CSFII) or DEEM® (1994-1996, 1998 CSFII) Monte Carlo estimate of dietary exposure. For DEEM® 1989-1992, CSFII, in parenthesis.
f/ lowest and highest values are in bold type.

A chronic, annual dietary exposure assessment used mean commodity residue levels combined with annualized consumption of tomato, potato and cotton. The estimated dietary exposures ranged from 0.001 to 0.012 μg/kg-d, (TAS EX-1®) or 0.001 to 0.013 μg/kg-d, (DEEM.®). The population sub-groups with the lowest and highest dietary exposure were nursing infants and children (1 – 6 yrs.), respectively, using TAS EX-1® and DEEM.® The respective MOS ranges were 1650 to 15,500 (TAS EX-1®) and 1550 to 16,700 (DEEM.®). These values are remarkably similar, especially considering that TAS EX-1® used the 1989-1992 CSFII database and DEEM,® the 1994-1996,1998 one. It should be noted that DPR uses percent crop-treated information when conducting a chronic dietary risk assessment for a pesticide, though not generally for an acute one.

Tolerance Assessment

An acute tolerance (MRL) assessment is routinely conducted by DPR to ensure that this level of residue will not result in adverse effects in consumers of treated commodities. It is the maximum amount of a pesticide residue that is legally allowed on that commodity *(14)*. By definition, the use of tolerance residue levels means that a point estimate, deterministic approach was used. DPR considers that acute, but not chronic, tolerance assessment is appropriate because it is highly improbable that an individual would consume a commodity with tolerance level residues on a chronic, annual basis. Similarly, it is considered by DPR to be inappropriate to use a percent crop-treated (%CT) adjustment on the tolerance assessment, whenever using a deterministic approach (Table VI). These tolerance assessments determined the MOS values at the 95th percentile of exposure.

For cotton, the exposure ranged from 0.016 – 0.052 μg/kg/day (TAS EX-4®) and 0.005-0.0.29 μg/kg/day (DEEM®). The sub-populations with the lowest and highest dietary exposure were seniors (55+ yrs.) and non-nursing infants (<1 yr.), respectively, using TAS EX-4® and seniors (55+ yrs.) and children (1 – 6 yrs.) using DEEM.® For potato, the exposure ranged from 0.340 - 0.892 μg/kg/day (TAS EX-4®) and 0.336 - 0.994 μg/kg/day (DEEM®). The sub-populations with the lowest and highest dietary exposure were females (13+ yrs.), pregnant, non-nursing and children (1 – 6 yrs.), respectively, using both TAS EX-4® and DEEM.® For tomato, the exposure ranged from 3.426 – 8.774 μg/kg/day (TAS EX-4®) and 3.189 - 8.620 μg/kg/day (DEEM®). The sub-populations with the lowest and highest dietary exposure were nursing infants (<1 yr.) and children (1 – 6 yrs.), respectively, using TAS EX-4® and seniors (55+ yrs.) and all infants using DEEM.®

The corresponding MOS values ranged from 10,300 – 56,900 (cottonseed), 301 – 892 (potato) and 34 – 94 (tomato) using TAS EX-4® and from 5,800 – 19,100 (cottonseed), 340 – 880 (potato) and 34 – 88 (tomato) using DEEM.®

Table VI. Margins of Safety (MOS) after Consumption of Commodites Registered for Methamidophos with Residues at U.S. EPA Tolerances[a/]

Population subgroup	(Acute) MOS at the 95th Percentile [b/]		
	Cottonseed[c/]	Potato[f/]	Tomato[g/h/]
US Pop. all seasons	$(30,600)^{d/}$ 11,300[c/]	$(597)^{d/}$ 580[c/]	$(65)^{d/}$ 63[c/]
Western Region	(25,000) 13,200	(571) 550	(62) 63
Pacific Region	14,000[c/]	560[c/]	65[c/]
Hispanics	(25,500) 14,700	(521) 600	(61) 50
Non-Hispanic Whites	(32,800) 11,300	(621) 600	(66) 65
Non-Hispanic Blacks	(26,700) 8,800	(532) 450	(65) 57
Non-Hispanic Other	(13,600) 12,700	(597) 470	(57) 61
All Infants	(10,400) 8,800	(406) 410	(37) **34**[1/]
Infants (nursing, <1 yr.)	(24,200) 9,100	(640) 540	**(94)**[2/] 37
Infants (non-nurs., <1 yr.)	**(10,300)**[1/] 9,100	(337) 400	(37) 46
Children (1-6 yr)	(12,500) **5,800**[1/]	**(301)**[1/] 340[1/]	**(34)**[1/] 35
Children (7-12 yr)	(21,100) 6,800	(456) 385	(54) 48
Females, 13-19 yr,NP, NN	(36,200) 11,400	(682) 620	(77) 62
Females, 13+ yr, P, NN	(51,300) 14,500	**(892)**[2/] **880**[2/]	(79) 78
Females, 13+ yr, N	(21,500) 10,300	(676) 600	(78) 79
Females, 20+ yr, NP, NN	(50,000) 16,600	(825) 780	(86) 82
Females (13-50 yr)	(44,000) 14,300	(768) 740	(82) 75
Males (13-19 yr)	(26,200) 10,300	(506) 400	(66) 72
Males (20+ yr)	(47,900) 13,700	(705) 700	(77) 76
Seniors (55+ yr)	**(56,900)**[2/] **19,100**[2/]	(828) 760	(84) **88**[2/]

Note: NN= non-nursing, NP = non-pregnant, P = pregnant, and N = nursing.
a/ Tolerances are: 0.1 ppm (cottonseed, potato), 1 ppm (tomato).
b/ MOS = NOEL / Acute Dietary intake
    NOEL of 0.3 mg/kg/day based on brain and plasma AChE inhibition and FOB effects in a rat neurotoxicity study (*12*).
c/ determined using the TAS EX-4® program.
d/ determined using the DEEM® program.
e/ highest and lowest MOS in bold.
h/ at the 99.9^{th} percentile (DEEM®), MOS ranged from 8 (females, 13-19 yr) to 50 (females, 13+nursing), based on exposures of 34.45 to 5.89 μg/kg/day, respectively.

Although chronic tolerance assessments are considered invalid by DPR, nonetheless USEPA conducts such investigations. Therefore, chronic tolerance evaluations were conducted for tomato for comparative purposes. Using the annualized mean dietary exposure (DEEM®), and the 1-yr. dog NOEL of 0.02 mg/kg/day, MOS values of 11 (children 1-6 yr) to 136 (nursing infants) were obtained. These corresponded to consumption of 0.15 – 1.80 µg/kg/day. All population groups except nursing infants had MOS values below 100 from chronic dietary exposure to tolerance level residues of methamidophos on tomato.

Discussion

Risk assessment is the process that is used to evaluate the potential for exposure and the likelihood that the toxic effects of a substance will occur under specific exposure conditions. There are limitations and uncertainties in estimating the potential risk to human health. The degree or magnitude of the uncertainty varies depending on the availability and quality of the data and the exposure scenarios being assessed (15). Specific areas of uncertainty associated with this risk assessment for methamidophos are described, as follows.

Hazard Identification

Acute toxicity tests measure the effects of a chemical after a single or brief period of exposure. Developmental toxicity tests, which are often utilized for acute risk assessment, did not show increased susceptibility of the developing organism to the effects of methamidophos relative to the dam. Maternal toxicity can also be used for acute risk assessment, for example, in cases where rapid body weight loss occurs during developmental toxicity tests. However, AChE inhibition in dams, which was determined at the study conclusion, is not considered an acute effect since it could have resulted from several days' dosings. Instead, the most sensitive acute endpoints were observed when methamidophos was evaluated in two acute neurotoxicity studies in the SD rat which, considered together, gave LOEL and NOEL values of 0.6 and 0.3 mg/kg/day, respectively. These values probably represent a fairly accurate determination of toxicity because they were based on both AChE inhibition (plasma, RBC and brain) and a FOB (behavioral) effect, (although the latter effect was only found in males). This helps to remove a lot of the uncertainty associated with the choice of a particular form of AChE for consideration as the endpoint for LOEL/NOEL determination(s). It should also be noted that AChE inhibition at the LOEL was ≥15%, was statistically significant and occurred in both sexes. However, it is unclear what level of inhibition of AChE should be considered adverse (see below). The rat reproductive toxicity and developmental neurotoxicity studies did not indicate greater sensitivity of the pups than adults to methamidophos. In the latter study, it was considered by both the registrants and the DPR reviewer, that 8% inhibition (p<0.05) of brain AChE in dams was toxicologically significant at the LDT (lowest dose tested) of 1 ppm. However, in pups at 1 ppm, brain AChE was not inhibited at PND4, and by only 3% (M) and 2% (F) at PND21 (n.s.).

In the evaluation of chronic toxicity, the most sensitive endpoint in the rat, mouse and dog, was also the inhibition of AChE. A statistically significant level of inhibition was recorded in all species at the LDT of 2 ppm in the diet. Therefore, an estimated NOEL (from the dog study), equivalent to 0.02 mg/kg/day was calculated, by dividing the LOEL by a default uncertainty factor (UF) of 3. This estimated NOEL was used for chronic risk characterization. The use of a default UF of 3 in estimating a NOEL from a LOEL is supported by BMD calculations. The lower bound on the dose giving a 10% reduction of brain AChE activity was just above 0.02, regardless of the program used, indicating that this value would be close to the "true" chronic NOEL. It should also be noted that there is currently much uncertainty concerning the most appropriate degree of inhibition of AChE to use for regulatory purposes. For example, it has been suggested that inhibition of (brain) AChE of 20% should be used to establish the LOEL/NOEL *(16)*. It should also be noted that, owing to its high hydrophilicity (log K_{ow} of -0.66), methamidophos would be anticipated to penetrate the blood brain barrier poorly, in the absence of a carrier-mediated specific uptake mechanism. Therefore, the brain AChE inhibition that was measured may have been predominantly localized in glial cells rather than having been synaptic. The presence of such extra-synaptic glial AChE has been demonstrated in the insect CNS using cytochemical methods, including electron microscopy *(17)*. The possible toxicological significance of the inhibition of brain, glial AChE is presently unclear, although it would not be anticipated to result in clinical signs to the same extent as would the inhibition of synaptic AChE. However, it may explain, at least in part, the relatively high inhibition of brain AChE that was reported in chronic and sub-chronic dietary studies in rodents and dogs, often as much as 90%, with no or only minimal clinical signs.

There is evidence that organophosphates may play a role in disrupting glial cell growth. It has been demonstrated that chlorpyrifos, chlorpyrifos-oxon and diazinon inhibit DNA synthesis in nerve cells, *in vitro (18)*. The effects were more pronounced in glial cells (C6) than in neuronal cells (C12). The disruption of DNA synthesis could, in turn, result in altered glial cell structure, with consequent changes in the properties of the blood-brain-barrier. However, counting against this theory to explain methamidophos' action is the finding that the parent organophosphates were more potent than their oxon metabolites, the form of methamidophos. It is also possible that methamidophos interacts with P-Glycoprotein transporters (PGTs) in the rat brain, also known as the multi-drug-resistance or MDR-protein mechanism. Such PGTs remove a variety of molecules from the brain, including the carbamate insecticide thiodicarb *(19)*. Disruption of the PGT system could result in higher levels of methamidophos in the brain being maintained than in the presence of an intact PGT system. However, in rodents such as the SD rat, the PGT system does not develop until the animal is 3 wks. of age *(20)*. The absence of the PGT system, which would be expected to be equivalent to its disruption, would therefore lead to greater levels of methamidophos in brain (for AChE inhibition) in neonatal rats than in adults. This is the opposite of the effects described in the DNT and reproductive toxicity studies, where the pups' brains appear to be less susceptible than adults,' thus diminishing the likelihood of a PGT explanation.

However, it should be mentioned that pre-weaning rats, receiving all of their methamidophos via the dams' milk, would be anticipated to receive very little of the organophosphate. This is because a compound with such high aqueous solubility, coupled with such low lipid solubility, would not be anticipated to partition into milk at very high concentrations. Perhaps this issue could be resolved by sampling rat dams' milk and measuring the concentration of methamidophos. Alternatively, measuring the pesticide in dams' and pups' blood and/or urine would also give a good indication of the actual level of exposure of pups relative to dams. Such measurements might enable a more precise estimate to emerge of pups' exposure and thus of their innate sensitivity relative to adults. Dosing pups *via* oral gavage could become necessary for hydrophilic compounds which do not partition into the dams' milk. It should also be noted that PND17 rats had the same MTD for methamidophos as did adults (PND70), whereas for chlorpyrifos, PND17 rats were ~5-fold more susceptible (*i.e.* lower MTD) than adults *(21)*. The explanation for these phenomena was the metabolic stability of methamidophos compared with chlorpyrifos.

Dietary Exposure

Methamidophos is an insecticide that is used on foliage and is therefore often detected in RACs (cotton and tomato) at harvest. Therefore, the residue values used for calculating possible dietary exposure are considered reasonable estimates rather than "worst-case" ones. For chronic exposure, the percentage of crop treated factor has been used, which will have the effect of reducing chronic dietary exposure and increasing the MOE values.

Dietary exposure computer models by TAS, EX-4® (acute) and EX-1® (chronic) have been used for many years, but are no longer supported. The calculations of exposure described here using TAS programs were quite similar to those using DEEM® programs, also designed by Novigen. For chronic dietary exposure, using the annualized mean dietary exposure, the results were almost identical. This is partly a result of the greater "stability" of values close to the mean than those at the 95th percentile, because outliers will have less impact on mean exposure estimates. The two DEEM® methods used for calculating acute dietary exposure, point estimate and Monte Carlo, have resulted in slightly different dietary exposure estimates and corresponding MOE values. The Monte Carlo method generally yielded lower dietary exposure estimates than the point estimate, the former being generally 20% to 40% of the latter. It is often considered that the Monte Carlo probabilistic simulation of acute dietary exposure is more appropriate than the point estimate, deterministic approach. Because the (Monte Carlo) exposure values are lower, this will have the effect of underestimating dietary exposure wherever the point estimate approach is more appropriate. On the other hand, by not using the percent crop-treated adjustment factor for acute exposure (unlike USEPA), this may have the effect of overestimating acute exposure in the present calculations. It should be noted that there is currently uncertainty about the most appropriate percentile of dietary exposure to use for acute risk determinations. In the current study, the 95th percentile was chosen since this was traditionally used by DPR with TAS EX-4® to calculate acute exposure. USEPA typically conducts acute risk assessments at the 99.9^{th} percentile of exposure, using a probabilistic approach. However, an analysis of commodity tolerances is generally conducted using a deterministic approach. As such tolerance levels of methamidophos on tomato would have decreased MOS values from 34 to 94 (95th percentile for children 1-6 yrs. and nursing infants, < 1 yr., respectively), to 8 to 50 (99.9^{th}. percentile for females, 13-19 yr. and females, 13+, nursing, respectively), using DEEM.® Other differences between DPR and USEPA in the conduct of acute dietary risk assessments are that DPR employs "user days" whereas USEPA uses "*per capita* days." The former means that only those persons who consumed a particular RAC on a specific day are included in the dietary exposure calculation; for the latter, all persons in the survey will be included in the calculation of the proportion of days on which a RAC is consumed, whether they are consumers or not. This difference has the effect of lowering theoretical dietary exposure when the latter method is used. Another difference between DPR and USEPA, which will tend to reduce exposure calculated by USEPA, is the use of %CT information. This is considered only for chronic and not for acute deterministic dietary exposure by DPR whereas USEPA may use this mitigation measure for both types of dietary exposure, when using probabilistic methods. The authors consider that it is inappropriate to use a stochastic variable like %CT for assessing *acute* deterministic dietary exposure.

The organophosphate insecticide acephate, which is the *N*-acetyl analog of methamidophos, is enzymatically converted to the latter, in both insects and mammals. Because acephate is inactive against AChE, it is considered to owe its toxicity to this conversion, *in vivo*. Acephate is widely used in California, on a range of ~20 crops, giving rise to potential indirect methamidophos exposure. It was also registered for Home & Garden uses, giving rise to additional potential exposure to methamidophos, but these uses have recently been canceled by the registrants. Because a risk assessment for acephate has not yet been completed, it has not yet been considered from the perspective of aggregate and/or combined exposure, under FQPA *(3)*. However, it should be recognized that dietary exposure to methamidophos, calculated here from just the use of methamidophos, will probably underestimate total methamidophos exposure from all sources.

Issues Related to the Food Quality Protection Act

Introduction

The Food Quality Protection Act of 1996 mandated USA EPA to "upgrade its risk assessment process as part of the tolerance setting procedures" *(3)*. The changes to risk assessment were based in part on recommendations from the National Academy of Sciences report *(22)*. The act required an explicit determination that tolerances were safe to children. US EPA was required to use an extra 10-fold safety factor to take into account both pre-/post natal developmental toxicity and the completeness of the database, unless US EPA determined, based on reliable data, that a different margin would be safe. In addition, US EPA must consider available information on: 1/ aggregate exposure from all non-occupational sources; 2/ effects of cumulative exposure to the pesticide plus others with a common mechanism of toxicity; 3/ effects of *in utero* exposure; 4/ the potential for endocrine disrupting effects.

Aggregate Exposure(s)

This refers to the possibility that an individual might be exposed to a particular chemical by more than one route. In the case of methamidophos, exposure is likely to be entirely *via* the oral route. There are no home and garden registrations for methamidophos in California. Therefore, dietary exposure will be the only likely route; methamidophos is unlikely to be found in potable water. Home and garden uses of the insecticide acephate, which is bioactivated to methamidophos, have recently been canceled. Although not mandated under FQPA, DPR has previously conducted, and will continue to conduct, aggregate exposure and risk estimations based on dietary and occupational exposure pathways, whenever appropriate. Because occupational exposure for any given pesticide is generally greater than dietary exposure, farm workers exposed to methamidophos could be especially at risk.

Cumulative Exposure(s)

There is a possibility that an individual could be exposed to multiple chemicals sharing the same mechanism of toxicity. An effort is to be made under FQPA to attempt to combine these "cumulative exposure(s)" to related chemicals *(23)*. In the case of methamidophos, such multiple chemical exposure(s) will include exposure to acephate, as well as other AChE inhibitors.

Pre-/Post Natal Sensitivity

Seven developmental toxicity studies (four rat, three rabbit) failed to show fetal or embryonic toxicity at doses of methamidophos less than those affecting dams. No evidence was forthcoming from these experiments that there was an increase in sensitivity among fetal/embryonic animals compared with adults. It is therefore unlikely that an additional factor will be required to protect against increased pre-/post natal sensitivity to methamidophos. The recently completed developmental neurotoxicity study of methamidophos also indicates that fetal/young rats are not more sensitive than adults. However, there must remain a degree of uncertainty about the precise dosage received by the pups in the reproductive toxicity and developmental neurotoxicity studies.

In a 2-generation (SD) rat reproductive toxicity study, reduced body weight was reported in adults and pups with the same LOEL and NOEL values. Because inhibition of AChE appeared to be more marked in adults than pups, it is therefore unlikely that methamidophos has adverse effects on reproduction.

Endocrine Effects

Endocrine effects caused by a pesticide are also to be addressed under FQPA. The main hormonal systems under consideration are male and female reproductive hormones and thyroid hormones. There are no indications that methamidophos may be toxic on any of these systems.

Tolerance assessment

The consumption of foods with residues at tolerance is considered a "worst-case allowable scenario" because tolerances are the maximum legal residues permitted on a crop for human consumption. In practice, only a very small proportion (<1%) of pesticide-treated crops have residues above tolerance *(24)*. Residues of methamidophos at the tolerance level resulted in MOS values that are considered adequate, for cotton and potato. However, for tomato this was not the case. The probable reason(s) for the USEPA calculations resulting in adequate MOS values, for all crops studied, are intriguing, since both Agencies use DEEM® software. There is no difference in the choice of acute toxicity endpoint or NOEL, thus suggesting that there are differences in the calculation of exposure. The use of percent crop-treated data by USEPA, along with the use of "*per capita*" rather than "user days," for acute exposure calculations, will both have the effect of reducing exposure. It is therefore possible that mitigation measures may be required for tomato in California.

Conclusions

- A dietary risk assessment for methamidophos has been completed.
- Acute risk assessment used a NOEL of 0.3 mg/kg-d, for brain AChE inhibition, from a rat oral gavage study.
- Chronic risk assessment used an estimated NOEL of 0.02 mg/kg-d, for brain AChE inhibition, from a 1-yr. dog, dietary study.
- Two computer models were used to estimate dietary exposure: TAS® and DEEM®
- Deterministic and probabilistic methods, at the 95th percentile of exposure, gave adequate MOS values (>100) for all population sub-groups.
- Two CSFII food consumption surveys were used, 1989-1992 and 1994-1998, both giving adequate MOS values.
- Acute and chronic exposures gave adequate MOE values, using measured residue levels.
- Residues at tolerance (0.1 ppm potato, cotton, 1 ppm tomato) gave adequate MOS values for potato and cotton.
- MOS values for tomato were inadequate (<100) for acute exposure at tolerance, for all population sub-groups.

References

1. USEPA, Interim Reregistration Eligibility Decision (IRED) for methamidophos; Case No 0043., **2002**.
2. USEPA. Federal Register **2002**, *67*(198), pp. 63423-4.
3. USEPA, **1996** FQPA at http://www.epa.gov/opppsps1/fqpa/
4. Zager, E.; Tarplee, B.; Wooge, W. *Pesticide Outlook* **2003**, *14*, 133-135.
5. TAS, **1996**. EX-4® (Acute) Detailed Distributional Dietary Exposure Analysis, Version 3.35. and EX-1® Chronic Dietary Exposure Analysis, Version 3.2. Technical Assessment Systems, Inc., Washington, D.C.
6. DEEM® Dietary Exposure Evaluation Model® Novigen Sciences, Inc., Washington, DC.
7. Magee, P.S. In *Insecticide Mode of Action;* Coats, J.R.; Ed. Academic Press: San Diego, CA, **1982**, pp 101- 161.
8. Hayes, R.H. **1984**. Study No. 81-174-01. Mobay Chem. Corp. DPR 315-061.
9. Gammon, D.W. *Pesticide Outlook* **2003**, *14*, 130-132.
10. Sheets, L.P., **2002**. Study Type 836. DPR 315-0167, #200767.
11. Crosley, J.; Lee, H., **1972** Chevron Report CDL: 223489-D DPR 108-128.
12. Sheets, L.P., **1994**. Miles Report 105053-1. DPR 315-139, #132008
13. Hamilton, B., **1993**. Report No. 105053. DPR 315-131. #127242.
14. C.F.R., **2002**. Title 40, Section 180.315 (methamidophos). U.S. Govt. Printing Office, Washington, DC.
15. Cochran, R.C. In *Encyclopedia of Agrochemicals;* Plimmer, J.R., Ed.; John Wiley & Sons, Inc: Hoboken, NJ, **2003**; Online Edition, pp 23.
16. Carlock, L.L. *et al. J. Toxicol. Environ. Hlth. B.* **1999**, *2*,105-160.
17. Smith, D.S.;Treherne, J.E. *J. Cell Biol.* **1965**, *26*, 445-465.
18. Qiao, D.; Seidler, F.J.; Slotkin, T.A. *Environ. Hlth. Perspec.* **2001**, *109*, 909-913.
19. Lanning et al., *Biochem. Biophys. Acta* **1996**, *1291*, 155-162.
20. Lankas, G.R.; Minsker, D.H.; Robertson, R.T. *Fd. Chem. Toxicol.* **1989**, *27*, 523-529.
21. Padilla, S.; Buzzard, J.; Moser, V.C. *Neurotoxicology* **2000**, *21*, 49-56.
22. *Pesticides in the Diets of Infants and Children;* National Academy Press: Washington, DC, **1993**, pp 386.
23. USEPA. Guidance on cumulative risk assessment of pesticide chemicals that have a common mechanism of toxicity. Office of Pesticide Programs, January 14, **2002**.
24. Lamont, M.; R.L. Epstein, Eds. Pesticide Data Program (PDP). Summary 2001. U.S.D.A. Agricultural Marketing Service, Washington, D.C. USDA, **2003**, pp. 115.

Chapter 14

Evaluation of Estrogen Receptor Binding Affinity of DDT-Related Compounds and Their Metabolites

Masahiro Miyashita[1,2], Takahiro Shimada[1,2],
Shizuka Nakagami[2], Norio Kurihara[3], Hisashi Miyagawa[1], and Miki Akamatsu[1,2]

[1]Graduate School of Agriculture, Kyoto University, Kyoto 606–8502, Japan
[2]CREST, Japan Science and Technology Agency, Saitama 332–0012, Japan
[3]Koka Laboratory, Japan Radioisotope Association, Shiga 520–3403, Japan

Methoxychlor, an analog of DDT, has been shown to exhibit estrogenic activity after metabolized to phenolic compounds. In this study, in order to investigate the structural requirements for increasing the estrogen receptor (ER) binding activity after metabolism, we evaluated the ER binding activity of a series of DDT-related compounds and compared it with that of their metabolite mixtures. Compounds that possess methoxy or ethoxy substituents on benzene rings exhibited enhanced ER binding activity after metabolic reaction with rat liver S9. These are likely to be enzymatically dealkylated into phenolic compounds similar to methoxychlor metabolites. In addition, we measured the ER binding activity of each enantiomer of mono-OH-MXC, and found that (S)-mono-OH-MXC, which is predominantly formed in rat and human livers, was 3-fold more potent than the (R) enantiomer.

There has been increasing concern regarding environmental chemicals that disrupt endocrine function in wildlife and humans (1). Several environmental compounds including industrial chemicals and pesticides have been shown to mimic or inhibit normal endogenous hormone action by binding to the estrogen receptor (ER) (2-4). In addition, some compounds are known to act as proestrogens, indicating that their metabolites have estrogenic activity. Methoxychlor (Figure 1), structurally similar to DDT, has been shown to exhibit estrogenic activity after being metabolized to phenolic compounds, although methoxychlor itself has negligible activity (5). Two demethylated metabolites are major estrogenic compounds: 2-(p-hydroxyphenyl)-2-(p-methoxyphenyl)-1,1,1-trichloroethane (mono-OH-MXC) and 2,2-bis(p-hydroxyphenyl)-1,1,1-trichloroethane (bis-OH-MXC; Figure 1). This fact indicates that the estrogenic activity of compounds might be underestimated unless the enhancing effects of metabolism on estrogenic activity are assessed. In this study, in order to investigate the structural requirements for increasing the ER binding activity after metabolism, we evaluated the ER binding activity of a series of DDT-related compounds and compared it with that of their metabolite mixtures obtained by treatment with rat liver S9 fraction. Moreover, we measured the ER binding activity of each enantiomer of mono-OH-MXC to further understand the binding mode of mono- and bis-OH-MXC to ER (6).

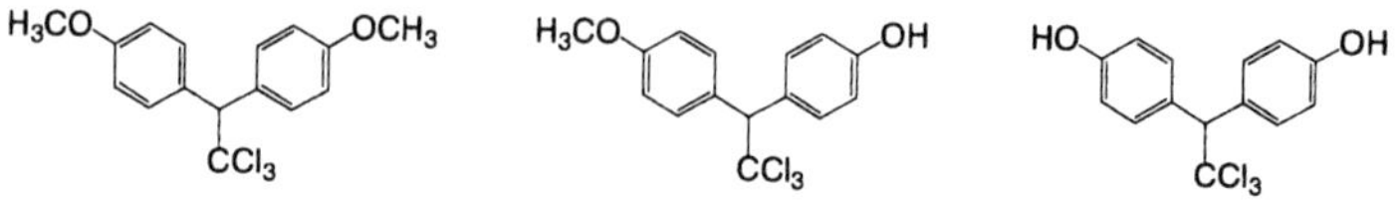

Methoxychlor (1)   Mono-OH-MXC   Bis-OH-MXC

Figure 1. Structure of methoxychlor and its metabolites

Comparison of the ER binding activity of DDT-related compounds with their metabolite mixtures

Test compounds used in this study are shown in Figure 2. Methoxychlor (**1**) was obtained from Wako Pure Chemical Industries, Ltd. (Osaka, Japan). Synthesis of all other compounds was previously reported (7). No contaminants in tested compounds were observed by thin-layer chromatography. Compounds were dissolved in dimethyl sulfoxide and diluted to 0.1 mM for the inhibition of ER binding assays (final concentration 38 μM). Metabolite mixtures of each test compound were obtained as follows. Each compound (final concentration 0.5 mM) was incubated for 60 min at 37°C with 5% rat liver S9 fraction (Kikkoman

Corp., Chiba, Japan) in 100mM Na-phosphate buffer (pH 7.2) containing 2 mM $NADP^+$, 2 mM glucose-6-phosphate, 10 mM KCl, 1 mM $MgSO_4$, and 4 mM HEPES. The incubation mixture was then diluted five fold with the buffer and used for the assay without any extraction or purification procedures. Inhibition of ER binding was measured using Ligand Screening System for Estrogen Receptor α kits (TOYOBO CO., LTD, Osaka, Japan) (8) according to the manufacturer's procedure. Diethylstilbestrol (DES, 113 nM), which was used as a positive control, gave 100% inhibition. The inhibition rate was calculated as the ratio between response of the test compound and the positive control after subtracting that of the negative control (1% DMSO). Each experiment was carried out in quadruplicate.

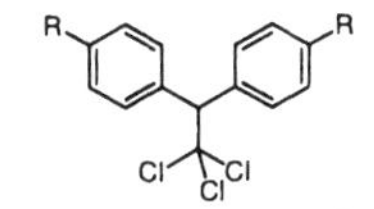

R= OCH_3 (1), OCH_2CH_3 (2), Cl (3), Br (4), F (5), CH_3 (6), Ph (7), NO_2 (8)

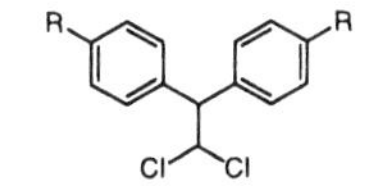

R= H (9), OCH_2CH_3 (10), Cl (11), Br (12), CH_3 (13)

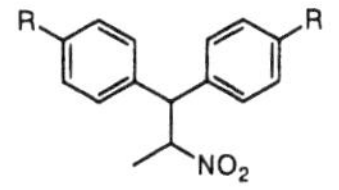

R= H (14), OCH_3 (15), Br (16), Cl (17), NO_2 (18), CH_3 (19)

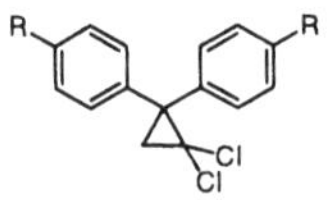

R= H (20), OCH_3 (21), OCH_2CH_3 (22), $O(CH_2)_2CH_3$ (23), $OCH(CH_3)_2$ (24), $O(CH_2)_3CH_3$ (25), Cl (26), Br (27), F (28), Br,F (29)

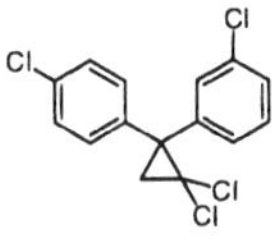

30

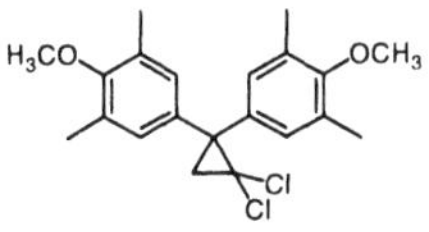

31

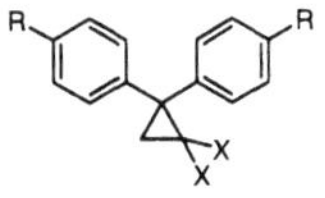

R=H, X= Br (32), R=Cl, X=Br (33), R=Cl, X=F (34), R= Cl, X=Cl,Br (35), R=OCH_2CH_3, X=Cl,Br (36), R=Cl, X=H,Cl (37)

Figure 2. Structure of DDT-related compounds

Inhibition rate of ER binding by test compounds and their metabolic mixture are shown in Figure 3. None of compounds except for **6** (54% inhibition) exhibited more than 50% inhibition of ER binding at the concentration of 38 μM without rat liver S9 treatment. DES, one of the strongest estrogenic compounds, had an IC_{50} of 16 nM in this assay. This indicates that all compounds tested in this study are at least 2,000 times less active than DES. On the other hand, compounds **2, 10, 15, 21, 22** and **36** in addition to methoxychlor (**1**) showed significantly enhanced ER binding activity after treatment with rat liver S9 fraction. All these compounds possess methoxy or ethoxy substituents on benzene rings. Since phenolic OH groups are known to be most important for

ER binding mimicking the 3- and/or 17β-OH in 17β-estradiol (Figure 6, A), the methoxy or ethoxy substituents of these compounds are likely to be enzymatically dealkylated and display enhanced ER binding activity. As previously reported, the inhibition of 17β-estradiol binding to ER by these phenolic metabolites is likely to be competitive (9). Bulky groups which are connected to the carbon atom bridging two benzene rings were diverse among these compounds. Although the hydrophobic interaction of ligands with ER is also important for the binding, the contribution of these bulky groups to the ER binding might be small. Compounds containing longer or branched O-alky, alkyl, nitro and halogen substituents on benzene rings showed little or no enhanced activity after rat S9 metabolism. This suggests that these groups are metabolically stable or, if metabolized, modified structures are not favorable for ER binding.

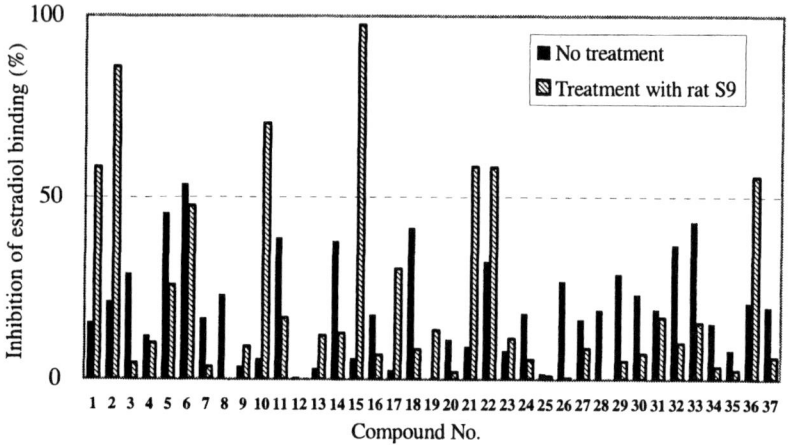

Figure 3. ER binding activity of DDT analogs and their metabolite mixtures.

Enantioselective recognition of mono-demethylated methoxychlor metabolites by estrogen receptor

Mono-OH-MXC is a chiral compound in which the carbon atom bridging two benzene rings is the chiral center (Figure 4). It is known that the enzymatic O-demethylation of methoxychlor into mono-OH-MXC in rat and human livers is enantioselective (10,11). In this metabolic reaction, the (S) enantiomer is predominantly produced by human liver microsomes at 77-87%. The activity of each enantiomer of mono-OH-MXC, however, has not yet been elucidated, and the hormonal activity of methoxychlor *in vivo* might be misinterpreted if each mono-OH-MXC enantiomer has distinct estrogenic activity.

(S)-, and (R)-mono-OH-MXC were prepared as described previously (12). Enantiomeric excess of each (S)- and (R)-mono-OH-MXC was determined to be >99% by chiral HPLC. Inhibition of ER binding for racemic, (S)-, and (R)-mono-OH-MXC and bis-OH-MXC was measured as described above and the concentration required for 50% inhibition (IC_{50}) was calculated (Figure 5).

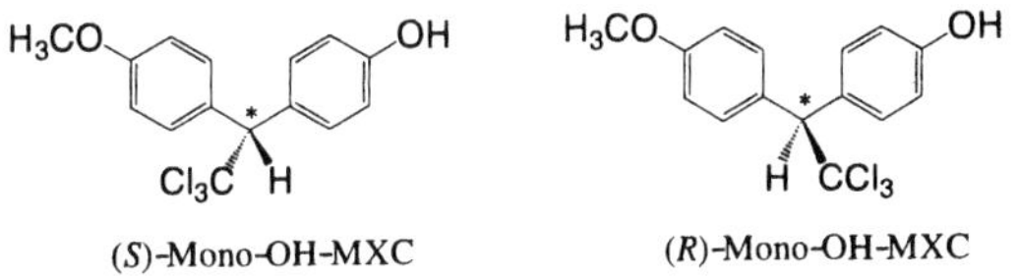

H₃CO‌‌‌‌‌‌‌‌‌‌‌‌‌‌‌‌‍ ...

(S)-Mono-OH-MXC   (R)-Mono-OH-MXC

Figure 4. Enantiomers of mono-demethylated methoxychlor metabolite

(S)-mono-OH-MXC showed 3-fold higher binding activity than that of (R) enantiomer (Table 1). The IC_{50} value of racemic mono-OH-MXC was in-between of those of (R)-, and (S)-mono-OH-MXC as expected. The activity of bis-OH-MXC was only 1.7-fold higher than that of (S)-mono-OH-MXC.

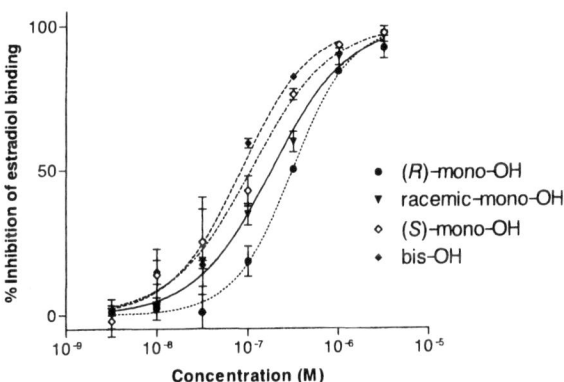

Figure5. Inhibition curves for ER binding by methoxychlor metabolites. Values represent the mean ±SEM of three separate experiments. Adapted from reference 6. Copyright 2004 Elsevier.

(S)-mono-OH-MXC is predominantly formed by enzymatic O-demethylation of methoxychlor using rat and human liver microsomes as previously described. Taking into account the result that (S)-mono-OH-MXC is 3 times more active than the (R) enantiomer, the estrogenic activity of methoxychlor after metabolic activation *in vivo* should be higher than estimated from the *in vitro* study using racemic mixtures.

Table 1. ER binding activity of demethylated methoxychlor metabolites

Compound	IC_{50} (μM)	Relative activity
Racemic mono-OH-MXC	0.26 ± 0.11	1
(S)-mono-OH-MXC	0.15 ± 0.05	1.7
(R)-mono-OH-MXC	0.47 ± 0.21	0.55
Bis-OH-MXC	0.09 ± 0.02	2.9

NOTE: IC_{50} values are given as mean ± S.D. of at least three independent experiments.
SOURCE: Reproduced from reference 6. Copyright 2004 Elsevier.

Binding mode of methoxychlor metabolites to estrogen receptor

Structure-activity relationship and X-ray crystallographic studies (*13,14*) have shown that the 3-OH and 17β-OH of estradiol (Figure 6) are critical for interaction with the receptor, which mainly serve as H-bond donors and acceptors in the receptor interaction. In particular, it has been demonstrated that the 3-OH is more important than the 17β-OH for binding. The 3-OH of estradiol was shown to interact with Glu 353 and Arg 394 of the estrogen receptor as a H-bond donor and acceptor, whereas the 17β-OH only formed one H-bond with His 524 as a H-bond donor (Figure 6, A). In this study, we found that the bis-OH-MXC had only 1.7-fold higher activity than (S)-mono-OH-MXC. This suggests that one hydroxy group of bis-OH-MXC is important for the receptor binding, mimicking the 3-OH of estradiol, but the contribution of the second hydroxy group to the binding is marginal because of its unfavorable spatial position for H-bond interaction (*13*) (Figure 6, B). The slightly lower activity of (S)-mono-OH-MXC than that of bis-OH-MXC might be due to the bulkiness of methoxy group as well as its H-bond characteristic that can act only as an acceptor (Figure 6, C). (S)-mono-OH-MXC had higher activity than the (R) enantiomer. Hydrophobic interactions between 7α- and 11β-substituents of estradiol and the receptor are also known to be important for the binding, particularly for nonsteroidal estrogens such as DES (15). The CCl$_3$ group of mono-OH-MXC could also form a hydrophobic interaction with the receptor, mimicking the 7α- or 11β- substituents of estradiol. With the configuration in which a hydroxy group of mono-OH-MXC forms the hydrogen bond with the receptor mimicking the 3-OH of estradiol, the CCl$_3$ group of the (S) enantiomer is likely to be better accommodated into the hydrophobic site of the receptor-binding pocket than that of the (R) enantiomer (Figure 6, D).

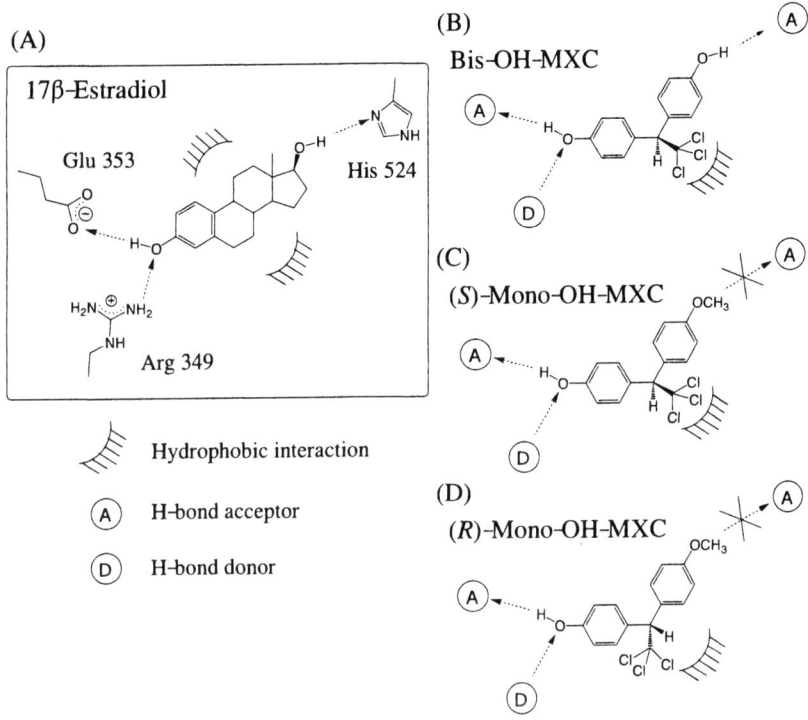

Figure 6. Hypothetical binding mode of demethylated methoxychlor metabolites.

Conclusion

In this study, we found that DDT-related compounds which possess methoxy or ethoxy substituents on benzene rings exhibited enhanced ER binding activity after metabolic reaction. This results would be useful for predicting the proestrogenicity of compounds before time-consuming syntheses. We also showed that (S)-mono-OH-MXC, which is predominantly formed in rat and human livers, was more potent than the (R) enantiomer. In order to accurately evaluate the estrogenic activity of methoxychlor after exposure in human, it is necessary to consider both the enantioselectivity of metabolism and of receptor-recognition.

Acknowledgements

We are thankful to Dr. Craig Wheelock of the University of California at Davis for reviewing this manuscript.

References

1. Colborn, T.; vom Saal, F. S.; Soto, A. M. *Environ. Health Perspect.* **1993**, *101*, 378-384.
2. Nelson, J. A. *Biochem. Pharmacol.* **1974**, *23*, 447-451.
3. Soto, A. M.; Justicia, H.; Wray, J. W.; Sonnenschein, C. *Environ. Health Perspect.* **1991**, *92*, 167-173.
4. Krishnan, A. V.; Stathis, P.; Permuth, S. F.; Tokes, L.; Feldman, D. *Endocrinology* **1993**, *132*, 2279-2286.
5. Bulger, W. H.; Muccitelli, R. M.; Kupfer, D. *Biochem. Pharmacol.* **1978**, *27*, 2417-2423.
6. Miyashita, M.; Shimada, T.; Nakagami, S.; Kurihara, N.; Miyagawa, H.; Akamatsu, M. **2004**, *54*, 1273-1276.
7. Nishimura, K.; Hirayama, K.; Kobayashi, T.; Fujita, T.; Holan, G. *Pestic. Biochem. Physiol.* **1986**, *25*, 153-162.
8. Koda, T.; Soya, Y.; Negishi, H.; Shiraishi, F.; Morita, M. *Environ. Toxicol. Chem.* **2002**, *21*, 2536-2541.
9. Bulger, W. H.; Muccitelli, R. M.; Kupfer, D. **1978**, *4*, 881-893.
10. Hu, Y.; Kupfer, D. *Drug Metab. Dispos.* **2002**, *30*, 1329-1336.
11. Kishimoto, D.; Oku, A.; Kurihara, N. *Pestic. Biochem. Phys.* **1995**, *51*, 12-19.
12. Ichinose, R.; Nakayama, M.; Kuramoto, J.; Kurihara, N. *Agric. Biol. Chem.* **1986**, *50*, 1331-1332.
13. Fang, H.; Tong, W.; Shi, L. M.; Blair, R.; Perkins, R.; Branham, W.; Hass, B. S.; Xie, Q.; Dial, S. L.; Moland, C. L.; Sheehan, D. M. *Chem. Res. Toxicol.* **2001**, *14*, 280-294.
14. Anstead, G. M.; Carlson, K. E.; Katzenellenbogen, J. A. *Steroids* **1997**, *62*, 268-303.
15. Shiau, A. K.; Barstad, D.; Loria, P. M.; Cheng, L.; Kushner, P. J.; Agard, D. A.; Greene, G. L. *Cell* **1998**, *95*, 927-937.

Metabolism

Chapter 15

Metabolism of Pesticides by Plants and Prokaryotes

Robert M. Zablotowicz[1], Robert E. Hoagland[1], and
J. Christopher Hall[2]

[1]Southern Weed Science Research Unit, U.S. Department
of Agriculture, Agricultural Research Service,
Stoneville, MS 38776
[2]Department of Environmental Biology, University of Guelph,
Guelph, Ontario N1G 2W1, Canada

The metabolism of pesticides by plants is a key factor in the susceptibility and tolerance of a species to a given pesticide, whereas metabolism by prokaryotes is often a key determinant in the environmental fate of that pesticide. Thus, understanding pesticide metabolism in both groups of organisms is crucial for efficient and environmentally sound pesticide management. The pathways of pesticide detoxification in plants have been traditionally divided into several metabolic phases, whereas bacterial biotransformations are characterized as either metabolic or cometabolic. There are common transformation mechanisms of many pesticides in both plants and bacteria; however, some prokaryotes can completely metabolize certain pesticides to mineral components (mineralization). The diversity of biotransformations in prokaryotic organisms for a given pesticide is also generally greater than in plants.

Modern crop production is highly dependent on pesticides to control various pests, e.g., insects, pathogens and weeds. Each year millions of tons of pesticides are used, with herbicides the most widely used class of agrochemicals. Tolerance of a crop to a pesticide, and susceptibility of the targeted pest, is the basis of agrochemical selectivity. Most modern pesticides are subject to extensive degradation by plants and microorganisms, however some can be somewhat recalcitrant, and thus residues in soil, water and food chain organisms is an environmental and health concern. Several recent books and reviews on pesticide metabolism by plants and microorganisms are available (*1, 2, 3, 4, 5*). Expanded knowledge regarding pesticide metabolic pathways has been attained using molecular biology and improved analytical instrumentation. Improved understanding of pesticide metabolism in crop plants and weeds, has and will continue to enhance efficient management of metabolically-based pesticide resistance. Advances in microbial pesticide metabolism has provided safe and efficient pesticide management strategies such as bioremediation, managing accelerated degradation, and minimizing environmental contamination. Novel degradative enzymes from plants and/or bacteria have and will continue to be exploited for production of transgenic herbicide resistant crops.

This review will focus on similarities and differences of pesticide metabolism by plants and prokaryotic organisms. Examples of pesticide biotransformations will be used to illustrate how these metabolic processes can affect efficacy, environmental fate, and persistence.

Physiological Processes in Plants and Prokaryotes and Potential for Pesticide Metabolism

There are some similarities regarding the enzymatic basis for metabolism of certain pesticides in plants and prokaryotic organisms. However, as outlined in Table I, there are fundamental differences between higher plants and prokaryotes that determine the potential for degradative processes among these broad divisions of organisms.

Prokaryotes (organisms lacking a nuclear membrane) represent a diverse assembly of unicellular organisms in the kingdoms, Archaea (archaeabacteria) and Bacteria (eubacteria). Structurally, prokaryotes are comprised of appendages (flagella, pilli, fribiae), a glycocalyx (extracellular capsule), a cell wall, periplasmic space, cell membrane, and cytoplasm containing ribosomes, granules, cytosol and a "naked" genome composed of a chromosome and plasmids (*6*). Although prokaryotic organelles contain a membrane, it is not composed of a lipid-protein bilayer as found in eukaryotic organelles. Plant cells possess a more complex cellular morphology including numerous organelles, e.g. mitochondria, chloroplasts, golgi bodies, vacuoles, cell membranes and a nuclear membrane. Plants are likewise differentiated into roots and shoots containing differentiated cellular structures, e.g., trichomes, root hairs, xylem, phloem that affect pesticide uptake and transport. Thus various transport and metabolic processes are more compartmentalized in plants on both a cellular and organ basis compared to prokaryotes. Plant genes are organized into a formal nucleus with a nuclear membrane and certain genes are also found in the mitochondria, chloroplasts and other plastids (7). In plants and prokaryotes, certain genes are constitutively expressed, while expression of other genes are inducible and regulated by various environmental factors. In plants, certain herbicide detoxifying glutathione S-transferases (GSTs) are constitutively expressed, while others are expressed upon exposure to chemicals such as safeners, pesticides or other xenobiotics.

Table I. Fundamental Differences between Plants and Prokaryotes

Characteristic	Higher Plants	Prokaryotes
Anatomy and Morphology	Multicellular, differentiated into various organs, and organelles	Unicellular, arranged in clusters, colonies
Nutrition	Primarily photo-autotrophic	Heterotrophic, chemotrophic, Photoautotrophic
Metabolism	Aerobic, O_2 sole terminal electron acceptor	Alternate terminal electron acceptors, or fermentation
Ring cleavage	None	Multiple aerobic and anaerobic pathways
Metabolic sites	Compartmentalized	Intra- and extracellular
Genetic structure	Genes present in nucleus, mitochondria and plastids	Genes present in chromosome and plasmids
Gene regulation	Transcription	Arranged in operons

Plants are generally photoautotrophs relying on energy from light to provide nutrition for cellular maintenace and development. Most prokaryotes are heterotrophs that require organic compounds. There are also some prokaryotes that also possess chemotrophic (e.g., hydrogen oxidizers, nitrifiers) or photoautotrophic (e.g., cyanobacteria) capabilities. Subsequently, many plants possess the metabolic potential for detoxification of xenobiotics, while certain prokaryotes actually utilize some pesticides as carbon sources. Although there are common mechanisms of pesticide biotransformations in both plants and bacteria, prokaryote pesticide metabolism is unique beause many pesticides can be completely degraded to mineral components, i.e., mineralization. This is based upon the prokaryotes ability to cleave various aromatic and heterocyclic ring strctures, and to oxidize these compounds as an energy source. Plants are strictly aerobic organisms, while many prokaryotes can utilize other compounds as terminal electron acceptors (O_2, NO_3^-, NO_2^-, Fe^{+2}, Mn^{+2}, SO_4^{-2}, CH_4) in lieu of oxygen to reoxidize NADH and other reduced electron carriers. Facultative aerobes and strict anaerobes also utilize fermentative pathways, e.g., lactic acid, ethanol, butyric acid, propionic acid, etc. for reoxidiation of NADH. Oxygen is actually toxic to many anaerobic bacteria since they lack superoxide dismutase for scavenging of toxic radicals.

Characteristics of Plant Pesticide Metabolism

For metabolism of a xenobiotic to occur in plants, a pesticide molecule must be absorbed via roots or through leaf cuticles and transported or translocated to active metabolic sites. As pointed out earlier, there are a variety of such sites within a given plant. Shimabukuro *et al.* (*8*) have categorized pesticide metabolism in plants into three phases. Oxidative, reductive, and hydrolytic mechanisms are characteristic of the initial phase of plant metabolism (Phase I). The primary enzymes responsible for oxidative Phase I transformations is cytochrome P450s that catalyzes reactions such as ring and alkyl hydroxylation, *N*-dealkylation, and sulfoxidation of pesticides. Esterases, lipases, aryl acylamidases, and nitrilases are the major Phase I hydrolytic enzymes. Phase II transformations are conjugative processes generally catalyzed by glutathione *S*-transferases (GSTs) or sugar, amino acid, or organic acid transferases. Phase III transformations encompass the formation of secondary conjugates (primarily via malonyl-CoA transferases), and bound residues (via peroxidases) within plant tissues.

Conjugation of pesticides in plants often involves utilization of existing enzymatic machinery, and is therefore called a co-metabolic process. Glucose (*O*-, *S*-, and *N*-glucosides, glucose ester, gentiobioside) and/or glutathione conjugations to pesticides primarily occur in plants, and are the main type of Phase II metabolites. The most common glucose conjugates are *O*-glucosides, because pesticide oxidation reactions (Phase I) form hydoxyl groups, which are suitable sites for glucose conjugation. UDP-glucosyl transferase, an enzyme involved in cellulose biosynthesis, mediates pesticide-glucose conjugation (*9*, *10*). Amino acid conjugation occurs less frequently.

Differences of 2,4-D (2,4-dichlorophenoxyacetic acid) conjugation in wheat (tolerant) versus some broadleaf weeds (susceptible) exemplifies how Phase II metabolism imparts herbicide selectivity. Many susceptible broadleaf weeds produce glucose ester metabolites, which are readily susceptible to hydrolysis, yielding phytotoxic 2,4-D. Conversely, 2,4-D-tolerant wheat rapidly produces amino acid conjugates and *O*-glucosides, which are stable, non-phytotoxic metabolites that are not easily hydrolyzed (*11*, *12*).

Glutathione (GSH, γ–L-glutamyl-L-cysteinylglycine), commonly present in the reduced form, is ubiquitously distributed in most aerobic organisms and at relatively high concentrations in most plant tissues. Non-enzymatic GSH conjugation can be important for the metabolism of several herbicides including fenoxaprop-ethyl (FE; ethyl(±)2-[4-[(6-chloro-2-benzoxazolyl)oxy]phenoxy] - propanoate), (*13, 14*) and the chloroacetamides (*15*). However, enzymatic conjugation of xenobiotics with GSH via GSTs is more common than non-enzymatic conjugation. GSTs are homo- or hetero-dimer, multifunctional enzymes, catalyzing the nucleophilic attack of the sulfur atom of GSH by the electrophilic center of some substrates (*16*). Some GSTs are constitutively expressed in certain tissues, but GST regulation can be modified by agrochemicals, including herbicide safeners and synergists (*17*).

GSTs and GSH have several roles in plants: i) the metabolism of secondary products (*18, 19*); ii) regulation and transport of both endogenous and exogenous compounds such as herbicides, which are often GS-X tagged for compartmentalization in the vacuole or cell wall (*20*); and iii) for protection against oxidative stress from herbicides, air pollutants (*21*), pathogen attack (*22*), and heavy metal exposure (*23*). GSH-conjugates and their terminal metabolites are often stored in the vacuole or bound to the cell wall (*24, 25*). GSH-conjugate pumps in the tonoplast membrane carry GSH-conjugates across the membrane (*16, 26-29*). In the vacuole, peptidases release the glutathionyl moiety (*25*).

GSTs in plants were first studied because of their ability to detoxify herbicides (*30, 16*). GST-based metabolism imparts herbicide tolerance in several plant species especially to the sulfonylureas, aryloxyphenoxypropionates, triazinone sulfoxides, and thiocarbamates herbicide families that are susceptible to GSH conjugation (*16, 31-35*). There is a positive correlation of both GSH levels and the activity of specific GST enzymes with the rate of herbicide conjugation and detoxification (*36-39*).

Metabolism of the proherbicide (i.e., a non-phytotoxic chemical that requires metabolic conversion to yield an active herbicide) FE, is an example of how Phase I and II metabolism, and use of a safener imparts selectivity between crop and weed species (Figure 1). FE is used to control many graminaceous weed species such as *Avena fatua, Echinochloa crus-galli, Digitaria ischaemum,* and *Setaria glauca* (*40*) in wheat which is moderately tolerant to this proherbicide. The primary basis for the selectivity of FE in grasses is due to differences in metabolism (*40-42*). FE is rapidly hydrolyzed to the herbicide fenoxaprop (F) in wheat, and in all the previously mentioned graminaceous weeds (*43*). In wheat,

F undergoes rapid, subsequent metabolism to 6-chloro-2,3-dihydrobenzoxazol-2-one (CDHB) as well as displacement of the phenyl group of F by GSH or cysteine, resulting in the production of the respective conjugates, and 4-hydroxyphenoxypropanoic acid (Figure 1). Little or no metabolism of F to these products occurs in *A. fatua, E. crus-galli, D. ischaemum,* and *S. glauca,* thus these species are more susceptible to FE.

Despite the tolerance of wheat to F, unacceptable yield losses may occur in wheat after application of FE. Stephenson et al. (*44*) showed that the commercial triazole safener, fenchlorazole-ethyl (FCE), which is used to protect wheat against the phytotoxic action of FE, also acts as a synergist for the herbicide in *E. crus-galli, D. ischaemum,* and *S. glauca,* but exhibits no synergy with FE when applied to *A. fatua.* In addition, FCE increased GSH levels in wheat, but had no effect on GSH levels in these weeds (*14*). Consequently, in the presence of FCE, more F was detoxified by conjugation with GSH and cysteine in wheat than in the absence of FCE, thereby safening it against damage by the herbicide. In *E. crus-galli, D. ischaemum,* and *S. glauca,* FCE increased the normal rate of conversion of FE to F, thus synergizing herbicidal activity. However, in *Avena fatua* this synergism does not occur since FCE is rapidly metabolized to water-soluble metabolites before it can mediate enhanced de-esterification of FE to phytotoxic F (*45*). Hall and Stephenson (*46*) pointed out that numerous chemical synergists and chemical safeners are capable of decreasing or increasing herbicide selectivity. However, FCE is unique. It may be the first chemical known to be an effective safener for a herbicide on certain crops (wheat), and also functions as a synergist on some important weeds. Further in-depth discussions on plant-based metabolism as the basis for herbicide selectivity, glutathione metabolism, and herbicide safeners is presented in subsequent chapters in this book.

Pesticides are often bound to plant cell walls (Phase III). Skidmore et al. (*47*) define a bound xenobiotic residues as "a residue associated with one or more classes of endogenous macromolecules. It cannot be dissociated from the natural macromolecule using exhaustive extraction or digestion without significantly changing the nature of the associated endogenous macromolecules". It appears that xenobiotics are randomly incorporated into different cell wall components (*48, 49*); however, little is known about the type of linkages involved in this binding. There is concern about the bioavailability of bound pesticides in plant residues. The ability of animals to release xenobiotics bound to plant residues is unknown. The toxicological nature and bioavailability of bound xenobiotic residues requires continued research to fully assess their impact on human health and the environment (*49*).

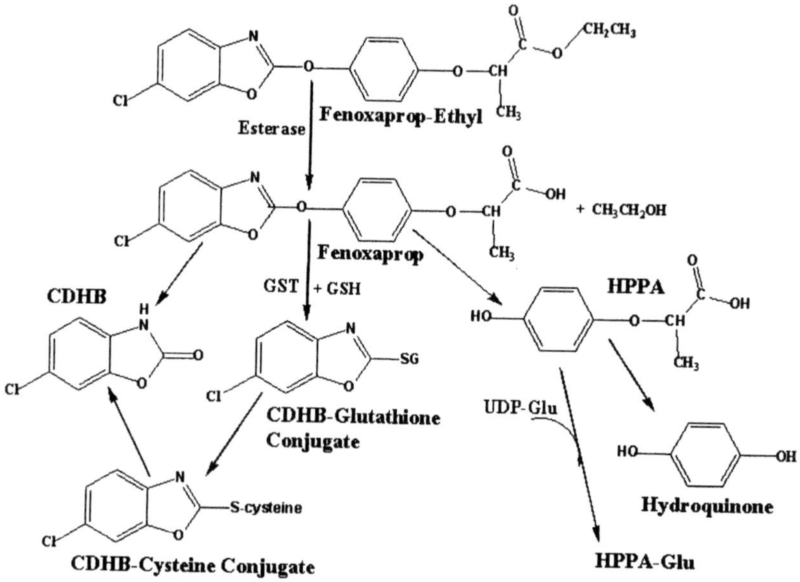

Figure 1. Metabolism of fenoxaprop-ethyl in plants and bacteria.

Characteristics of Prokaryotic Pesticide Metabolism

Bacterial biotransformation of xenobiotics is characterized as either metabolic (utilized as an energy or nutrient source) or co-metabolic (serves as a detoxification mechanism, without providing energy or nutrition) *(50)*. The metabolic potential to utilize pesticides as energy sources to support growth is governed by the ability of certain prokaryotes to cleave complex ring structures and to direct these metabolites into energy producing pathways. The diversity of metabolic mechanisms for dehalogenation, oxidation/reduction and cleavage of aromatic and heterocyclic ring structures by prokaryotic organisms demonstrates the remarkable range of pesticide transformations by these organisms.

Many of the metabolic transformations for pesticide detoxification that occur in plants are also found in prokaryotes, especially co-metabolic transformations. Glutathione conjugation of the chloroacteamide herbicides is observed in many species of gram negative bacteria including *Pseudomonas* sp. and various members of the Enterobacteriaceae *(51)*. Most of the degradative steps for FE metabolism in plants (de-esterification to F, and subsequent cleavage of the ether bond by gluathione conjugation) also occur in *Pseudomonas* sp. (Figure 1). A divergence is that HPPA is further metabolized to hydroquinone and propionic acid in *Pseudomonas* sp. (*52*), while HPPA is subject to glucosylation in plants. While sugar conjugation is the major conjugation mechanism for detoxifying phenols and anilines in plants, *O*-methyl and *N*-acetyl conjugates are more prevelant prokaryotes, e.g., *O*-methylation of chlorophenols (*53*) and acetylation of 3,4-dichloroaniline by *Pseudomonas* sp. (*54*).

Most plant dehalogenation processes are mediated by either GSTs or P450s. In prokaryotes, seven dehalogenation mechanisms for organic compounds have been outlined (*55*). With regard to bacterial pesticide biotransformation, dehalogenation, hydrolytic, thiolytic, oxidative, and reductive processes are most prevalent mechanisms (Table II). A wide range of bacterial dehalogenating enzymes have been discovered. The genes that encode these enzymes have been cloned. The amino acid sequences and structure-function of these enzymes have been elucidated. Finally, the regulation of gene expression and enzymatic activity have been determined. Selected examples are presented and discussed.

Table II. Prevalent Mechanisms for Prokaryotic Pesticide Dehalogenation

Mechanism	Reaction	Pesticicde Substrates
Oxygenolytic	R-Cl + NADH + H^+ $\Rightarrow$ R-OH + NAD + HCl	Pentachlorophenol (*56*)
Hydrolytic	R-Cl + H_2O $\Rightarrow$ R-OH + HCl	Lindane (*57*), Atrazine (*58*)
Reductive	R-Cl + $2H^+$ + $2e^-$ $\Rightarrow$ R-H + HCl	DDT, Aldrin (*59*)
Thiolytic	R-Cl + GSH $\Rightarrow$ R-SG + HCl	Alachlor (51), Pentachlorophenol (*56*)

Atrazine [6-chloro-*N*-ethyl-*N*-(1-methylethyl)-1,3,5-triazine-2,4-diamine] is one of the most extensively used herbicides in North America and is a major environmental concern as a potential water contaminant. Atrazine degradation by mineralizing species, e.g., *Pseudomonas* sp. strain ADP, is catalyzed via a series of hydrolytic reactions (Figure 2). The six genes encoding these processes in this bacterium are located on a single plasmid (*60*). The genes encoding the three initial enzymes (*atzA*, *atzB*, *atzC*) are constitutively expressed and genes encoding the final three enzymes (*atzD*, *atzE*, *atzF*) are organized in an operon, transcribed as a single transcript (*60*). The degradation process is initiated via hydrolytic dehalogenation by atrazine chlorohydrolase (AtzA), with secondary reactions, i.e., sequential removal of amino-ethyl and amino-isopropyl groups catalyzed via the AtzB and AtzC amidohydrolases, respectively (61). The triazine ring is then subject to cleavage by AtzD, with subsequent mineralization to CO_2 and NH_4 by the enzymes AtzE and AtzF. The three genes in the operon containing atzA, atzB, and atzC are flanked by insertion sequences providing mobility of these genes. Consequently these genes have been observed in several diverse species and are highly conserved (62). The findings that all genes for atrazine degradation occur on a single transmissible plasmid and insertion sequences are proximal to genes encoding for initial metabolism, suggest that either some or all the genes for atrazine mineralization may be readily lost or transferred to another prokaryote (60). These results help explain how accelerated degradation and loss of herbicidal efficacy can occur with repeated atrazine use.

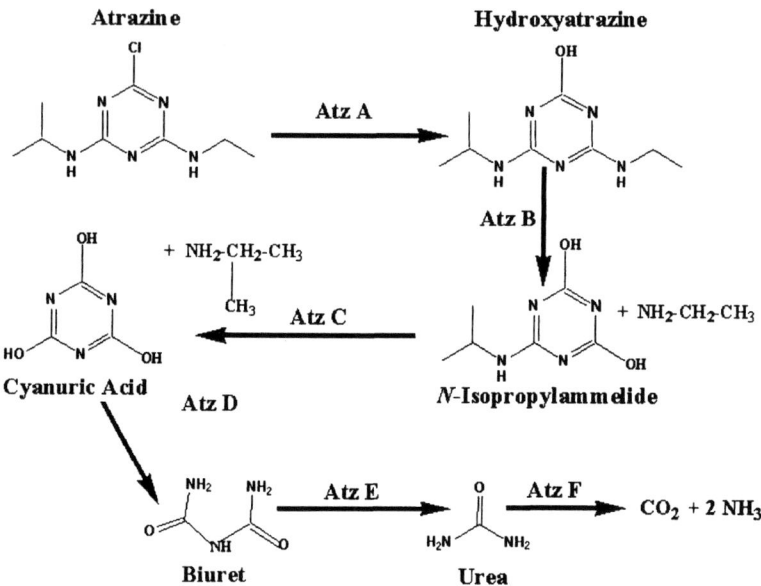

Figure 2. Pathway for atrazine metabolism in Pseudomonas strain ADP
(Adapted from Reference 60)

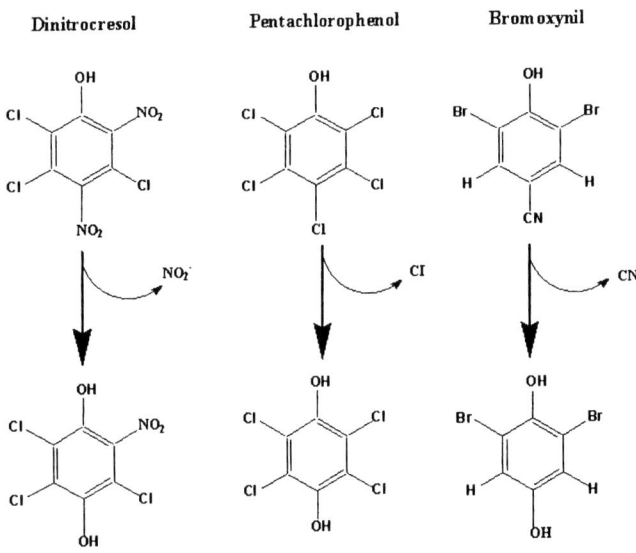

Figure 3. Pesticide substrates for pentachlorophenol flavin monooxygenase from Sphingomonas sp.

Several dehalogenation processes may be involved in the degradation of pesticides containing multiple halogen atoms. Dehalogenation may occur prior to, or after ring cleavage. The mineralization of pentachlorophenol by *Sphingomonas* sp. utilizes two dehalogenases (*56*). The initial oxidative removal of the *p*-chlorine is catalyzed by pentachlorophenol flavin monooxygenase (PcpA). Subsequent thiolytic removal of the two *m*-chlorines is catalyzed by the GST, tetrachlorohydroquinone dehalogenase (PcpB). The remaining chlorine is released from 6-chloro-1,2,4-benzenetriol after ring fission. The *Sphingomonas* pentachlorophenol flavin monooxygenase has a broad substrate specificity (Figure 3). In addition to the oxidative removal of the *p*-chlorine from pentachlorophenol, this enzyme liberates cyanide from the herbicide bromoxynil (3,5-dibromo-4-hydroxybenzonitrile) (*63*), or releases nitrite from the herbicide DNOC (2,4-dinitrocresol) (*64*). Three distinct dehalogenases are involved in mineralization of the insecticide lindane (1,2,3,4,5,6-hexachlorocyclohexane) by *Sphingomonas paucimoblis*: 1) LinA (a dehydrochlorinase) catalyzes the release of chlorines 1 and 4; 2) LinB (a chlorohydrolase) hydrolytically removes chlorines 2 and 5; and 3) LinD (a GST) catalyzes thiolytic removal of chlorines 3 and 6 (*57*). Both the PcpB and LinD reductive dehalogenases oxidize glutathione, without accumulation of the GSH conjugates of their substrates (*65*). Since these GSTs are involved in the mineralization of these pesticides, recycling GSH during the degradation process conserves energy and maintains GSH pools. These reductive GSTs are unique to prokaryotes and are not found in plants.

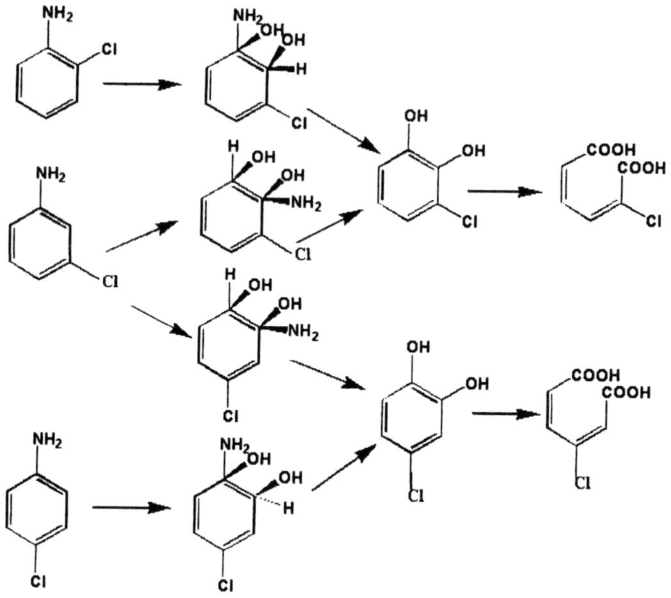

Figure 4. Dioxygenase-mediated cleavage of chloroanilines by Pseudomonas sp. Adapted from reference 67.

Aerobic prokaryotic fission of aromatic ring structures is catalyzed via two dioxygenases. The first reaction incorporates a molecule of oxygen into the substrate by dihydroxylation of the ring structure and requires either NADH or NADPH (*66*). Aromatic ring dioxygenases are multicomponent enzymes, including proteins that function in electron transport (reductase, ferredoxin), plus a hydroxylating oxygenase that incorporates molecular oxygen. Substrate specificity of the dioxygenase is mediated by the terminal iron-sulfur hydroxylase. Following hydroxylation, a second class of dioxygenases initiates ring cleavage. This can occur either between two hydroxyl groups (ortho or intradiol), or proximal to one of the hydroxyl groups (meta or extradiol).

Pathways for dioxygenase-mediated cleavage of chloroanilines by *Pseudomonas* sp. are outlined in Figure 4 (*67*). In this pathway, a molecular reorganization of the chlorine and amino groups occurs as a result of ring dihydroxylation and subsequent deamination. Following ring cleavage, the two resulting dicarboxylic acids enter the TCA cycle for further metabolism.

Prokaryote Genes for Creating Herbicide-Tolerant Crops

As outlined previously, prokaryotes can provide unique enzymes (especially hydrolytic) for transformation and deactivation of herbicides. Prokaryotic genes encoding for pesticide degrading enzymes have been used to transform various crop plants for resistance to several herbicides (Table III), but not all approaches have been commercialized. Transgenic crops for herbicide resistance have revolutionzed weed control and crop production in North and South America, but concern over genetic engineering of food crops has limited adoption of this technology in Asian and European markets. Among the successful genetically modified, herbicide-resistant crops that have been commercialized are cotton and potatoes transformed with the *bxn* gene from *Klebsiella ozaenae* (*71*). The basis of this resistance is the expression of a nitrilase that converts bromoxynil to the non-phytotoxic metabolite 3,5 dibromo-4-hydroxybenzoic acid. Phosphinothricin (PPT, L-2-amino-4-[hydroxy(methyl)phosphinyl]butyric acid) is the active ingredient of the natural product herbicide bialaphos, and the synthetic herbicide glufosinate (ammonium salt of PPT). Glufosinate tolerant crops including canola, corn, soybeans and rice have been developed based on their transformation with the *bar* and *pat* genes from various *Streptomyces* species (*69, 70*). The *bar* and *pat* genes encode *N*-acetyl transferases that inactivate either glufosinate or phosphinothricin via acetylation, thereby conferring resistance.

The sole basis for resistance to glyphosate [*N*-(phosphonomethyl)glycine] in most commercial transgenic crop plants is the insertion of a glyphosate-insensitive 5-enolpyruvylshikimic acid-3-phosphate synthase EC 2.5.1.19 (EPSPS) gene from an *Agrobacterium* strain CP4 allowing expression of a functional shikimic acid pathway (*75*). The *Ochromobacter antropi* (*Achromobacter* sp. strain LBAA) *gox* gene encodes glyphosate oxidioreductase which transforms glyphosate to glyoxylate and aminomethylphosphonate (*72*). This bioengineering strategy has limited utility because complete resistance to glyphosate was not achieved in *gox* transgenic crops (*75*). However *gox* is one of two resistance genes incorporated in glyphosate resistant canola and sugarbeet (*76*). The *tfdA* gene from *Alcaligenes eutrophus* (currently classified as *Ralstonia eutrophus*) encoding for the initial transformation of 2,4-D to 2-4-dichlorophenol and glyoxylate, has been successfully expressed in cotton, however no commercial lines have been released (*68*). All of these approaches have utilized a gene that inactivates a single herbicide or herbicide class. However another strategy is use of a gene that will encode for an enzyme that will detoxify a variety of pesticides have been suggested , e.g., cytochrome P450 (*77*).

Table III. Bioengineering Plants for Herbicide Resistance using Prokaryote Genes.

Herbicide	Enzyme	Gene Source	Reference
2,4-D	2,4-D dioxygenase	*Alcaligenes eutrophus* (*tfdA*)	(*68*)
Bialaphos (Glufosinate)	Acetyl transferase	*Streptomyces* sp. (*bar*) and (*pat*)	(*69, 70*)
Buctril	Nitrilase	*Klebsiella ozaenae* (*bxn*)	(*71*)
Glyphosate	Glyphosate oxidase	*Escherichia coli* (*gox*)	(*72*)
Dalapon	Dehalogenase	*Pseudomonas putida*	(*73*)
Phenmedipham	Phenmedipham hydrolase	*Arthrobacter* (*pcd*)	(*74*)

Conclusions

Advances in understanding the mechanisms and enzymes involved in plant and prokaryote pesticide metabolism can broaden our scientific knowledge of the fate of xenobiotics. Understanding the metabolism of pesticides in plants is essential to comprehend the mode of action of a given pesticide in various species (selectivity), its physiological role, and to predict its persistence in plants. This is essential for the safe use of pesticides as well as improving pesticide design. Knowledge of microbial metabolism of pesticides improves our ability to predict the environmental fate of a pesticide and develop management approaches to minimize risks to the environment. Microbial systems can be developed for the bioremediation of pesticide wastes and contaminants, and when exploited with living plants, i.e., phytoremediation, may provide more expansive benefits for contaminant removal. Much has been discovered thus far since the inception of use of synthetic organic compounds as pesticides. Promising gains in pesticide management have been accomplished via transgenic crops expressing prokaryotic genes.

Literature Cited

1. Hall, J.C.; Hoagland, R.E.; Zablotowicz, R.M.; Eds. *Pesticide Biotransformations in Plants and Microorganisms: Similarities and Divergences*, Hall, J.C.; Hoagland, R.E.; Zablotowicz, R.M., Eds.; ACS Symp. Ser. 777, 2001; 432 pp.
2. Hoagland, R.E.; Zablotowicz, R.M.; Hall, J.C. In *Pesticide Biotransformations in Plants and Microorganisms: Similarities and Divergences*, Hall, J.C.; Hoagland, R.E.; Zablotowicz, R.M., Eds.; ACS Symp. Ser. 777, 2001; pp 2-27.
3. Van Eerd, L.L.; Hoagland, R.E.; Zablotowicz, R.M.; Hall, J.C. *Weed Sci.* **2003**, *51*, 472-495.
4. Roberts, T. Ed. *Metabolism of Agrochemicals in Plants*. Wiley, New York. 2000; 300 pp.
5. Hoagland, R.E.; Zablotowicz R.M. In *Encyclopedia of Agrochemicals*. Plimmer, J.R. Ed.; Wiley Interscience, New York, Vol. III, pp 1034-1052.
6. White, D. *The Physiology and Biochemistry of Prokaryotes*, 2nd Edition, Oxford University Press, New York, 2000; pp 1-36.
7. Ferl, R.; Paul, A.-L.; In *Biochemistry and Molecular Biology of Plants* Buchannan, B.B.; Gruissem, W.; Jones, R.; Eds.; Amer. Soc. Plant Physiol.: Rockville, MD, 2000; pp 312-357.
8. Shimabukuro, R.H.; Lamoureux, G.L.; Frear, D.S. In *Biodegradation of Pesticides;* Matsamura, F; Krishna Muri, C.R., Eds.; Plenum: New York, 1982; pp 21-66.
9. Klambt, H.D. *Planta*, 1961, *57*, 339-353.
10. Mine, A.; Miyakado, M.; Matsunaka, S. *Pestic. Biochem. Physiol.* 1975, *5*, 566-574.
11. Andreae, W.A.; Good, N.E. *Plant Physiol.* 1957, *32*, 566-572.
12. Feung, C.; Hamilton, R.H.; Witham, F.H. *J. Agric. Food Chem.* 1971, *19*, 475-479.
13. Romano, M.L.; Stephenson, G.R.; Tal, A.; Hall, J. C. *Pestic. Biochem. Physiol.* 1993, *46*, 181-189.
14. Tal, A.; Hall, J.C.; Stephenson, G.R. *Weed Res.* **1995**, *35*, 133-139.
15. Scarponi, L.; Perucci, P.; Matinetti, L. *J. Agric. Food Chem.* **1991**, *39*, 2010-2113.
16. Marrs, K.A. *Ann. Rev. Plant Physiol. Plant Mol. Biol.* **1996**, *47*, 127-158.
17. Hatzios, K.K. In *The Metabolism of Agrochemicals in Plants*, Miyamoto, J.; Roberts, T.R., Eds.; John Wiley and Sons, Ltd., London, UK, 2000; pp 259-294.
18. Edwards, R.; Dixon, R.A. *Phytochemistry* **1991**, *30*, 79-84.
19. Marrs, K.A.; Alfenito, M.R.; Lloyd, A.M.; Walbot, V. *Annu. Rev. Plant Physiol. Plant Mol. Biol.* **1995**, *47*, 127-158

20. Hatzios, K.K. In *Pesticide Biotransformations in Plants and Microorganisms: Similarities and Divergences*, Hall, J.C.; Hoagland, R.E.; Zablotowicz, R.M., Eds.; ACS Symp. Ser. 777, 2001; pp 218-239.
21. Sharma, Y.K.; Davis, K.R. *Plant Physiol.* **1994**, *105*, 1089-1096.
22. Dudler, R.; Hertig.; Rebmann, G.; Bull, J.; Mauch, F. *Mol. Plant Microbe Interact.* **1991**, *4*, 14-18.
23. Kusaba, M.; Takahashi, Y.; Nagata, T. *Plant Physiol.* **1996**, *111*, 1161-1167.
24. Blake-Kalff, M.M.A.; Randall, R.A.; Coleman, J.O.D. In *Regulation of Enzymatic Systems Detoxifying Xenobiotics in Plants;* Hatzios, K.K. Ed.; NATO ASI Series, Kluwer Academic Publishers, Dodrecht, 1997; pp 245-259.
25. Schröder, P. In *Regulation of Enzymatic Systems Detoxifying Xenobiotics in Plants;* Hatzios, K.K. Ed.; NATO ASI Series, Kluwer Academic Publishers, Dodrecht, 1997; pp 233-244.
26. Gaillard, C.; Dufaud, A.; Tommasini, R.; Kreuz, K.; Amhrein, N.; Martinoia, E. *FEBS Letts.* **1994**, *352*, 219-221.
27. Li, Z-S.; Zhao, Y.; Rhea, P.A. *Plant Physiol.* **1995a**, *107*, 1257-1268.
28. Li, Z-S.; Zhao, Y.; Rhea, P.A. *Plant Physiol.* **1995b**, *109*, 177-185.
29. Martinoia, E.; Grill, E.; Tommasini, R.; Kreuz, K.; Amhrein, N. *Nature*, **1993**, *364*, 247-249.
30. Lamoureux, G.L.; Shimabukuro, R.H.; Frear, D.S. In *Herbicide Resistance in Weeds and Crops;* Caseley, J.C., Cussans, G.W., Atkin, R.K., Eds.; Butterworth-Heinemann, Oxford, 1991; pp 227-261.
31. Cole, D.J.; Cummins, I.; Hatton, P.J.; Dixon, D.; Edwards, R. In *Regulation of Enzymatic Systems Detoxifying Xenobiotics in Plants;* Hatzios, K.K. Ed.; NATO ASI Series, Kluwer Academic Publishers, Dodrecht, 1997; pp 139-154.
32. Lamoureux, GL., Rusness, D.G. In *Coenzymes and Cofactors. Glutathione: Chemical, Biochemical and Medical Aspects*; Dolphin, D., Poson, R., Avramovic, O.; Eds.; Wiley, New York, 1989; pp 153-196.
33. Lamoureux, G.L., Rusness, D.G. In *Sulfur Nutrition and Assimilation in Higher Plants*; De Kok, L.J., Stulen, I., Rennenberg, H., Brunold, C., Rauser, W.; Eds.; SPB Academic Publishing, The Hague, 1993; pp 221-237.
34. Timmermann, K.P. *Physiol. Plant.* **1989**, *77*, 323-342.
35. Zajc, A.; Neufeind, T.; Prade, L.; Reinemer, P.; Huber, R.; Bieseler, B. *Pestic Sci.* **1999**, *55*, 248-252.
36. Breaux, E.J. *Weed Sci.* **1987**, *35*, 463-468.
37. Breaux, E.J.; Patanella, J.E.; Sanders, E.F. *J. Agric. Food Chem.* **1987**, 35, 474-478.

38. Farago, S.; Kreuz, K.; Brunold, C. *Pestic. Biochem. Physiol.* **1993**, *47*, 199-205.
39. Boger, P., Sommer, A. In *Pesticide Biotransformations in Plants and Microorganisms: Similarities and Divergences*, Hall, J.C., Hoagland, R.E., Zablatowicz, R.M., Eds; ACS Symposium Series 777, 2001; pp 253-267.
40. Lefsrud, C.; Hall, J. C. *Pestic. Biochem. Physiol.* **1989**, *34*, 218-227.
41. Kocher, H.; Buttner, B.; Schmidt, E.; Lotzsch, K.; Schulz, *Brighton Crop Protection Conference – Weeds* **1989**, *2*, 495-500.
42. Yaacoby, T.; Hall, J.C.; Stephenson, G.R. *Pestic. Biochem. Physiol.* **1991**, *41*, 296-304.
43. Tal, A.; Romano, M.L.; Stephenson, G.R.; Schwann, A.L.; Hall, J.C. *Pestic. Biochem. Physiol.* **1993**, *46*, 190-199.
44. Stephenson, G.R.; Tal, A.; Vincent, N.A.; Hall, J. C. *Weed Technol.* **1993**, *7*, 163-168.
45. Romano, M.L.; Tal, A.; Yaacoby, T.; Stephenson, G.R.; Hall, J.C. *Brighton Crop Protection Conference - Weeds.* **1991**, *3*, 1087-1094.
46. Hall, J.C.; Stephenson, G.R. *Brighton Crop Protection Conference - Weeds.* **1995**, *1*, 261-268.
47. Skidmore, M.W.; Paulson, G.D.; Kuiper, H.A.; Ohlin, B.; Reynolds, S. *Pure Appl. Chem.* **1988**, *7*, 1423-1447.
48. Sandermann, H., Jr.; Arjmand, M.; Gennity, I.; Winkler, R.; Stuble, C.B.; Aschbacher, P.W. *J. Agric. Food Chem.* **1990**, *38*, 1877-1880.
49. Sandermann, H., Jr. In *Pesticide Biotransformations in Plants and Microorganisms: Similarities and Divergences*, Hall, J.C., Hoagland, R.E., Zablatowicz, R.M., Eds.; ACS Symposium Series 777, 2001; pp 119-128.
50. Alexander, M.A. *Biodegradation and Bioremediation,* Academic Press: San Diego, 1994; pp. 177-193.
51. Zablotowicz, R.M.; Hoagland, R.E.; Locke, M.A.; Hickey, W.J. *Appl. Environ. Microbiol.* **1995**, *61*, 1054-1060.
52. Hoagland, R.E.; Zablotowicz, R.M. *J. Agric. Food Chem.* **1998**, *45*, 4759-4765.
53. Dick, R.E.; Quinn, J.P. *FEMS Microbiol. Letts.* **1995**, *134*, 177-182.
54. Zablotowicz, R.M.; Locke, M. A.; Hoagland, R. E.; Knight, S. S.; Cash, B. *Environ. Toxicol.* **2001**, *16*, 9-19.
55. Fetzner, S.; Lingens, F. *Microbiol. Rev.* **1994**, *58*, 641-685.
56. Cassidy, M.B.; Lee, H.; Trevors, J.T.; Zablotowicz, R.M. *J. Ind. Microbiol. Biotechnol.* **1999**, *23*, 232-241.
57. Nagata, Y.T.; Miyauchi, K.; Takagi M. *J. Ind. Microbiol. Biotechnol.* **1999**, *23*, 380-390.
58. DeSouza, M.L.; Wackett, L.P.; Boundy-Mills, K.L.; Mandelbaum, R.T.; Sadowsky, M.J.; Appl. Environ. Microbiol. **1995**, 61, 3373-3378.

59. Barkovskii, A.L. In *Pesticide Biotransformations in Plants and Microorganisms: Similarities and Divergences*, Hall, J.C., Hoagland, R.E., Zablatowicz, R.M., Eds; ACS Symposium Series 777, 2001; pp 283-308.
60. Martinez, B.; Tomkins, J.; Wackett, L.P.; Wing, R.; Sadowsky, M.J. *J. Bacteriol.* **2001**, *183*, 5684-5697.
61. Sadowsky, M.J.; Wackett, L.P. In *Pesticide Biotransformations in Plants and Microorganisms: Similarities and Divergences*, Hall, J.C.; Hoagland, R.E.; Zablotowicz, R.M., Eds.; ACS Symp. Ser. 777, 2001; pp 268-282.
62. DeSouza, M.L.; Seffernick, J.; Martinez, B.; Sadowsky, M.J.; Wackett, L.P. *J. Bacteriol.* **1998**, *180*, 1951-1954.
63. Topp, E.; Xun, L.; Orser, C.S. *Appl. Environ. Microbiol.* **1992**, *58*, 502-507.
64. Zablotowicz, R.M.; Leung, K.T.; Alber, T.; Cassidy, M.B.; Trevors, J.T.; Lee, H.; Hall, J.C. *Can. J. Microbiol.* **1999**, *45*, 840-848.
65. Vuilleumier, S. In *Pesticide Biotransformations in Plants and Microorganisms: Similarities and Divergences*, Hall, J.C.; Hoagland, R.E.; Zablotowicz, R.M., Eds.; ACS Symp. Ser. 777, 2001; pp 240-252.
66. Gibson, D.T.; Parales, R.E. *Cur. Opin. Biotechnol.* **2000**, *11*, 236-243.
67. Häggblom, M.M. *FEMS Microbiol. Rev.* **1992**, *103*, 29-72.
68. Bayley, C.; Trolinder, N.; Morgan, M.; Quisenberry, J.E.; Ow, D.W.; Ray, C. *Theor. Appl. Genet.* **1992**, *83*, 645-649.
69. DeBlock, M.; Botterman, J.; Vanderwiele, M ; Dockz, J.; Thoen, C.; Gossele, V.; Movra, N.R.; Thompson, C.; van Montagu, J.; Leemans, J. *EMBO J.* **1987**, *6*, 2513-2518.
70. Broer, I ; Arnold, W.; Wohlleben, W.; Pühler, A. In *Proc. Braunschweig Symp. Applied Plant Molecular Biol.*, Braunschweig; Germany, 1989; pp 240.
71. Stalker, D.M.; McBride, K.E.; Malyj, L.B. *Science* **1988**, *242*, 419-423.
72. Barry, G.; Kishore, G.; Padgette, S.; Taylor, M.; Kolacz, K.; Weldon, M.; Re, D.; Eichholtz, D.; Fincher, K.; Hallas, L. In *Biosynthesis and Molecular Regulation of Amino Acids in Plants*, Singh, B.K.; Flores, H.E.; Shannon, J.C., Eds.; Amer. Soc. Plant Physiol.: Madison, WI, 1992; pp 139-145.
73. Wollaston, V.B.; Snape, A.; Cannon, F. *Plant Cell Reports*, **1992**, *11*, 627-631.
74. Streber, W.R.; Kutschka, U.; Thomas, F.; Pohlenz, H.D. *Plant Mol. Biol.* **1994**, *25*, 977-987.
75. Padgette, S.R., Re, D.B.; Barry, G.F.; Eichholtz, D.E.; Delannay, X.; Fuchs, R.L.; Kishore, G.M.; Fraley, R.T. In *Herbicide Ressistant Crops*; Duke, S. O., Ed.; CRC Press, Boca Raton, FL, 1996; pp 53-84.
76. U.S. Food and Drug Admininstration List of Completed Consultations on Bioengineered Foods. Online internet vm.cfsan.fda.gov/~ird/biocon.html.
77. Ohkawa, H.; Tsujii, H.; Ohkawa, Y. *Pestic. Sci.* **1999**, *55*, 867-874.

Chapter 16

Comparative In Vitro Metabolism of [^{14}C]Methoxychlor in Vertebrate Species Using Precision-Cut Liver Slice Technique

K. Ohyama

Metabolism Laboratory-I, Chemistry Division, Institute of Environmental Toxicology, Ibaraki 303–0043, Japan

Introduction

The organochlorine insecticide methoxychlor (MXC) [1,1,1-trichloro-2,2-bis(4-methoxyphenyl) ethane] is a structural analogue of the environmentally persistent compound DDT. Although MXC is biodegradable (*1-3*) and it exhibits relatively lower acute toxicity in mammals compared with DDT (*1*), various studies have shown that methoxychlor may elicit estrogenic responses in mammals (*4,5*) and also wildlife including fish (*6,7*). Such estrogenic effects of methoxychlor are believed to be caused by the action of its demethylated metabolites (*8-10*), which are major phase I metabolites catalyzed by the cytochrome P450 monooxygenase system. Therefore, it is important to understand the metabolism of methoxychlor in mammals and also wildlife species. Most of the studies, however, have focused on the phase I metabolic reaction (functionalization reaction) of methoxychlor, but little information is available for methoxychlor phase II metabolism (conjugation reaction).

Since the precision-cut tissue slice technique using an automated mechanical tissue slicer (*e.g.* Krumdieck Tissue Slicer) was introduced, it has been applied to drug and xenobiotic metabolism and toxicity studies in recent years (*11-13*). The slice of liver tissue contains many intact cell layers which maintain hepatic architecture and cell-to-cell interactions, therefore liver slices are expected to provide integrated profiles of phase I and phase II metabolites that would be more similar to *in vivo* metabolism than other *in vitro* models. This paper describes comparative *in vitro* metabolism of methoxychlor by precision-cut liver slices from rat, mouse, Japanese quail and rainbow trout.

In vitro metabolism of methoxychlor by precision-cut rat, mouse, Japanese quail and rainbow trout liver slices

The liver slices of male rats (Sprague-Dawley, 9 weeks old), male mice (CD-1, 18-19 weeks old), male Japanese quail (WE strain, 100-130 g body wt.) and juvenile rainbow trout (*Oncorhynchus mykiss*, 100-250g body wt.) were prepared using the Krumdieck Tissue Slicer (Alabama Research & Development) in ice chilled Krebs-Henselite buffer (containing 20mM fructose, pH 7.4) (*14*). The slices were cultured in Krebs-Henselite buffer containing 5 μM of [ring-U-^{14}C]methoxychlor (5.84 MBq/mg) in 12-well plates (*15*) with continuous gyratory shaking under air/CO_2 (95/5) atmosphere (1 slice/1 mL medium/well). Metabolites in the cultured medium and slice extracts were analyzed by a reversed phase radio-HPLC and LC/ESI/MS/MS. Validity of the test preparation was evaluated by monitoring a reaction of 7-ethoxycumarine metabolism (*12, 16*), forming 7-hydroxycoumarin and its sulfate and glucuronide conjugates, concurrently with each experimental run.

In precision-cut liver slice preparation from all test animal species, time-dependent metabolism of methoxychlor was observed, and quantitative and qualitative differences were detected (*17*). Representative HPLC profiles of methoxychlor metabolites after metabolic reaction are shown in Figures 1 and 2. In rat preparations (Figure 1), a bis-demethylated compound (bis-OH-MXC) and its *O*-glucuronide were the major metabolites. The glucuronide conjugate of mono-OH-MXC (mono-demethylated methoxychlor) was not found, even though trace amounts of the mono-OH-MXC was detected as the transient intermediate at shorter incubation times. Therefore, methoxychlor was appeared to be quickly metabolized to bis-OH-MXC by sequential *O*-demethylation, followed by subsequent *O*-glucuronidation. A very polar metabolite, bis-OH-MXC 4-*O*-sulfate 4'-*O*-glucuronide, was also formed as the rat-specific metabolite. On the other hand, mono-OH-MXC glucuronide was the major metabolite for mouse and Japanese quail, and bis-OH-MXC glucuronide was observed only as the minor one, so that mono-*O*-demethylation and subsequent *O*-glucuronidation may be considered as the main metabolic reactions for both

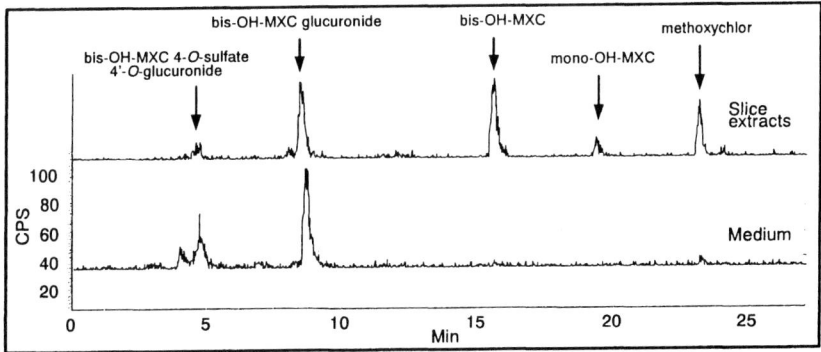

Figure 1. Representative radio-HPLC chromatograms of in vitro metabolism of [^{14}C]methoxychlor by precision-cut rat liver slices. The chromatograms show the radioactive components detected in the slice extracts after 1-hr incubation and in cultured medium after 4-hr incubation.

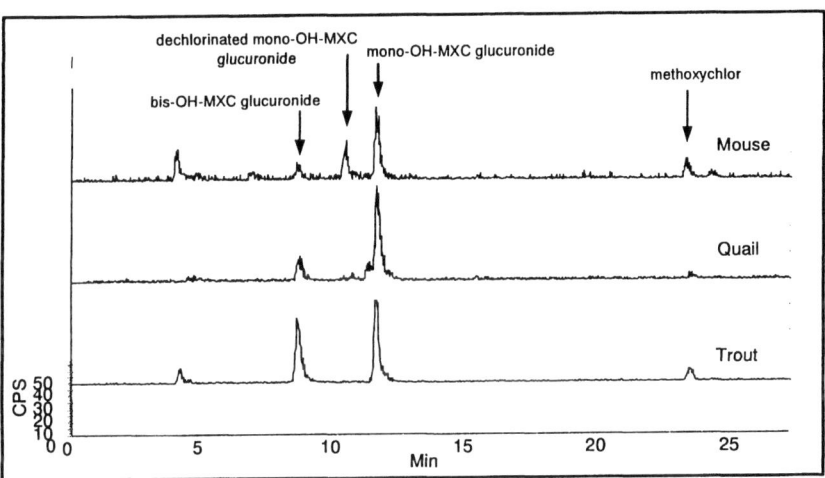

Figure 2. Representative radio-HPLC chromatograms of in vitro metabolism of [^{14}C]methoxychlor by precision-cut mouse, Japanese quail and rainbow trout liver slices. The chromatograms show the radioactive components detected in the cultured medium after metabolic reaction.

species. The reductively dehalogenated metabolite, dechlorinated mono-OH-MXC, was uniquely observed only in mouse preparation. In the case of trout, nearly equal amounts of mono- and bis-OH-MXC glucuronide were produced and unconjugated forms of those metabolites were detected only as minor products. It seems dual metabolic reactions, involving demethylation and glucuronidation, were equally contributed for metabolism of the intermediate metabolite, mono-OH-MXC, for this particular species.

Figure 3 shows the ratio of (mono-OH-MXC)/(mono- + bis-OH-MXC), where the values of mono- and bis- demethylated metabolites also include the amounts of their corresponding conjugates. After reaction in each animal preparation, more than 80% of demethylated metabolites were detected as a mono-demethylated form for mouse and quail, about 50% for trout, and less than 5 % for rat. The results indicate that O-demethylase enzyme(s) in rat can easily remove both methyl groups from the parent molecule, but only one methyl group can be preferably demethylated by the enzyme(s) in mouse and quail. Rainbow trout produces almost equal amounts of mono- and bis- demethylated metabolites. These facts suggest that there are species differences in the manner of phase I oxidative demethylation, and such differences are probably due to the

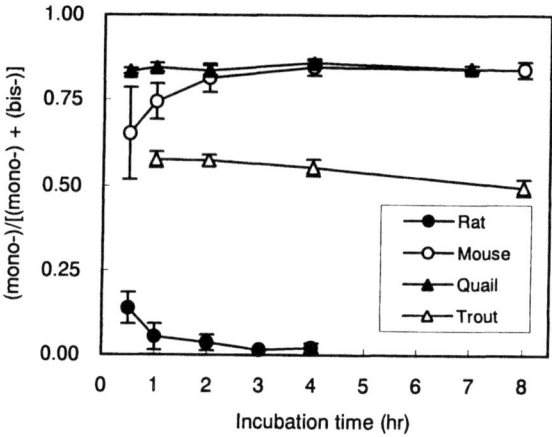

Figure 3. Ratio of mono- and bis-demethylated metabolites of methoxychlor formed by precision-cut rat, mouse, Japanese quail and rainbow trout liver slices. The amounts of corresponding glucuronide conjugate are also included in this analysis.

differences in the reactivity of the contributing P450s enzymes toward methoxychlor and its mono-demethylated metabolite.

Stereoselective formation of the mono-hydroxy metabolite of methoxychlor

Methoxychlor is a prochiral compound, and the mono-demethylation reaction would yield chiral metabolites, (R)- and (S)-mono-OH-MXC (Figure 4). Understanding the nature of the P450 enzymes contributing for the demethylation reaction of methoxychlor in different animal species, stereoselectivity of the initial demethylation step, which is the common metabolic reaction for all animal species tested, was further investigated.

Figure 5 shows the ratio of (R)- and (S)-mono-OH-MXC formed in methoxychlor metabolism. The amounts of corresponding conjugates are also included in this analysis. The enantiomers of the mono-demethylated metabolite were separated by chiral HPLC and identified by HPLC retention time comparison with authentic standards. Rat and mouse selectively formed (S)-mono-OH-MXC enantiomer at approximately 90 and 75 % yield, respectively. In contrast, quail and trout produced (R)-mono-OH-MXC at more than 85 % yield. The results indicate that the oxidative mono-demethylation reaction occurs stereo-selectively and different animal species exhibit different enantioselectivity, suggesting that contributing P450 can recognize the stereological conformation of methoxychlor and that there are species differences in such molecular recognizability.

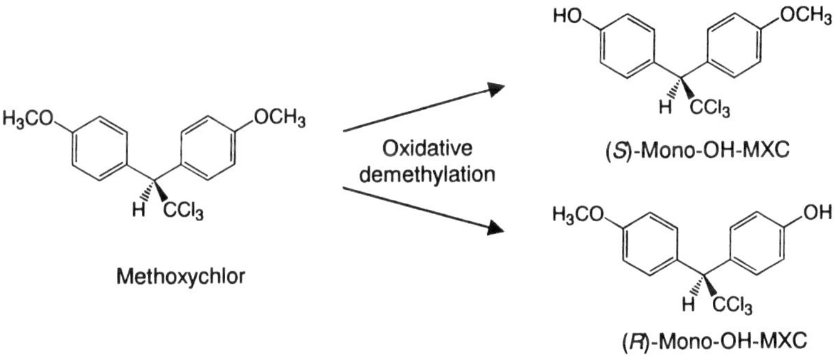

Figure 4. Stereochemistry of oxidative demethylation of methoxychlor.

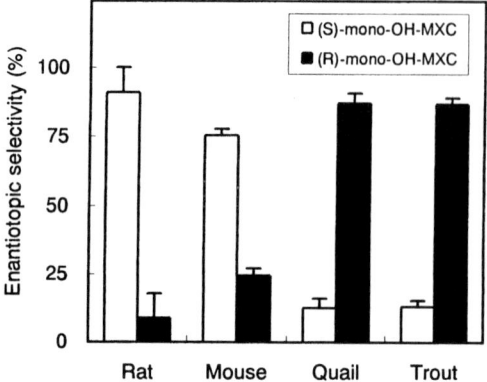

Figure 5. Enantiotopic selectivity in the oxidative mono-demethylation of methoxychlor by precision-cut rat, mouse, Japanese quail and rainbow trout liver slices. The amounts of corresponding conjugates are also included in this analysis.

Summary and conclusions

In the present study, precision-cut liver slices were used to investigate the metabolism of methoxychlor. The incubation of [^{14}C]methoxychlor with rat, mouse, Japanese quail and rainbow trout liver slices resulted in time-dependent metabolism, forming mono- and/or bis- demethylated metabolites. In addition, their corresponding phase II metabolites were subsequently formed as the final products. These metabolites were essentially those already reported as the major methoxychlor *in vivo* metabolites in mammals (*1-3*), therefore the precision-cut liver slice technique is shown to be a useful model to study biotransformations of xenobiotics. By comparing the metabolic profiles obtained from different animal species in detail, clear differences are observed, especially in the pattern of phase I metabolism. Rats showed intensive sequential *O*-demethylation forming bis-demethylated metabolites, in contrast, mouse and quail produce mostly mono-demethylated products. Rainbow trout can equally produce both mono- and bis-demethylated metabolites. It appears that the nature of the cytochrome P450 enzymes involved in the oxidative demethylation reaction are vary in different animal species. The enzymes in quail and mouse preferably remove only one methyl group and the enzyme in rat can easily remove both methyl groups. In addition, such a demethylation reaction proceeds enantioselectively and is also species dependent. The rodent species, rat and mouse, form (S)-mono-OH-MXC, preferentially, and quail and trout selectively produce the (R)-isomer.

Species differences in the metabolism of xenobiotics can be basically explained by the qualitative and quantitative differences of expressed enzymes contributing to the metabolic reactions. Among such enzymes, the cytochrome P450 monooxygenase system plays important roles in the oxidation of structurally diverse compounds, such as drugs and xenobiotics, as well as endogenous compounds. P450 enzymes consist of multiple isoforms, and such multiplicity causes structural and functional diversity of this enzyme. The mechanistic phase I metabolism of methoxychlor by P450s is already well studied and it is known that oxidative O-demethylation is the predominant reaction (18-20). Hu and Kupfer (21) indicated CYP1A2, 2C9, and 2C19 isoforms were likely the major enzymes contributing to methoxychlor metabolism by using cDNA expressed human P450s. Among those CYPs, CYP2C19 and 1A2 contributed to the catalytic reactions forming both mono- and bis-demethylated metabolites and 2C9 mainly contributed to an initial mono-demethylation path. Enantioselectivity of the mono-demethylation reaction by human liver microsomes and P450 isoforms has also been demonstrated (22). The (S)-mono-OH-MXC was the major enantiomer produced by CYP2C9, 2C19 and human liver microsomes, and CYP1A2 preferably forms (R)-mono-OH-MXC. Kishimoto et al. (23) demonstrated that rat CYP2C6 and 2A1 isolated from non-induced male rats, exhibited enantiotopic selectivity of methoxychlor O-demethylation forming (S)-mono-OH-MXC, selectively. They additionally showed, in contrast that rat 2B1 and 2B2, isolated from Phenobarbital-induced male rats, which also exhibited methoxychlor O-demethylase activity, had lower enantiotopic selectivity. These results indicate that there is a different substrate specificity in the different CYPs species contributing for methoxychlor O-demethylation and stereostructural preference may be involved in such substrate specificity. Differences in the substrate specificity of contributing P450s in different animal species may provide one possible explanation for the differences in methoxychlor metabolic profiles observed in the present study.

The metabolic pathways including stereochemistry of methoxychlor metabolism by liver slices obtained in this study are proposed in Figure 6. When the metabolic pathways are expressed in 2-dimentional molecular structures, mouse and quail show quite similar patterns, since the oxidative mono-O-demethylation and subsequent glucuronidation are the main metabolic pathways for both species. However, once the stereological structures are taken into account, the metabolic pathways in these two species turn out to be different, because of the enantioselectivity of the mono-demethylation reactions. Indeed, when the stereological structures are taken into consideration, it is shown that the metabolic pathways of methoxychlor from the four test animal species are all different. Superficially, the reactions involved in methoxychlor metabolism appear relatively simple, since oxidative demethylation and subsequent conjugation are the main processes in most cases. Even in such simple transformations, it has been shown that diverse metabolic reactions in different animal spices are involved.

From the toxicological point of view, the different metabolic profiles of methoxychlor observed in different animal species may need to be considered to understand the metabolism-induced methoxychlor toxicity in individual animal species. In the case of mouse and quail, only trace amounts of bis-OH-MXC, which is believed to be the most active metabolite causing methoxychlor induced estrogenic responses (*4, 8-10*), were detected, therefore, the toxic activity of the main metabolite, mono-OH-MXC and its *O*-glucuronide, also need to be considered. As for enantiotopic metabolites, it would be also important to understand if there is a difference in toxicological activity between (*R*)- and (*S*)-isomers. If marked toxicological difference is observed between these two isomers, the toxicity induced by stereoselective metabolism may need to be further considered.

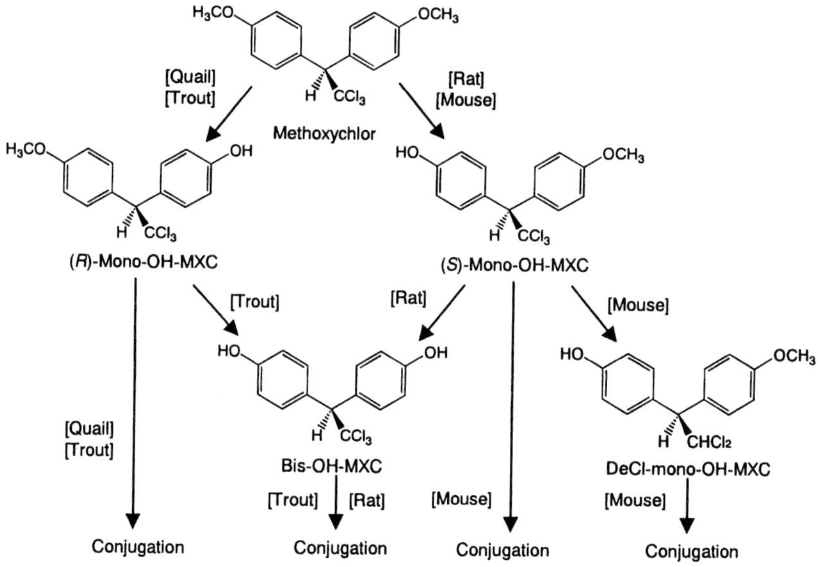

Figure 6. Proposed metabolic pathways of methoxychlor by precision-cut rat, mouse, Japanese quail and rainbow trout liver slices.

Additionally, all phase I metabolites were extensively conjugated with D-glucuronic acid and most metabolites detected after longer incubation, were conjugated forms in any test animal species, indicating that phase II reactions also play important roles for metabolism of methoxychlor. Therefore, contribution of phase II metabolism which may (or possibly may not) work as a detoxification process, needs to be taken into account to explain the *in vivo* toxicity of methoxychlor.

Acknowledgements

The author is grateful to Dr. Norio Kurihara kindly providing authentic (*R*)- and (*S*)-mono-hydroxy methoxychlor standards and for his helpful advice. The author acknowledges Dr. K. Kato, Dr. K. Sato and Mr. S. Maki for their valuable suggestions. I also would like to thank current and past colleagues at the Institute of Environmental Toxicology for their discussions and technical assistance. Part of this study was supported by the Ministry of Agriculture, Forestry and Fisheries of Japan.

References

1. Kapoor I. P.; Metcalf R. L.; Nystrom R. F.; Sangha G. K. *J. Agric. Food Chem.* **1970**, *18*, 1145-1152.
2. Davison K. L.; Feil V. J.; Lamoureux C. H. *J. Agric. Food Chem.* **1982**, *30*, 130-137.
3. Davison K. L.; Lamoureux C. H.; Feil V. J. *J. Agric. Food Chem.* **1983**, *31*, 164-166.
4. Bulger W. H.; Muccitelli R. M.; Kupfer D. *Biochem. Pharmacol.* **1978**, *27*, 2417-2423.
5. Gray L. E., Jr.; Ostby J.; Ferrell J.; Rehnberg G.; Linder R.; Cooper R.; Goldman J.; Slott V.; Laskey J. *Fundam. Appl. Toxicol.* **1989**, *12*, 92-108.
6. Hemmer M. J.; Hemmer B. L.; Bowman C. J.; Kroll K. J.; Folmar L. C.; Marcovich D.; Hoglund M. D.; Denslow N. D. *Environ. Toxicol. Chem.* **2001**, *20*, 336-343.
7. Ankley G. T.; Jensen K. M.; Kahl M. D.; Korte J. J.; Makynen E. A.; *Environ. Toxicol. Chem.* **2001**, *20*, 1276-1290.
8. Shelby M. D.; Newblod R. R.; Tully D. B.; Chae K.; Davis V. L. *Environ. Health Prespect.* **1996**, *104*, 1296-1300.
9. Gaido K. W.; Maness S. C.; Mcdonnell D. P.; Dehal S. S.; Kupfer D.; Safe S. *Mol. Pharmacol.* **2000**, *58*, 852-858.

10. Elsby R.; Maggs J. L.; Ashby J.; Paton D.; Sumpter J. P.; Park B. K. *J. Pharmacol. Exp. Ther.* **2001**, *296*, 329-337.
11. Ekins S. *Drug Metab. Rev.* **1996**, *28*, 591-623.
12. Steensma A.; Beamand J. A.; Walters D. G.; Price R. J.; Lake B. G. *Xenobiotica* **1994**, *24*, 893-907.
13. Singh Y.; Cooke J. B.; Hinton D. E.; Miller M. G. *Drug Metab. Dispos.* **1996**, *24*, 7-14.
14. Dogterom P. *Drug Metab. Dispos.* **1993**, *21*, 699-704.
15. Hashemi E.; Dobrota M.; Till C.; Ioannides C. *Xenobiotica* **1999**, *29*, 11-25.
16. Walsh J. S.; Patanella J. E.; Halm K. A.; Facchine K. L. *Drug Metab. Dispos.* **1995**, *23*, 869-874.
17. Ohyama K.; Maki S.; Sato K.; Kato. *Xenobiotica* **2004**, *unpublished*.
18. Dehal S. S.; Kupfer D.; *Drug Metab. Dispos.* **1994**, *22*, 937-946.
19. Stresser D. M.; Kupfer D. *Drug Metab. Dispos.* **1998**, *26*, 868-874.
20. Kurihara N.; Oku A. *Pestic. Biochem. Physiol.* **1991**, *40*, 227-235.
21. Hu Y.; Kupfer D. *Drug Metab. Dispos.* **2002**, *30*, 1035-1042.
22. Hu Y.; Kupfer D. *Drug Metab. Dispos.* **2002**, *30*, 1329-1336.
23. Kishimoto D.; Oku A.; Kurihara N.; *Pestic. Biochem. Physiol.* **1995**, *51*, 12-19.

Chapter 17

Herbicide Detoxification: Herbicide Selectivity in Crops and Herbicide Resistance in Weeds

Christopher Preston

Cooperative Research Centre for Australian Weed Management and School of Agriculture and Wine, University of Adelaide, PMB 1, Glen Osmond SA 5064, Australia

Crop selectivity is important in allowing weeds to be controlled by herbicides without damage to the crop. Often tolerant crops have more rapid and complete detoxification of a herbicide than occurs in sensitive crop species and weeds. A number of enzymatic systems are able to detoxify herbicides in plants. These include aryl acylamidases that cleave propanil, glutathione reductases that add the tripeptide glutathione to herbicides and cytochrome P450 monooxygenases that hydroxylate herbicides. *O*-Glycosyltransferases may further detoxify hydroxylated herbicides. Glycosylation and glutathione conjugation of herbicides are signals for extrusion of compounds from the cytoplasm, either into the vacuole or into the apoplast. Occasionally intensive use of herbicides will select weed populations with resistance to herbicides as a result of a greater capacity to detoxify herbicides. The types of enzymes responsible for detoxification of herbicides by weeds are the same as those responsible for herbicide selectivity in crops, but the regulation of these enzymes may be different.

Introduction

Herbicides have one almost unique characteristic among crop protection chemicals in that they are designed to control pest plants within a crop of plants. This property creates an immediate problem of how to control weeds without damage to the crop. Fortunately, many herbicides can be used to control weeds without damage to the crop, a property described as crop selectivity. The most common reason for crop selectivity is more rapid detoxification of the herbicide by crop plants (1). However, one consequence of crop selectivity is that weed species closely related to the crop can rarely be controlled; hence the interest in genetically modified herbicide-tolerant crops.

The biochemistry of crop selectivity has been well studied, but it is really only in the past decade, and with the advent of molecular tools, that we have started to obtain an understanding of the complexities involved. Detoxification of herbicides in plants is often a multi-step process and has more similarities to plant secondary metabolism than it has to plant tolerance to toxic compounds in the environment. The current understanding of the various processes involved in herbicide detoxification is summarized in Figure 1 (2,3).

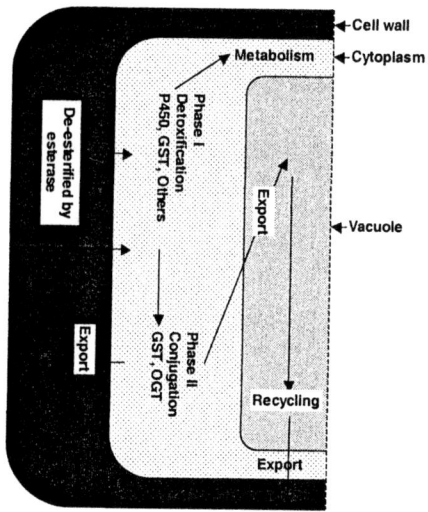

Figure 1. Pathways for the metabolism of herbicides in plant cells. Herbicides may need de-esterification mediated by esterases in the cell wall prior to transport into the cell. Once inside the cell, detoxification followed by conjugation and transport into the vacuole is the main pathway.

Enzymes involved in herbicide detoxification

As indicated in Figure 1, a number of different enzymes participate in the detoxification of herbicides in plants. Some herbicides are acted on by only a single enzyme, whereas for others several modifications occur in sequence. Not surprisingly, the same enzymes that are responsible for selectivity of herbicides in crops are also responsible for herbicide resistance in weeds.

A number of herbicides are formulated as carboxylesters when it is the acid component that is active. The most well known group is the grass herbicides of the aryloxyphenoxypropanoate group. Carboxylesterases of wide substrate specificity are located in the apoplast (4). Recently, it has become apparent that there are multiple carboxylesterases in the apoplast; however, the endogenous roles of these enzymes are unknown (4). Many acid herbicides enter cells via acid trapping (5). The herbicide is protonated in the acid apoplast to a more lipophilic form able to cross the membrane. In the neutral cytoplasm, the herbicide loses the proton, becomes charged and unable to return across the membrane. Therefore, carboxylesterases can be vital for effective entry into cells of acid herbicides formulated as esters. Lack of carboxylesterases could result in a failure of herbicides to be effective through exclusion from the cytoplasm. A failure to de-esterify chlorfenprop-methyl has been suggested as the mechanism of selectivity for this herbicide in wheat (6).

Aryl acylamidase catalyses the hydrolysis of the herbicide propanil and is responsible for providing selectivity to propanil in rice (Figure 2). The endogenous roles of aryl acylamidase are not known, but it is suspected to be involved in nitrogen metabolism (7). The number of genes encoding aryl acylamidase is not yet known and little is known about the regulation of this enzyme.

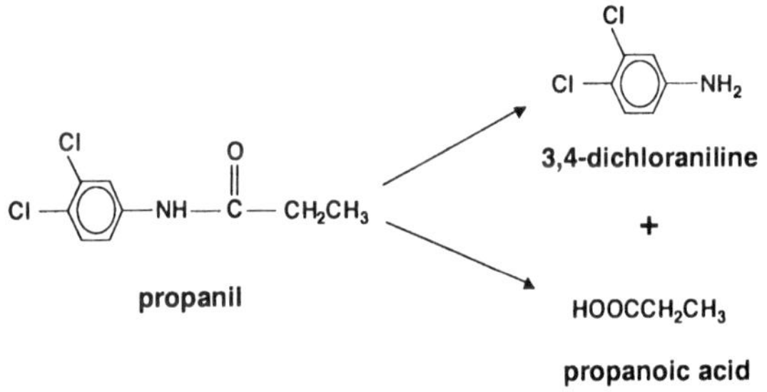

Figure 2. Cleavage of propanil by aryl acylamidase.

Populations of *Echinochloa crus-galli* and *E. colona* have evolved resistance to propanil in rice culture in several countries of the world. In the populations examined, resistance to propanil in these two weed species is the result of increased activity of aryl acylamidase *(8,9)*.

Cytochrome P450 monooxygenases (CYP) are membrane-bound enzymes that catalyze the addition of oxygen to hydrophobic compounds. CYPs are important enzymes in plant secondary metabolism; however, they are also important in the selectivity of a large number of herbicides, particularly in cereals *(10)*. CYPs catalyze a variety of reactions, but the most important ones for herbicide detoxification are hydroxylations and dealkylations (Figure 3). Ring hydroxylation reactions frequently render herbicides inactive; however, dealkylation reactions may only partially detoxify herbicides.

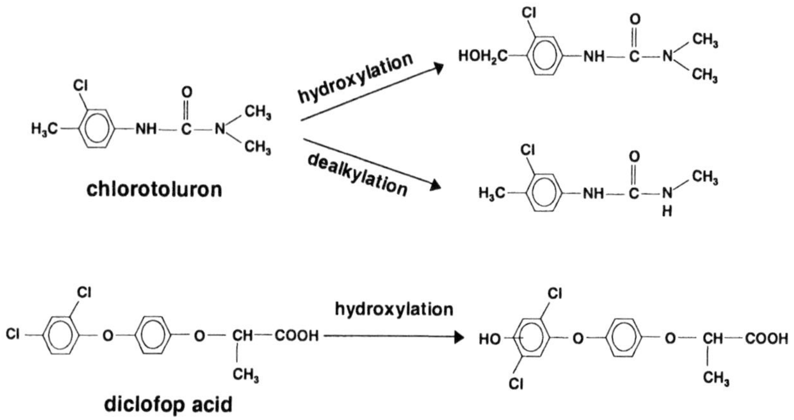

Figure 3. Examples of herbicide detoxification reactions catalyzed by cytochrome P450 monooxygenases.

There are over 180 open reading frames (ORF) for CYPs in the *Arabidopsis thaliana* genome *(11)* and there is increasing evidence that CYPs in plants have narrow substrate specificities. This situation would explain, at least in part, the diversity of patterns of herbicide selectivity between different crop species where CYPs are involved in herbicide detoxification.

Herbicide resistance has evolved in a number of weed species to many herbicides as a result of more rapid CYP-dependent detoxification. Some examples are provided in Table I. Evidence for increased CYP activity in these populations has come from the use of specific CYP inhibitors as synergists and from identification of the hydroxylated products of herbicide detoxification.

Table I. Herbicide Resistance in Weeds as a Result of Increased CYP-dependent Detoxification

Species	Selecting herbicide	References
Alopecurus myosuroides	Chlorotoluron	12
Lolium rigidum	Diclofop-methyl	13
Sinapis arvensis	Mecoprop	14
Avena sterilis	Diclofop-methyl	15
Phalaris minor	Isoproturon	16
Digitaria sanguinalis	Fluazifop-p-butyl	17
Stellaria media	Ethametsulfuron-methyl	18
Echinochloa phyllopogon	Fenoxaprop-p-ethyl	19

Glutathione transferases (GST) are soluble enzymes that catalyze the nucleophilic attack of the tripeptide glutathione on lipophilic, electrophilic compounds. GSTs are well characterized because of their important role in the detoxification of triazine and chloroacetanilide herbicides by maize (3). The endogenous roles of GSTs appear to be associated with plant responses to stress as a variety of stress conditions induce GSTs (20). GSTs are active as homodimers or heterodimers and individual enzymes have broad and overlapping substrate specificities in their reactions with herbicides (21). Typically, GSTs are involved in conjugation reactions such as with atrazine where glutathione is added to the herbicide; however, cleavage reactions, such as with fenoxaprop-p, also occur (Figure 4). GSTs occur as large gene families with 47 ORFs for GST in *Arabidopsis thaliana* (22), 42 different expressed sequences in maize and 25 in soybeans (21).

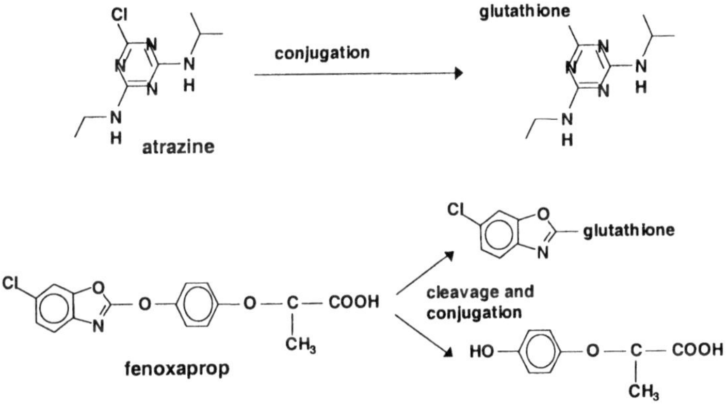

Figure 4. Conjugation and cleavage and conjugation reactions with glutathione with herbicides catalyzed by GST.

A few populations of *Abutilon theophrasti* have evolved resistance to triazine herbicides as a result of increased GST conjugation of the herbicide. One resistant population has an increase in GST with specific action against atrazine (23). It has also been suggested that multiple resistance in *Alopecurus myosuroides* may involve an increase in GST acting as glutathione peroxidase (24).

Herbicides detoxified by CYP are frequently glucosylated as well. UDP-glucose-dependent *O*-glucosyltransferases (OGT) catalyze this reaction. These enzymes add glucose to carboxyl or hydroxyl groups on herbicides (Figure 5). Little is known of OGTs; however, there appears to be multiple isozymes (25). While OGTs participate in the modification of herbicides, they do not appear to be important in either herbicide selectivity in crops or herbicide resistance in weeds.

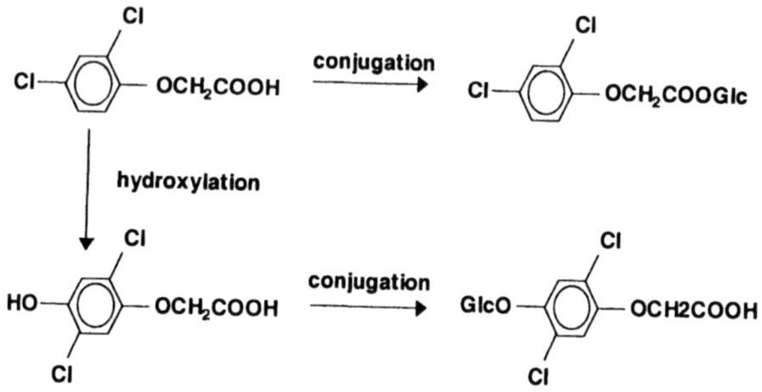

Figure 5. Conjugation of 2,4-D and hydroxy-2,4-D with glucose catalyzed by OGT.

Glutathione or glucose conjugation of herbicides provide substrates for export pumps. In plant cells, specific ATP-driven export pumps for both glutathione conjugates and sugar conjugates are located on the tonoplast membrane. These transporters are members of the ATP-binding cassette transporter family, that also includes multidrug resistance proteins. The herbicide conjugates are typically transported into the vacoule (26,27); however, there is also evidence that glucose conjugates of herbicides can be incorporated into the cell wall (28). Once inside the vacuole, the herbicide conjugates may be further modified.

Regulation of enzymes involved in herbicide detoxification

Most of our understanding of the regulation of the enzymes involved in herbicide detoxification has come from studies with safeners. Safeners are compounds applied to seeds or with herbicides that provide increased protection for cereal crops against the herbicide. Safeners typically act by increasing the rate of detoxification of herbicides (29). Safeners can also increase the activity of conjugating enzymes and conjugate transporters (27). This provides strong evidence that herbicide-detoxifying enzymes are co-ordinately regulated in plants. However, different safeners induce different GST isozymes, suggesting safener action is not a generalised response to stress, but a more specific response with different regulatory pathways (30).

It is tempting to conclude that the same members of the GST and CYP superfamilies are responsible for crop selectivity and herbicide resistance in weeds. However, this is probably unlikely to be the case. It is likely that any member of either of these superfamilies able to detoxify a herbicide could be recruited in the selection process leading to resistance. Indeed, in the cases of resistance in *L. rigidum* and *A. myosuroides* where increased herbicide detoxification is implicated, the weeds also become resistant to herbicides that will control cereals like wheat (12,31).

Multiple herbicide resistance has been intensely studied in *L. rigidum*. Some populations are resistant to many different herbicide chemistries through the increased activity of several detoxifying enzymes (32). It is possible that a single regulatory pathway has been up-regulated in these plants providing resistance to a large number of herbicides. It is known that CYPs will respond to a variety of chemical inducers as well as other stresses (33). When some of these inducers are applied to wheat plants, they increased the rate of detoxification of certain herbicides (Figure 6). For example, phenobarbital (PB) and the herbicide 2,4-D increased detoxification of diclofop acid in wheat and herbicide-susceptible *L. rigidum*. However, these inducers did not increase metabolism of diclofop acid in herbicide-resistant *L. rigidum*. Naphthalic anhydride (NA), PB and 2,4-D increased detoxification of chlorotoluron in wheat, but all inducers were inneffective on *L. rigidum*. NA, PB and ethanol increase detoxification of chlorsulfuron in wheat and both types of *L. rigidum*, but to different extents. PB was the best inducer in wheat, whereas ethanol was the best inducer in both populations of *L. rigidum*, and was better in resistant *L. rigidum* than it was in susceptible *L. rigidum*.

These experiments demonstrate that the pathways for inducing CYPs involved in herbicide detoxification in wheat and *L. rigidum* can be different. In addition, a different pattern of induction occurs in herbicide-resistant *L. rigidum* plants compared to susceptible plants.

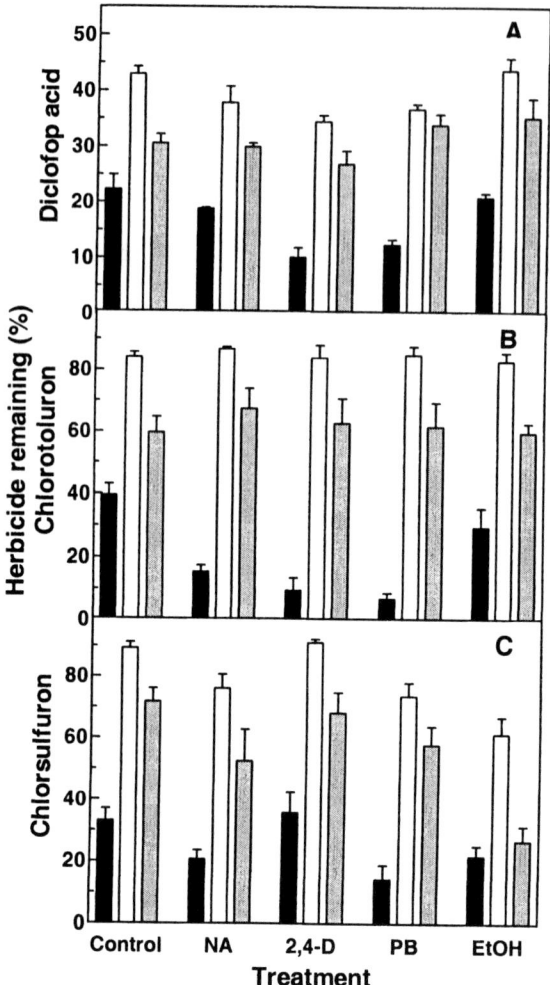

Figure 6. Effect of inducers on detoxification of herbicides in wheat (■), herbicide susceptible L. rigidum (□) and herbicide-resistant L. rigidum (▨). Bars are the mean amount (±SE) of herbicide remaining 24 h after treatment. The CYP inducers naphthalic anhydride (NA, 1 mM), 2,4-D (500 μM), phenobarbital (PB, 1 mM) and ethanol (EtOH, 3%) were applied in hydroponic culture 24 h prior to the addition of herbicides (Preston, C. unpublished data).

Conclusion

Herbicide selectivity in crops may involve a three-step process of detoxification, conjugation and transport into the vacuole. These three processes can be induced in a coordinated fashion by safeners. The major detoxifying enzymes occur in large gene families, which results in considerable variation in selectivity patterns. Our understanding of these enzymes and their regulation has been greatly assisted by new molecular techniques.

Some populations of herbicide-resistant weeds are resistant as a result of increased herbicide detoxification mechanisms. Where this has occurred, the same detoxification processes are involved as in crops. However, as the major detoxifying enzymes occur in large superfamilies in plants, the enzymes recruited for resistance in weeds are not necessarily the same as those responsible for herbicide detoxification in crops. This means that patterns of resistance in weeds do not necessarily follow the patterns of selectivity in crops and that regulation of the detoxification process may be different.

Acknowledgements

Some of the research reported here was supported by the Australian research Council. I would like to thank numerous colleagues for discussions that matured the ideas presented.

References

1. Cole, D.J.: *Pestic. Sci.* **1994**, *42*, 209-222.
2. Coleman, J.O.D.; Blake-Kalff, M.M.A.; Davies, T.G.E.: *Trends Plant Sci.* **1977**, *2*, 144-151.
3. Neuefeind, T.; Reinemer, P.; Bieseler, B.: *Biol. Chem.* **1997**, *378*, 199-205.
4. Haslam, R.;Raveton, M.; Cole, D.J.; Pallet, K.E.; Coleman, J.O.D.: *Pestic. Biochem. Physiol.* **2001**, *71*, 178-189.
5. Sterling, T.M.: *Weed Sci.* **1994**, *42*, 263-276.
6. Fedke, C.; Schmidt, R.R: *Weed Res.* **1977**, 17, 233-239.
7. Hirase, K.; Matsunaka, S.: *Pestic. Biochem. Physiol*, **1991**, *41*, 82-88.
8. Leah, J.M.; Caseley, J.C.; Riches, C.R.; Valverde, B.: *Pestic. Sci.* **1994**, *42*, 281-289.
9. Carey, V.F., III; Hoagland, R.E.; Talbert, R.E.: *Pestic. Sci.* **1997**, *49*, 333-338.

10. Barrett, M.: In *Herbicides and Their Mechanisms of Action*; Cobb, A.H. and Kirkwood,R.C., eds.; Sheffield Academic Press, Sheffield, UK., 2000; pp.25-37.
11. Nelson, D.R.: *Arch. Biochem. Biophys.* **1999**, *369*, 1-10.
12. Kemp, M.S.; Moss, S.R.; Thomas, T.H.: In *Managing Resistance to Agrochemicals: From Fundamental Research to Practical Strategies*; Green, M.B., LeBaron, H.M. and Moberg, W.K., eds.; American Chemical Society, Washington, DC, 1990; pp. 376-393.
13. Preston, C.; Powles, S.B.: *Pestic. Biochem. Physiol.* **1998**, *62*, 179-189.
14. Coupland, D.; Lutman, D.P.J; Heath, C.: *Pestic. Biochem. Physiol.* **1990**, *36*, 61-67.
15. Maneechote, C.; Preston, C.; Powles S.B.: *Pestic. Sci.* **1997**, *49*, 105-114.
16. Singh, S.; Kirkwood, R.C.; Marshall, G.: *Pestic. Sci.* **1998**, *53*, 123-132.
17. Hidayat, I.; Preston, C.: *Pestic. Biochem. Physiol.* **2001**, *71*, 190-195.
18. Veldhuis, L.J.; Hall, L.M.; O'-Donovan, J.T.; Dyer, W.; Hall, J.C.: *J. Agric. Food Chem.* **2000**, *48*, 2986-2990.
19. Fischer, A.J.; Bayer, D.E.; Carriere, M.D.; Ateh, C.M.; Yim, K.O.: *Pestic. Biochem. Physiol.* **2000**, *68*, 156-165.
20. Marrs, K.A.: *Annu. Rev. Plant Physiol. Plant Mol. Biol.* **1996**, *47*, 127-158.
21. McGonigle, B.; Keeler, S.J.; Lau, S-M.C.; Koeppe, M.K.; O'Keefe, D.P.: *Plant Physiol.* **2000**, *124*, 1105-1120.
22. Wagner, U.; Edwards, R.; Dixon, D.P.; Mauch, F.: *Plant Mol. Biol.* **2002**, *49*, 515-532.
23. Plaisance, K.L.; Gronwald, J.W.: *Pestic. Biochem. Physiol.* **1999**, *63*, 34-49.
24. Cummins, I.; Cole, D.J.; Edwards, R.: *Plant J.* **1999**, *18*, 285-292.
25. Brazier, M.; Cole, D.J.; Edwards, R.: *Phytochemistry* **2002**, *59*, 149-156.
26. Martinoia, E.; Grill, E.; Tommasini, R.; Kreutz, K.; Amrhein, N.: *Nature* **1993**, *364*, 247-249.
27. Galliard, C.; Dufaud, A.; Tommasini, R.; Kreutz, K.; Martinoia, E.: *FEBS Lett.* **1994**, *352*, 219-221.
28. Hidayat, I.; Preston, C.: *Pestic. Biochem. Physiol.* **1997**, *57*, 137-146.
29. Davies, J.; Caseley, J.C.: *Pestic. Sci.* **1999**, *55*, 1043-1058.
30. DeRidder, B.P.; Dixon, D.P.; Beussman, D.J.; Edwards, R.; Goldsbrough, P.B.: *Plant Physiol.* **2002**, *130*, 1497-1505.
31. Holtum, J.A.M.; Powles, S.B.: *Proc. Bri. Crop Prot. Conf. Weeds* **1991**, 1071-1078.
32. Preston, C.; Tardif, F.J.; Christopher, J.T.; Powles, S.B: *Pestic. Biochem. Physiol.* **1996**, *54*, 123-134.
33. Schuler, M.A.: *Crit. Rev. Plant Sci.* **1996**, *15*, 235-258.

Chapter 18

Identification of Thiolactic Acid Conjugated Metabolites of Fungicide Diethofencarb in Grape (*Vitis vinifera* L.) and the Mechanism of Their Formation in Plant and Rat

Takuo Fujisawa[1], Luis O. Ruzo[2], Yoshitaka Tomigahara[3], Toshiyuki Katagi[1], and Yoshiyuki Takimoto[1]

[1]Environmental Health Science Laboratory, Sumitomo Chemical Company, Ltd., Takarazuka, Hyogo 665–8555, Japan
[2]PTRL West, Inc., Hercules, CA 94547
[3]Environmental Health Science Laboratory, Sumitomo Chemical Company, Ltd., Osaka 554–8558, Japan

The carbamate fungicide diethofencarb [isopropyl 3,4-diethoxycarbanilate] is very effective for control of various fungal species, especially *Botrytis* sp., *Cercospora* sp. and *Venturia* sp., that are resistant to benzimidazole fungicide. The metabolism study in grape and rat was conducted using ^{14}C-diethofencarb. In grape, unique metabolite such as thiolactic acid attached to the 5-position of phenyl ring of 3-ethoxy-4-hydroxy carbanilate, oxidatively *O*-deethylated form of diethofencarb, *via* C-S linkage was detected. The thiolactic acid conjugate of 3-ethoxy-4-hydroxy carbanilate was also demonstrated to be involved in rat metabolism.

Introduction

Lamoureux et al. had firstly characterized thiolactic acid conjugate of pesticide in various plant species. Through the metabolism of PCNB (pentachloronitrobenzene) in peanut (1), thiolactic acid conjugate, S-(pentachlorophenyl)-2-thiolactic acid, was detected from the root in scarce amount (0.5% of the isolated radiolabelled compounds) while it could not be found from peanut cell suspension culture. With EPTC (S-ethyl dipropylthiocarbamate) using corn, cotton, and soybean suspension cell cultures (2), S-(N,N-dipropylcarbamoyl)-O-malonyl-3-thiolatic acid was characterized as a major metabolite which accounted for up to 33 % of the isolated radiolabelled compounds. Propachlor (2-chloro-N-isopropylacetanilide) was rapidly degraded to homoglutathione conjugate in the root and foliage of the soybean plant, and then further degraded to thiolactic acid S-oxide form (3). Fluorodifen herbicide (2,4'-dinitro-4-trifluoromethyldiphenyl ether) was rapidly metabolized by a spruce cell suspension culture and S-(2-O-glucosyl)-3-thiolactic acid conjugate was presented as major metabolite (4). Although the above studies have shown possible existence of thiolactic acid conjugates as plant metabolites, an ambiguity still remains for the definite determination of their chemical structure because it was conducted only from the results of mass spectrometric analysis.

In this study, extensive spectrometric analysis of plant extracts using liquid chromatography-mass spectrometry (LC-MS) and nuclear magnetic resonance (NMR) spectrometry in conjunction with direct chromatographic comparison with the synthetic standards were conducted to definitively identify thiolactic acid conjugate produced as plant metabolites. Also, metabolism study of diethofencarb in rat was conducted in detail and the involvement of the thiolactic acid conjugate in rat metabolism was thoroughly examined using *in vivo* and *in vitro* experiments.

1. Metabolism of diethofencarb in grape

Identification of metabolites

For the purpose of collecting enough amount of conjugated metabolites subjected to spectrometric analyses (MS, NMR), Pinot Noir grapes under the actual field conditions were used. Diethofencarb (1) labelled at the phenyl ring (8.88 MBq mg^{-1}) and 2-position of the isopropyl group (7.64 MBq mg^{-1}) was prepared in our laboratory (Figure 1). [^{14}C]-1 in the methanol/water solutions was applied to the grape bunches with a hand sprayer at a rate of 1000 g a.i. ha^{-1}, three times with an approximately 40 days interval between two subsequent applications. PHI (pre-harvest interval) is set as 14 days.

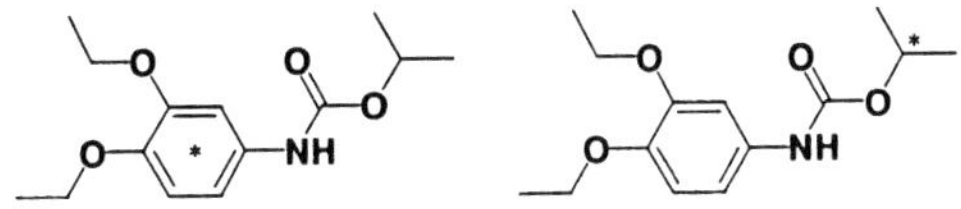

Figure 1.The chemical structure of diethofencarb and its radiolabelled positions.

* indicates label positions

^{1}H NMR spectra were measured with a Varian Unity 300 FT-NMR spectrometer. Liquid chromatography-atmospheric chemical ionization-mass spectrometry (LC-APCI-MS) in positive and negative ion modes was performed using a Hitachi M-1000 spectrometer. Unknown metabolites were isolated and purified from the grape fruits extracts with solid phase extraction method (Porapak Q) and HPLC. Isolated metabolites were analyzed by free form or derivatized (methylated, acetylated) form for identification. All of the non-radiolabelled reference standards **1-6** are synthesized in our laboratory and their chemical structures are shown in Figure 2.

Identification of metabolites 2 and 3

The chemical structure of **2** was easily confirmed by HPLC and TLC co-chromatographies of the isolated metabolite with the synthetic standard.

In the case of **3**, the corresponding HPLC fraction of extracts was collected and divided into two portions. One portion was incubated with cellulase (37 °C, over night) in 10 mM phosphate buffer at pH 5 and a new peak corresponding to **2** appeared *via* concomitant decrease of the peak due to **3**. To further investigate the chemical nature of **3**, the remained portion was subjected to LC-APCI-MS. The metabolite **3** was shown to have a molecular weight of 401. These results strongly suggest glucose, or equivalent, as the conjugation moiety. Definitive identification of **3** was achieved by synthesizing the glucose conjugate of **2** and carrying out direct comparison of the isolated **3** with this synthetic standard by HPLC and TLC co-chromatographies.

Identification of metabolites 4 and 5

The metabolite **5** was incubated with cellulase and a new peak corresponding to **4** appeared *via* concomitant decrease of the peak due to **5**. LC-APCI-MS analysis showed that the molecular weight of **5** was 521. With the methylated **5**, introduction of the two methyl groups were shown by a MS peak at *m/z* 548 [M-H]$^-$. When **5** was acetylated, a MS peak at *m/z* 730 [M-H]$^-$ was observed. In order to estimate the functional groups of **5** in detail, the successive derivatization (methylation followed by acetylation) was conducted. LC-APCI-

MS showed a MS peak at *m/z* 716 [M-H]⁻ and hence, the two methyl and four acetyl groups were added to the molecule. Assuming that the aglycone of **5** is **2**, one methyl group may be introduced *via* methylation of the activated phenol at 4-position, while the other one may be introduced from methylation of a carboxylic acid functionally associated with a conjugating moiety. The acetyl groups introduced are probably due to hydroxy functionalities of a monosaccharide which is also consistent with observed molecular weight. The NMR spectrum of the methylated derivative of **5** was investigated. Only two aromatic protons were observed at 6.95 and 7.05 ppm with the coupling constant of 2.0 Hz, indicating the presence of two aromatic protons in the *meta*-position and hence, some metabolic conversion at 5-position of the phenyl ring was most likely. Furthermore, the isopropyl and one ethoxy groups were found to remain intact. A large number of coupled resonance possibly due to glucopyranosyl protons appeared at 3.10-4.60 ppm, secondary hydroxyl groups, implying that **5** was a sugar conjugate. Taking account of all the spectrometric evidences together with the conjugation form of other pesticides (*1-4*), the chemical structure of **5** was proposed. Finally, the chemical identity of **5** as 3-{3-ethoxy-2-hydroxy-5-[(isopropoxycarbonyl)amino]phenylthio}-2-β-glucopyranosyloxy-propionic acid was definitely confirmed by both HPLC and TLC co-chromatographies with the reference standard.

Considering from the metabolic pathway, **4** was assumed to exist in grape as an intermediate metabolite of **5**. Therefore, the synthetic standard of **4** was used to confirm its existence in the extract of grape by HPLC and TLC co-chromatographies. As a result, the metabolite corresponding to **4** was detected in grape fruit.

Identification of metabolite 6

The metabolite **6** was incubated with cellulase, however, no degradation of **6** was observed. LC-APCI-MS spectrum of **6** showed the same molecular weight as **5** (*m/z* 520 [M-H]⁻), indicating isomeric structure. LC-MS analysis showed a MS peak of the methylated **6** at *m/z* 534 ([M-H]⁻) and that of methylated and acetylated derivative of **6** at *m/z* 744 ([M-H]⁻). The molecular weight of the latter derivative indicated the addition of five acetyl unites to the methylated molecule. These results together with the data on metabolite **5**, implied that **6** was a monosaccharide derivative, but only four acetyl groups was possibly to be introduced. Therefore, the metabolite should have an additional center sensitive to acetylation. Considering all the above spectrometric evidences, and taking chemical structure of **5** into account, **6** was proposed as 3-{3-ethoxy-2-β-glucopyranosyloxy-5-[(isopropoxycarbonyl)amino]phenylthio-2-hydroxy-

propionic acid}] which was definitely confirmed by both HPLC and TLC co-chromatographies with the synthetic standard.
Based on these results, the metabolic pathway of **1** in grape plants is proposed in Figure 2.

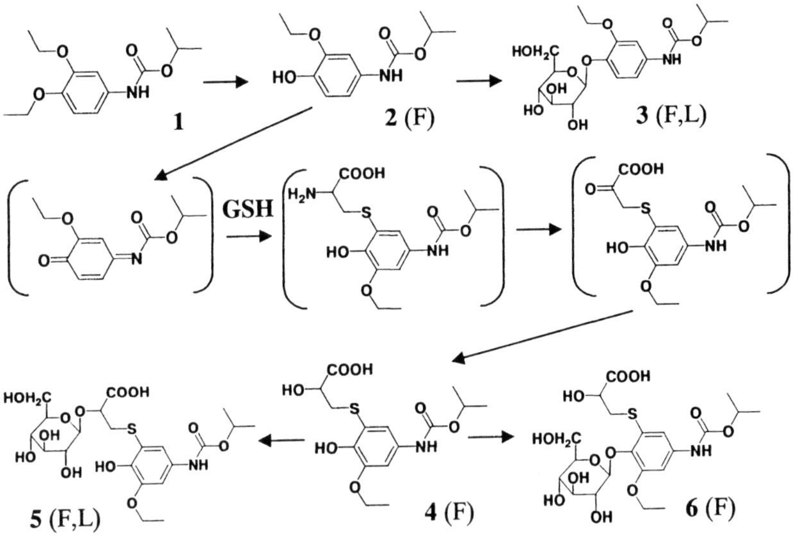

Figure 2. Proposed metabolic pathway of 1 in the grape plants.
F; Fruit treatment, L; Leaf treatment

Distribution of metabolites in grape

For investigation of metabolic profiles, grape vines (*Vitis vinifera* L., cv. Kyoho) with fruits setting at maturity stages were used. [^{14}C]-**1** in the acetonitrile solutions was topically applied at a rate of 500 g a.i. ha^{-1} to grape bunches (fruits) and leaves once with PHI 35 days.

Distributions of radioactivity and metabolites in the grape fruit are summarized in Table 1.

Within the whole grape, **1** remained as the major residue (52.8-60.9 % of total radioactive residue; TRR). Major metabolites detected were **3** (8.1-18.1 %TRR, 0.153-0.289 ppm), **5** (6.4-7.3 %TRR, 0.102-0.137 ppm), and **6** (8.7-13.5 %TRR, 0.165-0.216 ppm). Minor metabolites **2** and **4** were both detected below 1.7 %TRR. Meanwhile, the distribution of radioactivity from the treated leaves was different from that of the fruits (Table 2). Within the whole leaf, **1** was major residue (95.6-95.8 %TRR) and the amounts of metabolites **3** and **5** were less than 0.6 %TRR. These results indicated more penetration of **1** followed by metabolic degradation in fruits as compared with leaves.

Table 1 Distribution of radioactivities and metabolites of [^{14}C]-**1** with fruit treatment.

		[^{14}C-phenyl]-**1**		[^{14}C-isopropyl]-**1**	
		%TRR	ppm[1]	%TRR	ppm[1]
Fruit	1	52.75	0.844	60.87	1.149
	2	0.90	0.014	ND	ND
	3	18.06	0.289	8.11	0.153
	4	1.51	0.024	1.68	0.032
	5	6.39	0.103	7.25	0.137
	6	13.48	0.216	8.71	0.165
	others	6.30[a]	0.101[a]	8.68[b]	0.164[b]
	unextractable	0.61	0.010	4.68	0.088
Total		100.00	1.599	100.00	1.888

[1] ppm of **1** equivalent.
[a] Consisted of more than 6 metabolites, each metabolite below 3.70 %TRR (0.070 ppm).
[b] Consisted of more than 8 metabolites, each metabolite below 0.80 %TRR (0.013 ppm).
Source: Reproduced from ref. 18.

Table 2 Distribution of radioactivities and metabolites of [^{14}C]-**1** with leaf treatment.

		[^{14}C-phenyl]-**1**		[^{14}C-isopropyl]-**1**	
		%TRR	ppm[1]	%TRR	ppm[1]
Leaf	1	95.55	76.257	95.80	61.646
	2	ND	ND	ND	ND
	3	0.49	0.391	ND	ND
	4	ND	ND	ND	ND
	5	0.57	0.458	0.43	0.273
	6	ND	ND	ND	ND
	others	2.33[a]	1.856[a]	3.08[b]	1.969[b]
	unextractable	1.06	0.847	0.69	0.444
Total		100.00	79.809	100.00	64.332

[1] ppm of **1** equivalent.
[a] Consisted of more than 17 metabolites, each metabolite below 0.46 %TRR (0.367 ppm).
[b] Consisted of more than 8 metabolites, each metabolite below 0.55 %TRR (0.348 ppm).
Source: Reproduced from ref. 18.

2. Metabolism of diethofencarb in rats

Identification of metabolites

For metabolite identification, 8 male Crj:CD(SD) rats (7 weeks old) were treated orally with [^{14}C-phenyl]-1 at 300 mg/kg b.w./day for 2 days. Urine and feces were collected for 3 days after the first dose. Feces were not used following isolation. Urine was lyophilized and fractionated by solvent extractions (hexane, ethyl acetate, etc.). Eight metabolites were isolated from extracts by preparative TLC and preparative HPLC. Their chemical structures were determined by NMR (^{1}H, H-H COSY, HSQC, HMBC, etc.) and MS (ESI, EI) spectroanalyses. Their chemical structures are shown in Figure 3 (**8, 9, 10, 11, 13, 14, 15** and **16**).

Excretion of diethofencarb and metabolites in rats

Four male Crj:CD(SD) rats (7 weeks old) was treated orally with [^{14}C-phenyl]-1 at 10 mg/kg b.w. Urine and feces were collected for seven days. ^{14}C-Residue in residual carcass was also analyzed on the 7th day after administration. Identification of the metabolite was carried out by TLC co-chromatography with radiolabelled metabolites mentioned above and unlabelled authentic standards (**1, 2,** and **12**). The radiocarbon was rapidly eliminated from the body, and the ^{14}C-residue was low on 7 days after administration (^{14}C-Residue in residual carcass on the 7th day after administration was 0.1 % of the dose). Total ^{14}C-recoveries within 7 days after administration were 98.6 % of the dose (urine; 89.6 %; feces; 9.0 %). Pooled (0-2 day) urine and feces were analyzed for quantification of metabolites. **1** was detected only in feces (0.3 % of the dose). More than 16 metabolites were detected and quantified. The conjugates (sulfate and/or glucuronide) of **2** and a mixture of conjugates (sulfate and/or glucuronide) of **11, 12,** and **13** were the major metabolites, accounting for 55.1 % and 18.9 % of the dose, respectively. S-containing metabolites such as free and conjugates (sulfate and/or glucuronide) of **8, 14, 15,** and **16** were detected, accounting for 0.9 % or less of the dose. Other metabolites accounted for 2.8 % or less of the dose.

The major metabolic reactions were 1) *O*-deethylation of 4-ethoxy group; 2) cleavage of the carbamate linkage, 3) acetylation of the amino group; 4) conjugation of the resultant phenols with sulfuric acid and/or glucuronic acid. The minor reactions were 1) oxidation of the isopropyl group; 2) cyclization of the oxidized isopropyl carbamate group; 3) oxidation of the 4-ethoxy group; 4) conjugation of the phenols with glutathione; 5) γ-glutamyltranspeptidation and depeptidation of the glutathione to form cysteine conjugate; 6) *N*-acetylation of the cysteine; 7) cleavage of C-S linkage of the cysteine followed by *S*-methylation; 8) oxidation of the *S*-methyl group. Proposed metabolic pathways of **1** in rat are shown in Figure 3.

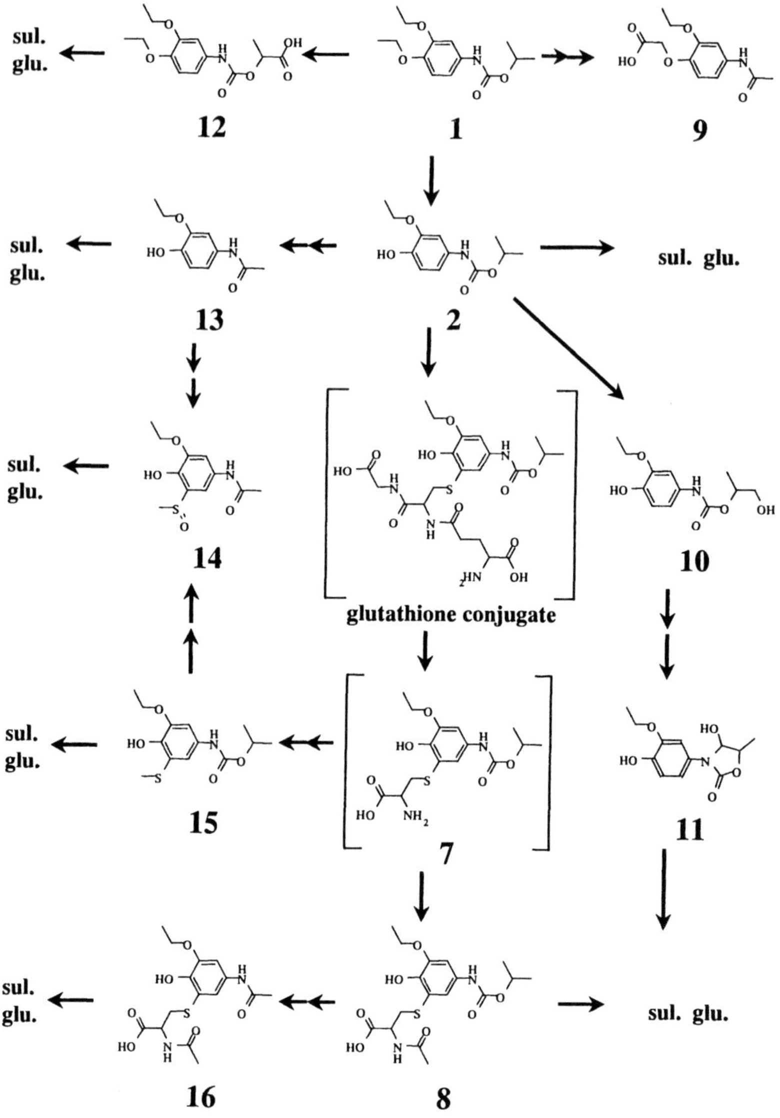

sul. glu. : sulfate and/or glucuronide

Figure 3. Proposed metabolic pathways of 1 in rats.

In vitro *metabolism of* **7**

Three Crj:CD(SD) male rats were sacrificed by bleeding from the abdominal artery under ansesthesia with diethyl ether. Liver was taken from these three rats. The liver was homogenized at 4 °C with 10 mM potassium phosphate buffer (pH 7.5) containing 1.15 % KCl, 10 µM (4-amidinophenyl)-methanesulphonyl fluoride and 50 µM pyridoxal 5'-phosphate, and then the homogenates was centrifuged at 9000×g for 20 min. The supernatant was subsequently centrifuged at 105000×g for 60 min to separate cytosol and microsomal fractions. The reaction mixtures (n=3) contained 100 mM Tris-acetate buffer (pH7.0), 2 mg of rat liver cytosol protein (final conc.;4 mg/mL), 5 mM α-ketoglutaric acid, 1 mM NADH, 1 mM NADPH, 1 % (v/v) dimethylsulfoxide and 1 µmol of [^{14}C-phenyl]-**7** (final conc.; 2 mM), in a volume of 0.5 mL. Control experiment (n=3) were performed by excluding rat liver cytosol from the reaction mixtures. The reaction mixtures were incubated at 37 °C for 16 hours. The metabolites were analyzed by TLC and quantified by the scraping-LSC method. **4** was only detected in the reaction mixture containing rat liver cytosol fraction. In 16 hours of reaction, 5.3 % of **7** was biotransformed to **4**. The formation rate of **4** in rat liver cytosol was 1.67 nmol/hr/mg protein.

The metabolic pathway of **7** to **4** in rats was proposed. The metabolic reactions were 1) transamination of amino group of cysteine; 2) reduction of ketone group of pyruvic acid. Above-mentioned step 1) and 2) were catalyzed by cysteine conjugate transaminase and 3-mercaptopyruvic acid *S*-conjugate reductase, respectively. Proposed metabolic pathway of **7** to **4** in rat (*in vitro*) are shown in Figure 4.

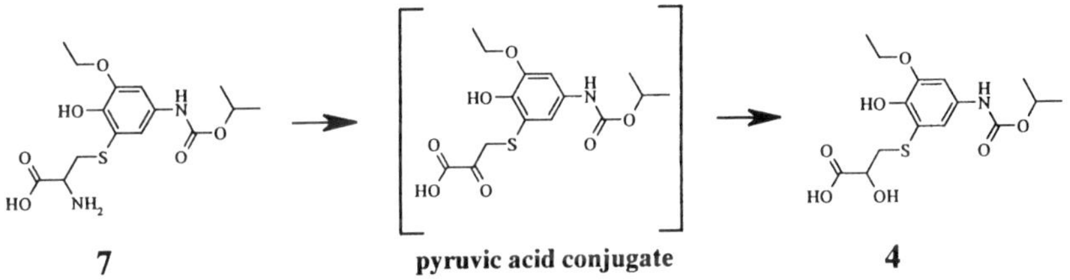

7 → pyruvic acid conjugate → **4**

Figure 4. Proposed pathways of 7 to 4 in rats.

3. Discussion

The mechanism on formation of the thiolactic acid conjugate was postulated from the various works conducted with acetaminophen (*N*-acetyl-*p*-aminophenol). Acetaminophen is a widely used analgesic medicine which has *p*-

aminophenol as a fundamental chemical structure. The *p*-aminophenol moiety is considered to undergo two-electron oxidation and it is easily transformed to a highly reactive benzoquinoneimine. The *in vitro* oxidation of acetaminophen has been extensively investigated using various enzymes such as cytochrome P-450, prostaglandin H synthase [PHS], and peroxidase (*5-11*) and the three possible routes to generate benzoquinoneimine from *p*-aminophenol by enzymatic oxidation was postulated. Firstly, acetaminophen is oxidized with peroxidase or PHS *via* one-electron oxidation to produce *N*-acetyl-*p*-benzosemiquinoneimine and the resulting semiquinoneimine disintegrates to give acetaminophen and *p*-benzoquinoneimine. Secondly, acetaminophen is oxidized with P-450 and PHS directly *via* two-electron oxidation to form benzoquinoneimine. Thirdly, acetaminophen is oxidized with P-450 to form hydroxylated intermediate (*N*-acetyl-*N*-hydroxy-*p*-aminophenol), subsequently followed by dehydration to form benzoquinoneimine. The main metabolite of 1 at the early stage of metabolism is most likely to be 2, possessing the common *p*-aminophenol structure to acetaminophen. Therefore, 2 would be similarly oxidized to form the corresponding benzoquinoneimine as an intermediate.

Additionally, Finley *et al.* (*12*) had discussed the specificity of reacting positions on benzene ring of *N*,*N*'-quinonediimine against a nucleophilic substitution. It was proposed that the 1-, 3- and 5-positions of *N*,*N*'-quinonediimine ring was highly susceptible to nucleophilic substitution. Actually, it has been reported in metabolism of acetaminophen (*13-16*) that the 3- and 5-positions of quinoneimine is highly reactive to the nucleophilic attack by glutathione (GSH).

In summary, 1 primary underwent *O*-deethylation at 4-position of the phenyl ring to form 2. Successively, 2 was likely to be converted *via* two-electron oxidation to the corresponding benzoquinoneimine whose 5-position of the ring was attached by cysteine *via* C-S linkage. Cysteine is the most probable nucleophile, since it is known to be formed by catabolism of the corresponding conjugates of GSH, a major ingredient in grape fruit (*17*). Actually, acetylcysteine conjugate of 2 was detected as rat metabolites to support our assumption. Then, the amino group of cysteine was considered to be transformed to the pyruvate structure as an intermediate by cysteine conjugate transaminase and finally the pyruvate form to the thiolactic acid by 3-mercaptopyruvic acid *S*-conjugate reductase (Fig 4).

References

1. Lamoureux, G. L.; Gouot, J. M.; Davis, D. G.; Rusness, D. G. *J. Agric. Food Chem.* **1981**, 29, 996-1002.
2. Lamoureux, G. L.; Rusness, D. G. *J. Agric. Food Chem.* **1987**, 35, 1-7.

3. Lamoureux, G. L.; Rusness, D. G. *Pestic. Biochem. Physiol.* **1989**, 34, 187-204.
4. Lamoureux, G. L.; Rusness, D. G.; Schröder, P.; Rennenberg, H. *Pestic. Biochem. Physiol.* **1991**, 36, 291-301.
5. Hinson, J. A.; Monks, T. J.; Hong, M.; Highet, R. J.; Pohl, L. R. *Drug Metab. Dispos.* **1982**, 10, 47-50.
6. Morgan, E. T.; Koop, D. R.; Coon, M. J. *Biochem. Biophys. Res. Commun.* **1983**, 112, 8-13.
7. Albano, E.; Rundgren, M.; Harvison, P. J.; Nelson, S. D.; Moldéus, P. *Mol. Pharmacol.* **1985**, 28, 306-311.
8. Potter, D. W.; Miller, D. W.; Hinson, J. A. *Mol. Pharmacol.* **1985**, 29, 155-162.
9. Potter, D. W.; Hinson, J. A. *Mol. Pharmacol.* **1986**, 30, 33-41.
10. Potter, D. W.; Hinson, J. A. *J. Biol. Chem.* **1987**, 262, 966-973.
11. Potter, D. W.; Hinson, J. A. *J. Biol. Chem.* **1987**, 262, 974-980.
12. Finley, T. K.; Tong J. L. K.; Patai, S., Eds.; Interscience publishers, London, 1970; pp 663-729.
13. Gemborys, M. W.; Gribble, G. W.; Mudge G. H. *J. Med. Chem.* **1978**, 21, 649-652.
14. Miner, D. J.; Kissinger, P. T. *Biochem. Pharmacol.* **1979**, 28, 3285-3290.
15. Buckpitt, A. R.; Rollins D. E.; Nelson, S. D.; Franklin, R. B.; Mitchell, J. R. *Anal. Biochem.* **1977**, 83, 168-177.
16. Pascoe, G. A.; Calleman, C. J.; Baille, T. A. *Chem.-Biol. Intractions* **1988**, 68, 85-98.
17. Park, S. K.; Boulton, R. B.; Noble, A. C. *Food Chem.* **2000**, 68, 475-480.
18. Fujisawa, T.; Ichise-Shibuya, K.; Katagi, T.; Ruzo, L. O.; Takimoto, Y. Metabolism of Fungicide Diethofencarb in Grape (Vitis vinifera L.): Definitive Identification of Thiolactic Acid Conjugated Metabolites. *J. Agric. Food Chem.* **2003**, 51, 5329-5336.

Chapter 19

The Role of Plant Glutathione S-Transferases in Herbicide Metabolism

Dean E. Riechers[1], Kevin C. Vaughn[2], and William T. Molin[2]

[1]Department of Crop Sciences, University of Illinois, Urbana, IL 61801
[2]U.S. Department of Agriculture, Agricultural Research Service, Southern Weed Science Research Unit, Stoneville, MS 38776

Glutathione S-transferase (GST) enzymes catalyze the conjugation of reduced glutathione to pesticide substrates, leading to their irreversible detoxification. Glutathione conjugation of xenobiotics is a very well studied Phase II detoxification reaction in plants, and is presumed to be requisite for their transport into the vacuole (Phase III) and possible further catabolism within the vacuole (Phase IV). GST-catalyzed glutathione conjugation is thus critical for removing xenobiotics from the cytosol and preventing them from interacting with their target sites. In plants, the expression of GST genes is regulated by many stimuli, including biotic and abiotic stresses and exposure to xenobiotics, thus indicating an important role for GST proteins in various detoxification processes. Chemicals called herbicide safeners, which protect grass crops from herbicide injury, stimulate herbicide metabolism by increasing the activity of GST enzymes that detoxify herbicides. GST proteins are soluble and their subcellular localization has usually been presumed to be cytosolic. However, immunocytochemical studies in our lab have shown that GST proteins are localized in the cytosol and inside the vacuoles of epidermal and sub-epidermal cells in herbicide safener-treated coleoptiles [Planta 217:831-840]. The majority of immunoreactive GST protein was located in the outer two cell layers of safener-treated coleoptiles, indicating that these cell types may be involved in a novel form of herbicide detoxification mechanism that involves vacuolar accumulation of GST protein and xenobiotic-glutathione conjugates.

Glutathione *S*-transferases (GSTs; EC 2.1.5.18), also termed glutathione transferases, are critical enzymes in the metabolism of herbicides and detoxification of other xenobiotics in plants. They mainly function as Phase II detoxification enzymes, where glutathione conjugation of electrophilic molecules leads to their irreversible detoxification and subsequent transport to the vacuole (Phase III) (*1*). In crop plants, glutathione conjugation is responsible for the selectivity of several major herbicide families (*1*), and in some cases, herbicide resistance has developed in weed biotypes due to increased GST and associated activities (*2*). The biochemical characteristics of plant GSTs, classification system, and proposed nomenclature of genes and isozymes have been recently reviewed (*3-9*) and will not be discussed in detail here. The purpose of this article is to highlight some recent research findings from the authors' laboratories and emphasize some areas of research that clearly deserve more attention, such as tissue-specific expression of GSTs and subcellular localization patterns before and after herbicide stress. The possible relevance of these areas with respect to herbicide metabolism in plants, and regulation by chemicals termed herbicide safeners, will also be discussed.

Functions and Regulation of Plant Glutathione *S*-Transferases

Herbicide Selectivity in Major Cereal Crops and Soybean

One biochemical function of GSTs that is very well defined in the scientific literature is their role in herbicide metabolism in crops. GSTs are the predominant detoxification enzymes in maize and cereal crops that are responsible for the metabolism of triazine and acetamide herbicides, as well as certain graminicides such as fenoxaprop-ethyl in wheat (*10-14*), leading to herbicide selectivity between the crop and problem weed species (*15*). Herbicide-detoxifying GSTs have been well characterized in maize (*5,6,8,16-*

18), but have also been identified and partially characterized in wheat, rice, grain sorghum, and soybean (*11,13,19-25*) to varying extents.

Function and Role of GSTs in Plant Response to Environmental Stresses

Glutathione *S*-transferase (GST) gene expression has been shown to be induced following exposure to many stresses, including heat shock, cold, high salt, and UV light exposure; biotic stresses such as pathogen attack and fungal elicitors, and abiotic stresses such as heavy metals, herbicides, and safeners; and phytohormone treatments such as ethylene, auxins, abscisic acid, methyl jasmonate, and salicylic acid (reviewed by *3,6,7*). Induction of GST expression by so many diverse stimuli implies that plant GSTs are critical in plant response to stress, either by participating in the signal transduction process and/or detoxifying harmful compounds produced in response to or as a result of a given stress. It is likely that GST gene expression is induced by conditions that lead to oxidative stress (*26*). The encoded GST proteins play an important but poorly understood role in plant response to stress, possibly through the central role of antioxidant function. GST enzymatic activity could involve direct glutathione conjugation to toxic electrophilic molecules (such as herbicides), or glutathione-dependent peroxidase activity, using glutathione as reductant for the detoxification of toxic oxygen species, oxygen radicals, and lipid peroxides formed during or after plant stress (*4,7*).

Herbicide Safeners Enhance Herbicide Selectivity by Inducing GSTs and Other Enzymes

Safeners are chemical compounds that increase the tolerance of certain grass crops (*e.g.*, maize, grain sorghum, wheat, rice) to herbicides (*27*). Crops can be protected from herbicides that are frequently used for selective control of annual grass weeds in maize and grain sorghum. Herbicide safeners protect the crop plant by increasing herbicide metabolism and detoxification pathways (*12,27-30*). The increase in metabolism results from an increase in the activity of herbicide detoxification enzymes, such as glutathione *S*-transferases, cytochrome P450-dependent monooxygenases, and glucosyl-transferases (*1,11,22,25,31,32*).

In addition to increasing the expression of cytochrome P-450 and GST enzymes (Phase I and II), safeners and herbicides have also been shown to increase the enzymatic activity of a vacuolar transporter of xenobiotic-glutathione conjugates (Phase III) (*33-34*). Once inside the vacuole, xenobiotic-glutathione conjugates are degraded by various peptidase enzymes that sequentially cleave the amino acids glycine then glutamate from the glutathione moiety (potentially called Phase IV) (*35-37*). The effect of safeners on peptidase or other glutathione conjugate catabolic activities, and their turnover rates within the vacuole, has not been reported. Thus, a fundamental question that needs to be addressed is whether the entire herbicide detoxification pathway is coordinately regulated by safeners (*38*). It is becoming more apparent that the goal of glutathione-mediated herbicide detoxification in plants is to remove the glutathione conjugate from the cytosol and transport it into the vacuole for further metabolic processing (*3,39*). Whether or not safeners induce expression of all members of this detoxification pathway in grass crops, or selectively induce only certain components of the pathway, is not known.

Despite the widespread agronomic use of safeners and information about their effects on GST and P-450 enzymatic activity, there is virtually no information on the precise molecular mechanism of safener induction. Safeners (applied as seed treatments, foliar applications, or as soil drenches at low micromolar concentrations) induce the expression of plant defense genes, such as GSTs and cytochrome P-450s, yet they are not toxic to the plant and actually confer protection from herbicide injury. This implies that safeners are tapping into pre-existing signaling pathways for detoxification of endogenous toxins or xenobiotics. Safeners are non-toxic to crop plants (at the labeled rate of usage as a seed treatment; or when applied as soil drenches at low micromolar concentrations, usually less than 10 µM) yet extremely effective inducers of gene expression, and thus present an excellent tool to understand fundamental aspects of GST proteins, glutathione conjugation, vacuolar sequestration, and their respective roles in xenobiotic detoxification in plants. Molecular information regarding safener-mediated induction of GST gene expression in crops is very limited. A safener-binding protein and its activity have been characterized in maize seedlings, along with its gene expression patterns (*40,41*). Another area that is especially lacking in information is the identification and characterization of important regulatory sequences present in the promoters of safener-responsive GST genes, and the transcription factors that bind to these DNA sequences, in agronomically important grasses such as maize, sorghum, rice, and wheat (*42*).

Regulation of GST Gene Expression in Plants by Safeners

A current model for safener-induced GST gene expression is that safeners cause a status of oxidative stress in the cell, possibly by perturbing glutathione homeostasis, or status of the reduced glutathione to oxidized glutathione ratio GSH:GSSG (*39,43,44*). It is important to note that safeners are conjugated to reduced glutathione in corn (*45,46*), although it is not known if this is non-enzymatic or GST-catalyzed conjugation with glutathione. Therefore, glutathione conjugation of safeners may lead to decreased levels of total glutathione or a decrease in the ratio of reduced glutathione to oxidized glutathione. Such a perturbation of glutathione homeostasis in the plant cell is potentially the signal that induces the cell to respond by producing transcription factors that bind to the promoters of GST genes, leading to transcriptional activation of the gene (*47*). The encoded GST enzymes then protect essential cellular components from damage by detoxifying reactive oxygen species and lipid peroxides that are produced during or after the oxidative stress. This could occur through either direct conjugation to glutathione or via glutathione-dependent peroxidase activity utilizing glutathione as reductant (*3,4,7*). A signal transduction pathway that is responsive to oxidative stress has been proposed to confer inducible GST gene expression following treatment with safeners (*5,6,48*), but information about the molecular components of this signal transduction pathway are extremely limited.

Glutathione Conjugation is Requisite for Export of Detoxified Herbicides From the Cytosol to the Vacuole

It has been established that glutathione-herbicide conjugates formed in the cytosol are transported to the tonoplast, actively pumped into the vacuole by a GS-X pump, and degraded once inside the vacuole by carboxypeptidases and other peptidase enzymes to recycle the amino acids of glutathione (glutamate and glycine) as part of the pathway of herbicide detoxification in monocot crops (*35-37,49*). Do safeners regulate the expression for all proteins/enzymes involved in the herbicide detoxification pathway, starting with glutathione conjugation of herbicides by GSTs, and ending with degradation of the glutathione-herbicide conjugate (or GST proteins) in the vacuole? It is possible that safeners protect grass crops from herbicide injury by coordinately regulating the expression of proteins that comprise an entire detoxification pathway that:

1. Conjugate herbicides containing an electrophilic site with glutathione (GST proteins).
2. Facilitate the transport of the glutathione-herbicide conjugates to the vacuole (GSTs or other proteins).
3. Transport the glutathione-herbicide conjugate into the vacuole (tonoplast ABC transporters).
4. Catabolize the glutathione-herbicide conjugate once inside the vacuole (*e.g.*, carboxypeptidases, other peptidases).

In support of this theory, one report showed that the transport activity of glutathione conjugates into isolated vacuoles was induced by safener treatment in barley (*33*), and transcripts for specific ABC transporter genes were increased up to 40-fold following treatment of *Arabidopsis* seedlings with xenobiotics, including the herbicide primisulfuron (*34*). In addition, the expression of a multidrug resistance associated protein homologue (MRP) was induced by the safener cloquintocet-mexyl in combination with phenobarbital in wheat, with the protein being localized to the tonoplast (*38*).

The safener-induced, tau class GST proteins *Tt*GSTU1 and *Tt*GSTU2 from *Triticum tauschii* are most likely involved in conjugation of herbicides with glutathione (*42,50*), but they may also play a role in transporting the glutathione-herbicide conjugate to the tonoplast, where it is actively pumped into the vacuole by a GS-X pump, similar to the multidrug resistant ATP binding cassette (ABC) transporters from humans and yeast (*39,49,51-53*). This additional "ligandin" function has been speculated to occur with other plant GSTs such as the tau class or type III maize Bz2 protein (*54*) and the phi class or type I petunia AN9 protein (*55,56*), as well as plant GST proteins that may bind flavonoids, porphyrins, or auxins (*57-60*).

Safener-enhanced Glutathione Conjugation Prevents Chloroacetamide Herbicide Effects on Plant Cell Membranes

Members of the chloroacetamide herbicide family (such as propachlor, alachlor, and metolachlor) have been used many years for weed control in maize and soybeans, and their mechanism of selectivity between crops and weeds has been shown to be increased GST-mediated herbicide detoxification in the crop (*12,61*). Despite their long time use and agronomic importance, there have been very few qualitative studies of the effects of these herbicides at the ultrastructural level to help determine their mechanism of herbicidal action, or to determine the effect of safeners in preventing their phytotoxic effect. Most of the earlier studies that were conducted involved treatment of seedlings with high concentrations of herbicide and documenting effects on plant senescence; however, effects on growth and development that were observed gave little insight into the biochemical site of action of these compounds (early work reviewed in *61*). Recent biochemical studies have indicated that the chloroacetamide herbicides inhibit an elongase enzyme involved in very long chain fatty acid biosynthesis in plants (*62,63*), suggesting a phytotoxic effect due to an inability to synthesize the necessary lipids and cell membranes needed for actively dividing cells. This is consistent with preliminary observations we have noted and described below in our ultrastructural studies with *Triticum tauschii* shoots (Figure 1). In addition to helping identify the specific effect of chloroacetamide herbicides on plant cell membranes and lipids, cytological studies can provide information on the interactions between safeners and the

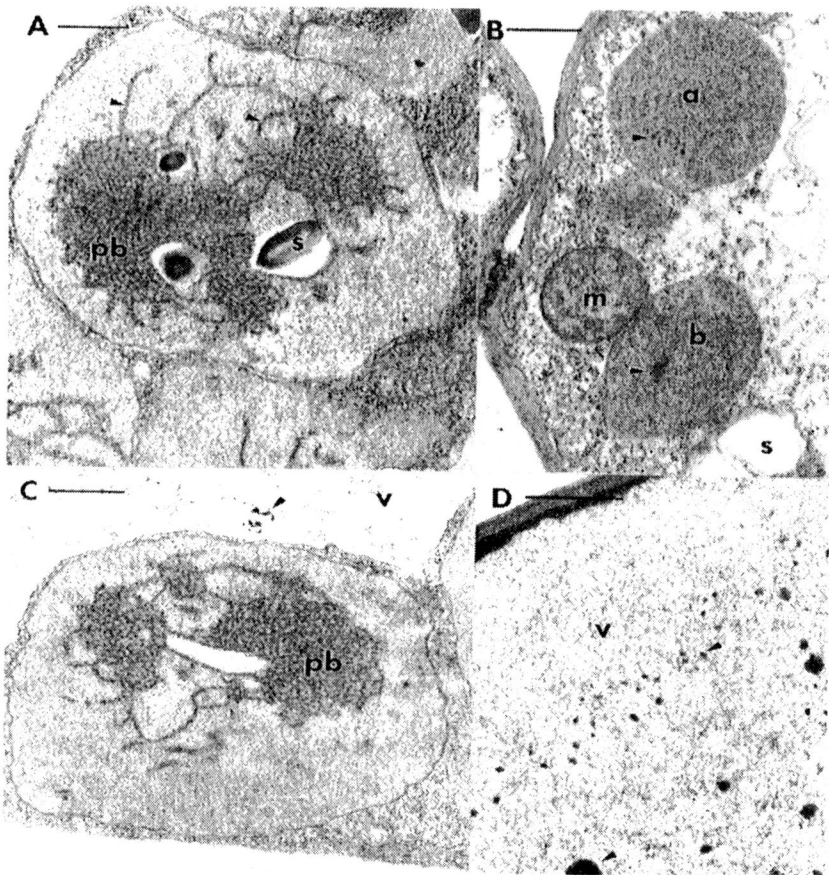

Figure 1. Electron micrographs of Triticum tauschii seedlings grown for 5 days in the dark. **A.** Control cells reveal well developed etioplastids that contain prominent prolamellar bodies (pb) and prothylakoids (arrowheads). **B.** Treatment with 1 µM alachlor results in severe inhibition of etioplastid development. In the etioplastids (marked a and b), virtually no internal structure is noted and they are only slightly larger than mitochondria (m). **C.** Treatment of 10 µM fluxofenim with the alachlor treatment renders the etioplastid ultrastructure back to that of the control. **D.** Vacuole (v) of epidermal cell after treatment with safener only reveals a large accumulation of osmiophilic material with some larger accumulations (arrowheads). Vacuoles in control tissue contain no such inclusions. Bars = 0.5 µm; s = starch.

group of chloroacetamide herbicides at the cellular level, and the involvement of GST proteins in the overall detoxification process.

Preliminary cytological experiments (Figure 1) have revealed the usefulness of cytochemical analyses in determining the cellular effects of the chloroacetamide herbicide alachlor and the effect of safeners. Dark grown, untreated seedlings of *Triticum tauschii* have characteristic etioplasts with prominent prolamellar bodies (PLBs) and prothylakoids (Figure 1A). Treatment of these seedlings with 1 µM of the chloroacetamide herbicide alachlor results in shorter plants that contain etioplasts lacking both the PLB and the prothylakoids (Figure 1B), structures that are enriched in galactolipids. The etioplast size in these treated plants is also greatly reduced over the untreated controls (Figure 1B). When seedlings are treated with 10 µM of the safener fluxofenim with the same herbicide treatment as in Figure 1B, the etioplasts have an ultrastructure virtually indistinguishable from the control (Figure 1C), indicating that the toxic effects associated with alachlor treatment are reversed (or prevented) by the safener fluxofenim. In addition, there are striking changes in cellular physiology and vacuolar ultrastructure as a consequence of safener treatment. Epidermal cells (Figure 1D), and to a lesser extent mesophyll cells (Figure 1C), accumulate masses of osmiophilic material in the vacuole that consist of an unusual and as of yet unidentified morphology. It is known that glutathione conjugates accumulate in the vacuole (*1*), so we presume that these particles are an ultrastructural manifestation of these conjugates. The very interesting findings that have resulted from these initial cytological studies demonstrate the utility of microscopy and cytochemical approaches in documenting and monitoring the relative efficacy of safener treatments.

Possible Physiological Functions of Plant GST Proteins Implied by Their Localization

Although GSTs metabolize herbicide molecules to inactive compounds, it is obvious that these enzymes must have physiological functions other than their role in herbicide metabolism. Determining the subcellular location and tissue distribution of GST proteins may help to uncover some of their natural physiological functions. The subcellular location of GSTs has usually been presumed to be cytosolic (*4,8*). This has been largely based on a lack of protein-targeting sequences detected in their genes and encoded proteins, or differential centrifugation properties of the GST proteins from crude protein extracts, where they are usually reported as being soluble and not membrane-bound. However, safener-inducible GSTs involved in dimethenamid metabolism were recently shown to be localized predominantly to the epidermal and outer sub-epidermal cellular layers of wheat coleoptiles rather than in developing leaves, and within these cell layers, to the cytosol and vacuoles following safener treatment (*50*). Now that we know the distribution of the constitutive and inducible GST isozymes in *Triticum tauschii* seedlings (*50*), some possible functions and roles for these enzymes in other metabolic pathways are implied.

The GST distributions in etiolated *Triticum* shoots that we observed (*50*) are strikingly correlated with the distribution of anthocyanins, for example. Anthocyanins occur primarily in epidermal and sub-epidermal tissues and are often more abundant in non-photosynthetic tissues than photosynthetic ones. Similar to herbicides, anthocyanins are generally cytotoxic molecules that must be compartmentalized in the vacuole so as to keep them away from other sensitive cellular processes. Even the vacuole frequently relegates anthocyanins to separate cellular compartments in some plant species, termed the anthocyanophore (*64,65*), to keep these pigments away from other vacuolar processes. Furthermore, Walbot and colleagues have shown convincingly that the Bz2 locus in maize encodes for GST that is involved in moving the pigment to the vacuole, with the mutant bz2 protein arresting movement of the anthocyanin to the vacuole (*54*). Surprisingly, the GST encoded by the Bz2 locus does not appear to metabolize the anthocyanin, but simply assists its movement to the vacuole (*54*). It is possible that an interference in flavonoid biosynthesis by chloroacetamide herbicides is caused by a competition between the herbicides and the flavonoids for carrier GST molecules (*66*). Thus, the presence of the enzyme machinery that allows the cell to deal with one of its own toxic molecules might be recruited in the case of an applied herbicide to deal with a similarly phytotoxic xenobiotic.

Another possible common denominator is found between the distribution of GSTs in *Triticum* seedlings and the distribution and metabolism of the plant hormone phytochrome. In oat seedlings, phytochrome is found predominantly in the epidermal and sub-epidermal tissue and is randomly distributed throughout the cytoplasm under dark conditions (*67,68*). Upon exposure to red or white light, the phytochrome aggregates into complexes and is tagged by ubiquitin so that the protein is marked for vacuolar movement and eventual degradation. Thus, these cell layers would not only have a system requiring a GST "ligandin" mechanism for anthocyanin biosynthesis and intracellular transport, but also a high level of ubiquitin-tagging activity to quickly convert large quantities of phytochrome to an inactive form. Such a tagging would be required to move the GST protein (which lacks any such signal sequence) to the vacuole.

Safeners May Function by Exploiting Pre-existing Biochemical Pathways

Until our GST localization studies in *Triticum tauschii* (*50*), it was unclear as to why a safener could induce a detoxification system that effectively metabolized herbicide molecules even though the plant had certainly not been exposed to herbicide molecules of any type previously throughout its evolutionary history. Now, there are at least two biochemical systems, involving flavonoid metabolism and phytochrome degradation, that occur in the same tissue and would require the same types of cellular processes for their function. Thus, what the safener has apparently done is to tap into an existing system for a new metabolic problem for the emerging seedling.

The localization of GST in both the vacuolar and cytoplasmic compartments was a very surprising result of our studies in etiolated *Triticum* shoots (*50*). However, Tagu et al. (*69*) report a similar cytoplasmic and vacuolar localization in roots of *Eucalyptus globulus* as well, including a report of small amounts of dense structures in the vacuole. Could these dual sites of localization reflect an unusual property of the GST enzyme, which by sequence analysis strongly indicates a cytoplasmic localization? A possible clue to these results is an interesting observation from mammalian cell studies, where dyes conjugated to GST could be transported across membranes even though this enzyme has no protein transduction domains (*70*). If plant GSTs have similar properties, this could explain how GSTs could be involved in the transmembrane passage of an enzyme despite the lack of any obvious signal or targeting sequences (*50,71,72*).

Tissue-specific GST Expression: Implications for Xenobiotic Metabolism

Prior to our report on the tissue-specific expression of GSTs in the coleoptile of wheat seedlings (*50*), crude protein, total RNA, or cell-free extracts for determination of GST activities, GST mRNA levels, or glutathione levels were typically made from intact seedlings, roots or shoots with little consideration of tissue-specific localization (*10,11,13,14,16,22-25,32,42,73-75*). Safener-inducible GSTs can now be characterized in a functional and anatomical framework, and questions about the compartmentalization of metabolic processes in safener-responsive grass species can be addressed. Henceforth, expression of GSTs in species that contain constitutive and inducible GST isoforms probably should be considered in a tissue-specific context.

Many aspects of safener action and xenobiotic metabolism might be further clarified if considered in light of tissue-specific GST induction. For example, a re-examination of the safener rates needed to initiate induction, and the magnitude of responses among GSTs following induction, may be prudent because of the disproportionate protein content between leaf tissues and coleoptiles in grass shoots (*50*). Similarly, glutathione concentrations and synthesis, and mechanisms for balancing redox states may be disproportionately distributed among tissues or cell layers based on GST abundance rather than protein contents. Elucidating the number, distribution and coordinate regulation of safener-induced GST isoforms and other detoxifying mechanisms across cell and tissue types would surely provide new insights into GST function and regulation. The coordinate induction of the processes of xenobiotic activation and vacuolar loading may also be localized within the tissues containing induced GSTs. Questions also emerge on whether tissue tolerance or susceptibility to various herbicide classes varies among tissues and is based on substrate specificity among induced GST isoforms. Additionally, botanical specificity (*i.e.*, monocot crop responsiveness and lack of a response in dicots) of the safener response may be related to tissue-specific expression of herbicide-detoxifying GSTs in the coleoptile of grass crops, and/or the lack of an essential component of the overall detoxification pathway in dicots (*50*).

The literature regarding effects of safeners on GST activities is further complicated by differences in species, plant age, and degree of tissue dissection. Several inducible GST isozymes were identified in 7-day-old safener (fenchlorazole-ethyl) treated wheat shoots, and only the inducible forms could metabolize fenoxaprop-ethyl (*11*). In 5-day-old wheat shoots in which coleoptile and leaves were separated, the constitutive expression of GSTs was localized to the coleoptile and the inducible forms occurred in both coleoptile and leaves (*50*). The diversity of isoforms present was not determined so it is not known whether there were differences in isoforms induced by fluxofenim or cloquintocet–mexyl treatments or in coleoptiles and leaves. In 3-day-old fluxofenim-treated grain sorghum shoots without separation of leaves and coleoptiles, two GSTs were purified; one had high activity with 1-chloro-2,4-dinitrobenzene (CDNB) and low activity with metolachlor, and the other had high activity with metolachlor and low activity with CDNB (*23*). In 3-day-old benoxacor-treated maize shoots without separation of leaves and coleoptiles, three inducible isoforms had activity with metolachlor and low activity with CDNB (*14,16*). Each of these studies would have been more informative if coleoptiles and leaves had been separated, GSTs fractionated into their various isoform components, and had substrate specificities been determined for individual GST isoforms that were resolved.

Localization of phase II detoxification mechanisms (via glutathione or glucose conjugation) to the outer layers of the coleoptile would allow for xenobiotic conjugation and detoxification at the point near uptake. Xenobiotic phase II conjugation reactions should increase water solubility and allow for dispersion of glutathione or glucose conjugates within the cytosol, thereby decreasing their concentration at lipophilic surfaces or intercalation within membranes (*76*). The flurazole-glutathione conjugate was more water-soluble than flurazole based on its retention on reversed phase HPLC columns (*45*). Following conjugation, glutathione conjugates are sequestered in vacuoles, where further catabolism occurs, such as cleavage of amino acids, and oxidation of xenobiotics (*1,35-37,76*). Since inducible GSTs accumulated in the epidermal and sub-epidermal cell layers of wheat coleoptiles (*50*), co-induction of vacuolar loading mechanisms might also be expected to occur in these cells and tissues as well. Other constitutive and inducible metabolic enzymes that co-exist or are co-induced by safener or herbicide treatments, such as cytochrome P-450 monooxygenases or ABC transporters, may also be localized with respect to which isoforms of GSTs are present.

Considerations of the Physical Properties and Fate of Safeners in Relation to Activity

Knowledge of the chemical and physical properties of xenobiotic compounds involved in GST induction and subsequent metabolism by GSTs, and consequences of their application to plant systems, may provide greater insights into their possible interactions with plants. Application of safeners to seeds, or herbicides and safeners to seeds or seedlings as a drench, or in tank-mixes to emerged seedlings, exposes plant surfaces to high xenobiotic concentrations and may establish a gradient in xenobiotic concentration from the outer epidermis to the innermost leaves. For xenobiotics with low water solubility and high lipophilicity, absorption and concentration of xenobiotics at the waxy surface of the coleoptile would be expected to be high, and indeed, plant responses to such levels might be expected to be rapid and excessive. Induction of GSTs in the epidermal and sub-epidermal layers of the coleoptile is conceptually consistent with xenobiotics for which tissue permeation is likely to be low and confined to outer tissue layers of the coleoptile.

Listed in Table 1 are the water solubilities and octanol/water partition coefficients ($K_{o/w}$) of herbicides that undergo conjugation with glutathione. For comparison, the polar, charged herbicide glyphosate has a solubility of 900,000 mg/L and a $K_{o/w}$ of 0.0006 to 0.0017, whereas trifluralin (a lipophilic herbicide) has a solubility of 0.3 mg/L and a $K_{o/w}$ of 118,000. Hence, the herbicides that undergo glutathione conjugation (as well as the safeners listed in Table 1) are more like trifluralin than glyphosate, and would likely associate with surfaces, membranes or membrane systems rather than accumulate in the cytosol. In addition, if one considers that the membrane fraction of a cell represents about 0.25 percent of the cell volume, then by association with membrane systems, xenobiotics with low water solubility may be concentrated over four hundred times at the membrane surface.

Table 1. Solubilities and $K_{o/w}$ of Herbicides that Undergo Detoxification by Conjugation with Reduced Glutathione

Herbicides and Safeners	Solubility (mg/L)	$K_{o/w}$
Atrazine	33	481
Alachlor	200	794
Dimethenamid	1174	141
Metolachlor	488	794
Cloquintocet-mexyl	0.8	126,000
Fluxofenim	300	7943

NOTE: The safeners cloquintocet-mexyl and fluxofenim are included for comparison purposes.
SOURCE: Data taken from *Herbicide Handbook*, Weed Science Society of America, (77).

Both fluxofenim and cloquintocet-mexyl (applied as a soil drench to seeds) induced GSTs with dimethenamid-conjugating activity, indicating that the tissues responding to safener treatment were quite specific (50). Fluxofenim was a better inducer of coleoptilar GSTs than cloquintocet-mexyl, inducing GST activity fifty-fold compared to twenty-fold (50). Reasons for the differences in safener efficacy are not clear at this time. Cloquintocet-mexyl is registered as a postemergence safener in wheat that is applied as a tank mix with the herbicide, and protects all types of wheat from injury from clodinafop-propargyl (77,78). Its activity as a safener is achieved in wheat leaves by enhancement of hydroxylation and ether cleavage of clodinafop-propargyl rather than glutathione conjugation (78). Use of cloquintocet-mexyl as a soil drench safener may limit its uptake, or place it in contact with a tissue in which it has limited activity (50).

From a water solubility standpoint, fluxofenim is much more water soluble than cloquintocet-mexyl (Table 1) and may simply penetrate, permeate, and saturate the coleoptilar tissues more readily when applied as a soil drench than cloquintocet-mexyl. The rate of safener metabolism, its longevity and cellular fate as an inducing species, and the concentration gradients that may form across tissue layers, may also impact its effectiveness. Unfortunately, little is known of the fate of safeners in plant tissues. Flurazole was rapidly metabolized in corn and sorghum shoots to its glutathione conjugate, similar to the chloroacetanilides, as well as to the thiazolecarboxylate and 2-hydroxy-thiazolecarboxylate metabolites (45). Benoxacor metabolism was examined in maize Black Mexican Sweet cell cultures, and both mono- and di-glutathione conjugates of benoxacor were identified (46). In wheat, cloquintocet-mexyl may simply not induce all or the same forms of GSTs as fluxofenim in coleoptiles and new leaves (50). Interestingly, an elution profile of GSTs from maize shoots treated with either benoxacor or flurazole showed differences in amounts of GST activities (*14*), and presumably different amounts and/or forms of GST isozymes induced. However, no explanation was provided for the differences in GSTs induced by the two safeners examined in this study.

Conspectus

Future research examining GST protein localization and tissue-specific expression patterns of individual GST isozymes and isoforms should help to clarify their precise roles in herbicide metabolism and detoxification pathways, and may also provide important clues as to their natural physiological and biochemical functions in plants. One important question that needs to be experimentally addressed is whether GST protein localization is determined by its gene sequence (or encoded amino acid sequence), or if subcellular localization is a direct result or consequence of the detoxification processes that they catalyze. Clearly, in *Triticum* seedlings the localization pattern and cellular distribution of inducible GSTs differed following herbicide safener treatment (*50*), whereas in *Eucalyptus* roots the localization pattern did not change following infection by ectomycorrhiza (*69*). In soybean hypocotyls, extracellular (apoplastic) GSTs were detected but only following treatment with the known GST inducer 2,3,5-triiodobenzoic acid (*72*). It will be of great interest in future studies to localize GST proteins following both herbicide and herbicide safener treatment (applied both separately and together), or following other xenobiotic or abiotic stresses.

Acknowledgments

We thank Qin Zhang and Dr. Fangxiu Xu for excellent technical assistance, and Dr. Stephen Jones, Dr. Patrick Fuerst, and Dr. Evans Lagudah for helpful discussions concerning the research. Research in the laboratory of D.E.R. is supported by a grant from the USDA/NRICGP, No. 2003-00777. The authors would also like to acknowledge the mentoring, friendship, and research collaborations of Dr. Kriton K. Hatzios.

Literature Cited

1. Kreuz, K.; Tommasini, R.; Martinoia, E. *Plant Physiol.* **1996,** *111,* 349-353.
2. Cummins, I.; Cole, D. J.; Edwards, R. *Plant J.* **1999,** *18,* 285-292.

3. Dixon, D. P.; Cummins, I.; Cole, D. J.; Edwards, R. *Curr. Opin. Plant Biol.* **1998**, *1*, 258-266.
4. Dixon, D. P.; Lapthorn, A.; Edwards, R. *Genome Biol.* **2002**, *3*, reviews3004.1-3004.10. URL http://genomebiology.com/2002/3/3/reviews/3004.1.
5. Droog, F. *J. Plant Growth Regul.* **1997**, *16*, 95-107.
6. Marrs, K. A. *Annu. Rev. Plant Physiol. Plant Mol. Biol.* **1996**, *47*, 127-158.
7. Edwards, R.; Dixon, D. P.; Walbot, V. *Trends Plant Sci.* **2000**, *5*, 193-198.
8. Edwards, R.; Dixon, D. P. In *Herbicides and Their Mechanisms of Action;* Cobb, A. H.; Kirkwood, R. C.; Eds.; Sheffield Academic Press: Sheffield, UK, 2000; pp 38-71.
9. Wagner, U.; Edwards, R.; Dixon, D. P.; Mauch, F. *Plant Mol. Biol.* **2002**, *49*, 515-532.
10. Edwards, R.; Cole, D. J. *Pestic. Biochem. Physiol.* **1996**, *54*, 96-104.
11. Cummins, I.; Cole, D. J.; Edwards, R. *Pestic. Biochem. Physiol.* **1997**, *59*, 35-49.
12. Riechers, D. E.; Fuerst, E. P.; Miller, K. D. *J. Agric. Food Chem.* **1996**, *44*, 1558-1564.
13. Riechers, D. E.; Irzyk, G. P.; Jones, S. S.; Fuerst, E. P. *Plant Physiol.* **1997**, *114*, 1461-1470.
14. Fuerst, E. P.; Irzyk, G. P.; Miller, K. D. *Plant Physiol.* **1993**, *102*, 795-802.
15. Neuefeind, T.; Reinemer, P.; Bieseler, B. *Biol. Chem.* **1997**, *378*, 199-205.
16. Irzyk, G. P.; Fuerst, E. P. *Plant Physiol.* **1993**, *102*, 803-810.
17. Irzyk, G. P.; Fuerst, E. P. In *Regulation of Enzymatic Systems Detoxifying Xenobiotics in Plants;* Hatzios, K. K.; Ed.; NATO ASI Series, Kluwer Academic Publishers: Dordrecht, the Netherlands, 1997; pp 155-170.
18. Jepson, I.; Lay V. J.; Holt, D. C.; Bright, S. W. J.; Greenland, A. J. *Plant Mol. Biol.* **1994**, *26*, 1855-1866.
19. Andrews, C. J.; Skipsey, M.; Townson, J. K.; Morris, C.; Jepson, I.; Edwards, R. *Pestic. Sci.* **1997**, *51*, 213-222.
20. Cummins, I.; O'Hagan, D.; Jablonkai, I.; Cole, D. J.; Hehn, A.; Werck-Reichhart, D.; Edwards, R. *Plant Mol. Biol.* **2003**, *52*, 591-603.
21. Thom, R.; Cummins, I.; Dixon, D. P.; Edwards, R.; Cole, D. J.; Lapthorn, A. J. *Biochemistry* **2002**, *41*, 7008-7020.
22. Gronwald, J. W.; Fuerst, E. P.; Eberlein, C. V.; Egli, M. A. *Pestic. Biochem. Physiol.* **1987**, *29*, 66-76.
23. Gronwald, J. W.; Plaisance, K. L. *Plant Physiol.* **1998**, *117*, 877-892.
24. Deng, F.; Hatzios, K. K. *Pestic. Biochem. Physiol.* **2002**, *72*, 24-39.
25. Riechers, D. E.; Yang, K.; Irzyk, G. P.; Jones, S. S.; Fuerst, E. P. *Pestic. Biochem. Physiol.* **1996**, *56*, 88-101.
26. Polidoros, A. N.; Scandalios, J. G. *Physiol. Plant.* **1999**, *106*, 112-120.
27. Davies, J.; Caseley, J. C. *Pestic. Sci.* **1999**, *55*, 1043-1058.
28. Fuerst, E. P.; Gronwald, J. W. *Weed Sci.* **1986**, *34*, 354-361.
29. Fuerst, E. P.; Lamoureux, G. L.; Ahrens, W. L. *Pestic. Biochem. Physiol.* **1991**, *39*, 138-148.
30. Fuerst, E. P.; Lamoureux, G. L. *Pestic. Biochem. Physiol.* **1992**, *42*, 78-87.

31. Cole, D. J. *Pestic. Sci.* **1994**, *42*, 209-222.
32. Brazier, M.; Cole, D. J.; Edwards, R. *Phytochemistry* **2002**, *59*, 149-156.
33. Gaillard, C.; Dufaud, A.; Tommasini, R.; Kreuz, K.; Amrhein, N.; Martinoia, E. *FEBS Lett.* **1994**, *352*, 219-221.
34. Tommasini, R.; Vogt, E.; Schmid, J.; Fromentau, M.; Amrhein, N.; Martinoia, E. *FEBS Lett.* **1997**, *411*, 206-210.
35. Wolf, A. E.; Dietz, K.-J.; Schröder, P. *FEBS Lett.* **1996**, *384*, 31-34.
36. Cole, D. J.; Edwards, R. In *Metabolism of Agrochemicals in Plants;* Roberts, T.; Ed.; John Wiley and Sons: New York, NY, 2000; pp 107-154.
37. Beck, A.; Lendzian, K.; Oven, M.; Christmann, A.; Grill, E. *Phytochemistry* **2003**, *62*, 423-431.
38. Theodoulou, F. L.; Clark, I. M.; He, X.-L.; Pallett, K. E.; Cole, D. J.; Hallahan, D. L. *Pest. Manag. Sci.* **2003**, *59*, 202-214.
39. Foyer, C. H.; Theodoulou, F. L.; Delrot, S. *Trends Plant Sci.* **2001**, *6*, 486-492.
40. Walton, J. D.; Casida, J. E. *Plant Physiol.* **1995**, *109*, 213-219.
41. Scott-Craig, J. S.; Casida, J. E.; Poduje, L.; Walton, J. D. *Plant Physiol.* **1998**, *116*, 1083-1089
42. Xu, F.-X.; Lagudah, E. S.; Moose, S. P.; Riechers, D. E. *Plant Physiol.* **2002**, *130*, 362-373.
43. May, M. J.; Vernoux, T.; Leaver, C.; van Montagu, M.; Inze, D. *J. Exp. Bot.* **1998**, *49*, 649-667.
44. Foyer, C. H.; Rennenberg, H. In *Sulfur Nutrition and Sulfur Assimilation in Higher Plants;* Brunold C. et al.; Eds.; Paul Haupt: Bern, Switzerland, 2000; pp 127-153.
45. Breaux, E. J.; Hoobler, M. A.; Patanella, J. P.; Leyes, G. A. In *Crop Safeners for Herbicides;* Hatzios, K. K.; Hoagland, R. E.; Eds.; Academic Press: New York, NY, 1989; pp 163-175.
46. Miller, K. D.; Irzyk, G. P.; Fuerst, E. P.; McFarland, J. E.; Barringer, M.; Cruz, S.; Eberle, W. J.; Fory, W. *J. Agric. Food Chem.* **1996**, *44*, 3335-3341.
47. Pastori, G. M.; Foyer, C. H. *Plant Physiol.* **2002**, *129*, 460-468.
48. Daniel, V. *CRC Crit. Rev. Biochem.* **1993**, *25*, 173-207.
49. Coleman, J. O. D.; Blake-Kalff, M. M. A.; Emyr Davies, T. G. *Trends Plant Sci.* **1997**, *2*, 144-151.
50. Riechers, D. E.; Zhang, Q.; Xu, F.; Vaughn, K. C. *Planta* **2003**, *217*, 831-840.
51. Tommasini, R.; Vogt, E.; Fromentau, M.; Hortensteiner, S.; Matile, P.; Amrhein, N.; Martinoia, E. *Plant J.* **1998**, *13*, 773-780.
52. Rea, P. *J. Exp. Bot.* **1999**, *50*, 895-913.
53. Martinoia, E.; Klein, M.; Geisler, M.; Sanchez-Fernandez, R.; Rea, P. A. In *Vacuolar Compartments, Annual Plant Reviews;* Robinson, D. G.; Rogers, J. C.; Eds.; Sheffield Academic Press: Sheffield, UK, 2000; Vol. 5, pp 221-253.
54. Walbot, V.; Mueller, L. A.; Silady, R. A.; Goodman, C. D. In *Sulfur Nutrition and Sulfur Assimilation in Higher Plants;* Brunold C. et al.; Eds.; Paul Haupt: Bern, Switzerland, 2000; pp 155-165.

55. Alfenito, M. R..; Souer, E.; Goodman, C. D.; Buell, R.; Mol, J.; Koes, R.; Walbot, V. *Plant Cell* **1998**, *10*, 1135-1149.
56. Mueller, L. A.; Goodman, C. D.; Silady, R. A.; Walbot, V. *Plant Physiol.* **2000**, *123*, 1561-1570.
57. Bruce, W.; Folkerts, O.; Garnaat, C.; Crasta, O.; Roth, B.; Bowen, B. *Plant Cell* **2000**, *12*, 65-79.
58. Debeaujon, I.; Peeters, A. J. M.; Leon-Kloosterziel, K. M.; Koornneef, M. *Plant Cell* **2001**, *13*, 853-871.
59. Lederer, B.; Böger, P. *Biochim. Biophys. Acta* **2003**, *1621*, 226-233.
60. Bilang, J.; Sturm, A. *Plant Physiol.* **1995**, *109*, 253-260.
61. Fuerst, E. P. *Weed Technol.* **1987**, *1*, 270-277.
62. Böger, P.; Matthes, B.; Schmalfuss, J. *Pest. Manag. Sci.* **2000**, *56*, 497-508.
63. Schmalfuss, J.; Matthes, B.; Knuth, K.; Böger, P. *Pestic. Biochem. Physiol.* **2000**, *67*, 25-35.
64. Nozzolillo, C.; Ishikura, N. *Plant Cell Rep.* **1988**, *7*, 389-392.
65. Pecket, R. C.; Small, C. J. *Phytochemistry* **1980**, *19*, 2571-2576.
66. Molin, W. T.; Anderson, E. J.; Porter, C. A. *Pestic. Biochem. Physiol.* **1986**, *25*, 105-111.
67. Speth, V.; Otto, V.; Schafer, E. *Planta* **1987**, *171*, 332-338.
68. Vierstra, R. D.; Langan, S. M.; Haas, A. L. *J. Biol. Chem.* **1985**, *260*, 12015-12021.
69. Tagu, D.; Palin, B.; Balestrini, R.; Gelhaye, E.; Lapeyrie, F.; Jacquot, J.-P.; Sautiere, P.-E.; Bonfante, P.; Martin, F. *Plant Physiol. Biochem.* **2003**, *41*, 611-618.
70. Namiki, S.; Tomida, T.; Tanabe, M.; Iino, M.; Hirose, K. *Biochem. Biophys. Res. Commun.* **2003**, *305*, 592-597.
71. Jakoby, W. B.; Keen, J. H. *Trends Biochem. Sci.* **1977**, *2*, 229-231.
72. Flury, T.; Wagner, E.; Kreuz, K. *Plant Physiol.* **1996**, *112*, 1185-1190.
73. Farago, S.; Kreuz, K.; Brunhold, C. *Pestic. Biochem. Physiol.* **1993**, *47*, 199-205.
74. Riechers, D. E.; Kleinhofs, A.; Irzyk, G. P.; Jones, S. S. *Genome* **1998**, *41*, 368-372.
75. Wu, J.; Cramer, C. L.; Hatzios, K. K. *Physiol. Plant.* **1999**, *105*, 102-108.
76. Alfenito, M.; Walbot, V. In *Regulation of Enzymatic Systems Detoxifying Xenobiotics in Plants;* Hatzios, K. K.; Ed.; NATO ASI Series, Kluwer Academic Publishers: Dordrecht, the Netherlands, 1997; pp 197-208.
77. *Herbicide Handbook;* Weed Science Society of America: Lawrence, KS, 2002; 8th Edition.
78. Kreuz, K.; Gaudin, J.; Stingelin, J.; Ebert, E. *Z. Naturforsch.* **1991**, *46c*, 901-905.

Resistance and Management

Chapter 20

Sodium Channel Point Mutations Associated with Pyrethroid Resistance in the Head Louse, *Pediculus humanus capitis*

Takashi Tomita[1], Noboru Yaguchi[2], Minoru Mihara[3], Noriaki Agui[1], and Shinji Kasai[1]

[1]Department of Medical Entomology, National Institute of Infectious Diseases, Toyama 1–23–1, Shinjuku-ku, Tokyo 162–8640, Japan
[2]Toshima City Ikebukuro Health Center, Higashi-Ikebukuro 1–20–9, Toshima-ku, Tokyo 170–0013, Japan
[3]Japan Environmental Sanitation Center, Yotsuya-Kamimachi 10–6, Kawasaki, Kanagawa 210–0828, Japan

The problem of pyrethroid-resistance in the head lice is growing worldwide and an insensitive sodium channel is suspected as the major mechanism of this resistance. We established an acute toxicity test with a limited number of louse sample to discriminate pyrethroid resistance and analyzed an ORF encoding for the *para*-orthologous sodium channel from an insecticide-susceptible strain of the body louse. Phenothrin-susceptible and -resistant head louse colonies from Japan were individually analyzed for point mutations of the sodium channel cDNA; The resistant head lice shared 23 base substitutions homozygously, in which four resulted in amino acid substitutions: D11E in the N-terminal inner-membrane segment; M850T in the outer-membrane loop between DII S4 and S5; T952I and L955F in DII S5.

Head louse infestations in school children are a difficult public health problem in many countries, irrespective of the socioeconomic status of those infested (1). Disease transmission by head lice has not been reported, however, the putative conspecific body louse, Pediculus humanus humanus, transmits typhus, trench fever, and relapsing fever infested (1) and the potential vectorial role of head lice should be considered. Pediculicides used worldwide have changed from DDT and lindane, to carbaryl and malathion, and subsequently to the pyrethroids, in the interests of human safety and increased efficacy. In the last two decades pyrethroids such as permethrin and phenothrin were the most popular pediculicides where available (1). However, control failures with pyrethroids due to resistance is a concern in many developed and developing countries (2, 3, 4, 5, 6, 7). The voltage-sensitive Na^+ channels contain integral membrane proteins responsible for the generation of action potentials in excitable cells and it is the neuronal target of DDT and pyrethroid insecticides (8).

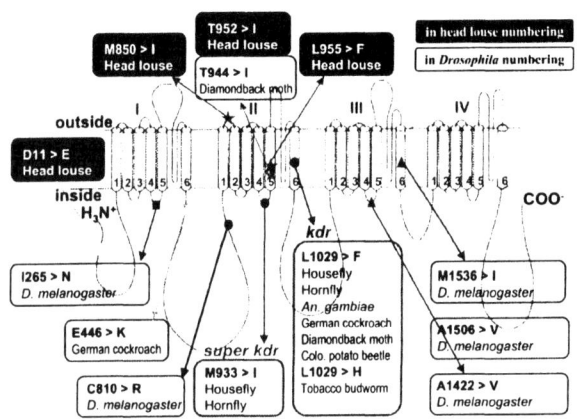

Figure 1. Resistance-associated structural changes of para-orthologous Na^+ channel in insects.

Resistance showing reduced sensitivity of the Na^+ channel was first described as knockdown-resistance (kdr) in the house fly, Musca domestica (9). Kdr is a major mechanism of DDT- and pyrethroid-resistance in a number of insects with medical and agricultural importance (10, 11). A point mutation in para-orthologous voltage-sensitive Na^+ channel α-subunit gene is the mechanism of kdr. Typical kdr mutations are kdr (L1014F) occurring at the trans-membrane segment 6 of domain II (DII S6) and super-kdr (M918T) appearing at DII S4-5 intracellular loop in the house fly (12, 13). The kdr-type (and -subtype) mutations were found in resistant strains of another 7 species (13,

14, 15, 16, 17, 18, 19) and seem to be ubiquitous among pyrethroid-resistant insects, while the *super-kdr*-type mutation was identified in a resistant strain of the horn fly, *Haematobia irritans (18)*. However, phenotypically *kdr*-like (that is atypical) mutations which are also specifically associated with pyrethroid-resistance but non-homologous to either *kdr* or *super-kdr* site have been identified. Those mutations were, for example, found at four sites in *Drosophila melanogaster (20, 21)*, at two homologous sites each in two Heliothine lepidopterans *(22)*, and at two sites in the German cockroach, *Blattella germanica (23)*. The roles of those mutations for sensitivity-reducers have not been experimentally confirmed. Another *kdr*-like mutation, V421M (numbering in *D. melanogaster para*) occurring at DI S6, from the tobacco budworm, *Heliothis virescens (19)*, was functionally expressed by *in vitro* mutagenesis and heterologous expression and its contribution to nerve insensitivity was electrophysiologically exemplified *(24)*. Resistance-associated sodium channel structural changes in insects are shown in Figure 1.

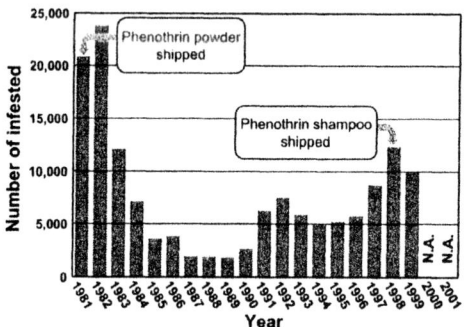

Figure 2. Fluctuation of head lice infestations reported by the Ministry of Health and Welfare.

Reduced Na^+ channel sensitivity in pyrethroid-resistant head lice was first suggested by Hemmingway *et al. (25)*. In pyrethroid-resistant head louse colonies from the US and the UK, resistance-specific two concomitant amino acid substitutions were commonly identified as *kdr*-like mutations at DII S5 of para Na^+ channel *(26)*. Their analysis was based on a partial cDNA sequence of as many as 132 amino acid residues including DII S5 and S6. Analysis of the whole coding sequence is a prerequisite for the development of molecular diagnosis or the study of the genetic origins of *kdr*-like mutations in head lice.

DDT was successfully used for controlling head lice in Japan after WWII and pediculosis greatly decreased in the 1950s. The incidence increased dramatically in the 1970s, and government surveillance initiated in 1981 recorded 24,000 cases in 1982 *(27)*. With phenothrin as the sole registered pediculicide since 1981, cases declined to ca. 2,000 annually in the late 1980s. However, there was an increase to 10,000 cases in 1999 when surveillance ended (Figure 2). The official incidence estimates seem to have underestimated the actual levels, thought to be 10-fold higher, in light of the amount of phenothrin powder and shampoo shipped in 1999 (520,000 units; Sumitomo Pharmaceuticals Co., Ltd., personal communications). The efficacy of phenothrin pediculicides has not been re-evaluated and the cause of the recent increases in pediculosis in Japan is still not known. We started to evaluate the susceptibility of head lice to phenothrin in Japan in 2001 and first identified the presence of the resistant head louse colonies in Japan.

Susceptibility to Phenothrin

In order to establish a simplified bioassay method with head lice, knockdown ratios in response to phenothrin concentrations was analyzed, in changing treatment hours by the method of continuous contact of lice with insecticide-impregnated filter paper *(28)*. We used the long-established insecticide-susceptible strain of body louse, NIID, as a reference. Figure 3A shows that phenothrin effectively knocked down adult body lice after 3 hours treatment. Similar result was obtained from 1st instar nymphs, though knock down speed was relatively low (Figure 3B).

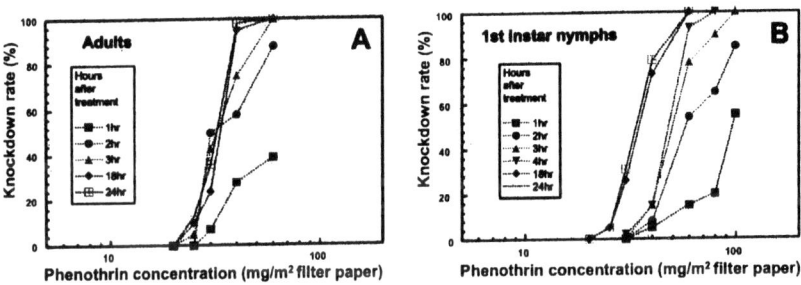

Figure 3. Knockdown rates in response to phenothrin concentrations using adult (A) and 1st instar (B) body lice.

Using the rapid toxicity of phenothrin to adult body lice, we established a short-time bioassay method for head lice. Since KC_{99} of the susceptible body lice (after 3 hours treatment) was 54 mg/m^2 (Table 1), we set a diagnostic phenothrin concentration at 100 mg/m^2 filter paper. When a louse is alive at 100 mg/m^2, the louse is evaluated as a resistant. Since the KC_1 of the susceptible body lice was 20 mg/m^2, a resistance-ratio (RR) of the louse alive at 100 mg/m^2

can be estimated at least as 5-fold as compared to the susceptible strain. This simplified method enabled us to estimate minimum resistance-ratios with small numbers of head lice.

Table 1. Susceptibility of insecticide-susceptible body lice to phenothrin

Stage of lice	Treatment	Regression line	KC^a value (95% CL) in mg/m^2
1st instar nymphs	3 hrs	$P^b = 8.43X^c - 14.4$	$KC_1 = 27$ (22-31)
			$KC_{50} = 52$ (48-55)
			$KC_{99} = 97$ (86-117)
	24 hrs	$P = 12.0X - 18.3$	$KC_1 = 22$ (18-24)
			$KC_{50} = 34$ (32-26)
			$KC_{99} = 53$ (47-64)
Adults	3 hrs	$P = 10.8X - 16.5$	$KC_1 = 20$ (17-22)
			$KC_{50} = 33$ (32-35)
			$KC_{99} = 54$ (48-66)
	24 hrs	$P = 15.1X - 22.5$	$KC_1 = 22$ (19-23)
			$KC_{50} = 31$ (30-32)
			$KC_{99} = 44$ (40-51)

[a] KC, knockdown concentration, a theoretical concentration at a knockdown percentage. [b] P, probit. [c] X, logarithm of phenothrin concentration in mg/m^2

We investigated the phenothrin-susceptibility of the head lice collected in Japan, by the simplified bioassay method. Head lice were obtained from dermatologists in Tokyo metropolitan area (Tokyo, Saitama and Kanagawa prefectures) and used for bioassay. Diagnostic phenothrin concentrations were set as 100, 400, 800, 1,600 and 3,200 mg/m^2. Three out of ten colonies survived at these diagnostic concentrations (>= 100 mg/m^2) so that they are judged as resistant colonies (Figure 4). The two of three resistant colonies (R1 and R2) were collected from schoolchildren living in the same city. These two colonies may be originated from same ancestor. Another resistant colony (R3) was collected from a 88 years old British female. She noticed the head lice infestation just after 1-month trip from UK suggesting high possibility of resistant head lice transported from outside of the country. Since 7/10 colonies showed high susceptibility to phenothrin, it seems that recent increasing of head lice infestations is not only due to development of insecticide resistance.

239

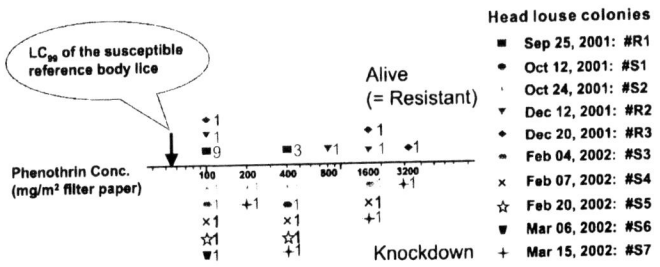

Figures above and below the scale that shows phenothrin concentration indicate the numbers of lice tested at respective concentrations

Figure 4. Phenothrin susceptibilities of head louse colonies.

Resistance-Associated Mutations

To determine the lice sodium channel cDNA sequence, we conducted two rounds of primer walking using the long established body louse strain, NIID. The primary round was based on RT-PCR using mainly degenerate primers and some previously reported GSPs and then RACEs. As a result, 6,521 bases of cDNA (GenBank #AB090951) including an ORF encoding 2,086 amino acids were first determined by Tomita *et al.* (29).

Adults from the NIID body louse strain and the five head louse colonies (S1, S2, R1-3; see Figure 4) were individually analyzed for the sodium channel cDNA. The analysis covered the base# 100-6401 segment including the complete coding sequence, except for R3 lice that mainly covered the domain II. All the lice tested were deduced to be homozygous for the sodium channel gene from the results of direct sequencing. Totally 24 base substitutions were identified among the NIID body lice and the head louse colonies tested (Figure 5). There were two base substitutions (0.016%) between S1 (or S2) and NIID lice, and 23 base substitutions (0.36%) between R1 (or R2) and NIID lice, in the 6302 bases analyzed. Only allelic levels of differences were present between the two sibling louse species. The partially analyzed cDNA sequences from phenothrin-resistant R3 lice completely coincided with those from R1 and R2 lice, as they were classified by phenothrin-susceptibility. 20 out of 24 bases were synonymous substitutions. The other four resulted in amino acid substitutions and they all associated with the resistant head lice. No amino acid substitution was detected in susceptible S1 and S2 head lice.

The four amino acid substitutions associated with phenothrin-resistant R1 head lice were (i) D11E in the N-terminal inner-membrane segment, (ii) M850T in the outer-membrane loop between the trans-membrane segments 4 and 5 of domain II, and (iii) T952I and (iv) L955F in the trans-membrane segment 5 of domain II (Figure 1). The former two substitutions were first identified. The latter two substitutions were the same as those commonly discovered from pyrethroid-resistant head lice in the US and UK (25). Interestingly, the resistant head louse colony, R3, was collected in Japan from a British woman who was long stayed in Japan. She retuned her homeland to spend for a month, and soon after she came back to Japan she realized head irritation. This colony completely shared 9 base substitutions (and as a consequent also M850I, T952I, and L955F) with R1, within currently available cDNA sequence (base# 1780-3451). The genetic origin descended to R1 and R3 should be further studied.

Colony #	RR	Asp 11 Glu	Val 143	Ala 235	Leu 328	Asp 363	Ala 442	Arg 461	Arg 481	Ser 489	Ser 518	Arg 596	Arg 697
NIID	-	GAT	GTA	GCC	CTG	GAT	GCC	AGA	AGA	TCA	TCT	CGC	AGG
S1	-	...	...	...	...	...	...	...	...	...	...	...	...
S2	-	...	...	...	...	...	...	...	...	...	...	...	...
R1	>= X20	..A	..T	..T	T..	..C	..T	..G	..G	..T	..A	..T	..A
R2	>= X80	..A	..T	..T	T..	..C	..T	..G	..G	..T	..A	..T	..A
R3	>= X160	..A	-	-	-	-	-	-	-	-	-	..T	..A

Colony #	RR	Thr 766	Val 805	Met 850 Ile	Thr 952 Ile	Leu 955 Phe	Ser 1012	Leu 1046	Gly 1120	Gly 1231	Phe 1275	Ala 1376	Phe 1749
NIID	-	ACA	GTT	ATG	ACA	CTT	TCT	TTG	GGT	GGA	CCG	GCC	TTT
S1	-	..G	...	...	...	...	...	...	...	...	...	...	..C
S2	-	..G	...	...	...	...	...	...	...	...	...	...	..C
R1	>= X20	..G	..A	..T	.T.	T..	..A	C..	..A	..T	..A	..T	...
R2	>= X80	..G	..A	..T	.T.	T..	..A	C..	..A	..T	..A	..T	...
R3	>= X160	..G	..A	..T	.T.	T..	..A	C..	-	-	-	-	-

Figure 5. Base substitutions in louse Na^+ channel cDNA.

The resistance-associated four amino acid substitutions of the louse sodium channel are located to some extent in evolutionarily conserved regions among insect species. D11E may have the least significance for pyrethroid-insensitivity, because of the common acidic nature and evolutionary vulnerability between Asp and Glu. It is difficult to evaluate the significance of M850T, which is the first report for a resistant-associated substitution put in outer-membrane loop among the insect species studied, even though the M850 position is conserved at least in 7 susceptible insects. The T952I and L955F positions are located in the highly conserved DIIS5 trans-membrane segment. As for L955F, an analogous contribution to insensitivity is speculated by Lee et al (25) due to its close proximity to the DIIS6 Leu position in which the *kdr*-type Leu to Phe substitution occurring in many pyrethroid-resistant pests. The T952I substitution was shared by pyrethroid-resistant diamondback moth (15) and thus the substitution may be potentially involved in pyrethroid-insensitivity of head lice sodium channel. We designed TaqMan probes in order to preliminary demonstrate molecular diagnosis of pyrethroid-resistance associated Na^+ channel gene of head lice (Figure 6). The results successfully discriminated possible three genotypes for T952I.

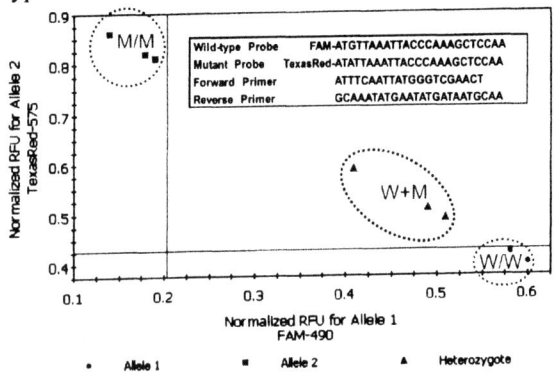

Figure 6. Discrimination of Na^+ channel genotypes.

Perspective

The present study was dependent on the evaluation of insecticide-susceptibility with live lice and the subsequent individual-genotyping of the Na+ channel gene. Such work is costly and time-consuming because live lice have to be bioassayed on the day of collection due to low viability after removal from the host, limited availability of human volunteers, and the lack of effective alternative blood feeding systems. Cumulative information on resistance-associated representative Na^+ channel gene haplotypes by such a basic study and experimental confirmations of insensitive-responsible mutations are prerequisite to make molecular diagnosis of pyrethroid-resistance possible from dead lice. PCR-based high throughput systems for single nucleotide polymorphisms (SNP)-typing would facilitate further studies on the origins or migrations of the lice resistant Na+ channel genes worldwide as well as on the frequency distribution of presumed recessive resistant genes. Such works should also be a remedy for resistant infested dermatology patients against overuse of inefficacious pediculicides when the resistant diagnostic system is timely linked to a medical examination. Recent reports of pyrethroid-resistant in lice demonstrate the need for the development of novel pediculicides.

Conclusion

Pyrethroid-resistance of head lice was discriminated with phenothrin at 100 mg/m^2 by the method of continuous contact of insecticide-impregnated filter paper for 3 hours. We first identified pyrethroid-resistant head lice colonies in Japan. Pyrethroid pediculicides seemed to be out of use for the resistant colonies, however, the resistance has not been an overwhelming majority in Tokyo metropolitan area. The complete coding sequence of *para*-orthologous Na$^+$ channel was determined in the head louse and resistance-associated four amino acid substitutions were commonly identified in the Na$^+$ channel from the three resistant colonies tested.

Acknowledgements

We thank Tsuchiya, T., Mizushima, J., and a number of dermatologists for providing live lice, Ishii, N., for coordinating lice-sampling from medical facilities, and Sumitomo Chemical Co., Ltd. for providing pediculicide-shipping information and phenothrin.

References

1. Gratz, N. G.: WHOPES, CTD, WHO, Geneva, Switzerland, **1997**.
2. Coz, J.; Combescot-Lang, C.; Verdier, V: *Bull. Soc. Fra. Parasitol.* **1993**, *11*, 245-252 (in French).
3. Mumcuoglu, K. Y.; Hemingway, J.; Miller, J.; Ioffe-Uspensky, I.; Klaus, S.; Ben-Ishai, F.; Glaun, R: *Med. Vet. Entomol.* **1995**, *9*, 427-432.
4. Burgess, I. F.; Brown, C. M.; Peock, S.; Kaufman, J.: *Brit. Med. J.* **1995** *311*, 752.
5. Rupes, V.; Moravec, J.; Chmela, J.; Ledvinka, J.; Zelenkova, J.: *Centr. Eur. J. Pub. Health* **1995**, *3*, 30.
6. Picollo, M. I.; Vassena, C. V.; Casadio, A. A.; Massimo, J.: *J. Med. Entomol.* **1998**, *35*, 814-817.
7. Pollack, R. J.; Kiszewski, A.; Armstrong, P.; Hahn C.; Wolfe, N.; Rahman, H. A.; Laserson, K.; Telford III, S. R.; Spielman, A: *Arch. Pediatr. Adlesc. Med.* **1999**, *153*, 969-973.

8. Gordon, D.; Moskowitz, H.; Zlotkin, E.: *Arch. Insect. Biochem. Physiol.* **1993**, *22*, 41-53.
9. Farnham, A. W.: *Pestic. Sci.* **1977**, *8*, 631-636.
10. Oppenoorth, F. J.: Biochemistry and genetics of insecticide resistance, pp. 731-773. In G.A. Kerkut and L.I. Gilbert [eds], Comprehensive insect physiology. Pergamon, Oxford, **1985**.
11. Soderlund, D. M.: pp 21-56. *In* Sjut V. [ed], Molecular mechanisms of resistance to agrochemicals. Springer Verlag, Berlin, **1997**.
12. Williamson, M. S.; Martinez-Torres, D.; Hick, C. A.; Devonshir, A. L.: *Mol. Gen. Genet.* **1996**, *252*, 51-60.
13. Miyazaki, M.; Ohyama, K.; Dunlap, D. Y.; Matsumura, F.: *Mol. Gen. Genet.* **1996**, *252*, 61-68.
14. Lee, S. H.; Dunn, J. B.; Clark, J. M.; Soderlund, D. M.: *Pestic. Biochem. Physiol.* **1999**, *63*, 63-75.
15. Martinez-Torres, D.;, Devonshire, A. L.; Williamson, M. S.: *Pestic. Sci.* **1997**, *51*, 265-270.
16. Martinez-Torres, D.; Chandre, F.; Williamson, M. S.; Darriet, F.; Berge, J. B.; Devonshire, A. L.; Guillet P.; Pasteur, N.; Pauron, D.: *Insect Mol. Biol.* **1998**, *7*, 179-184.
17. Martinez-Torres, D.; Foster, S. P.; Field, L. M.; Devonshire, A. L.; Williamson, M. S.: *Insect Mol. Biol.* **1999**, *8*, 339-346.
18. Guerrero F. D.; Jamroz, R. C.; Kammlah, D.; Kunz, S. E.: *Insect. Biochem. Mol. Biol.* **1997**, *27*, 745-55.
19. Park, Y.; Taylor M. F.: *Insect Biochem. Mol. Biol.* **1997**, *27*, 9-13.
20. Pittendrigh, B.; Reeman, R.; ffrench-Constant, R. H.; Ganetzky, B.: *Mol. Gen. Genet.* **1997**, *256*, 602-610.
21. Martin, R. L.; Pittendrigh, B.; Liu, J.; Reenan R.; ffrench-Constant, R.; Hanck, D. A.: *Insect Biochem. Mol. Biol.* **2000**, *30*, 1051-1059.
22. Head, D. J.; McCaffery, A. R.; Callaghan, A.: *Insect Mol. Biol.* **1998**, *7*, 191-196.
23. Liu, Z.; Valls, S. M.; Dong, K.: *Insect. Mol. Biol.* **2000**, *30*, 991-997.
24. Lee, S. H.; Soderlund, D. M.: *Insect Biochem. Mol. Biol.* **2001**, *31*, 19-29.
25. Hemmingway, J.; Miller, J.; Mumcuoglu, K. Y.: *Med. Vet. Entomol.* **1999**, *13*, 89-96.
26. Lee, S. H.; Yoon, K.-S.; Williamson, M. S.; Goodson, S. J.; Takano-Lee, M.; Edman, J. D.; Devonshir, A. L.; Clark, J. M.: *Pestic. Biochem. Physiol.* **2000**, *66*, 130-143.
27. Agui, N.: *Seikatsu To Kankyo,* **1999**, *44*, 18-22. (in Japanese)
28. Kasai, S.; Mihara, M.; Takahashi, M.; Agui, N.; Tomita, T.: *Med. Entomol. Zool.* **2003**, *54*, 31-36.
29. Tomita, T.; Yaguchi N.; Mihara M.; Takahashi M.; Agui, N.; Kasai, S.: *J. Med. Entomol.* **2003**, *40*, 468-474.

Chapter 21

Molecular Characterization of Resistance to Acetolactate Synthase Inhibitors in *Lindernia micrantha*: Origin and Expansion of Resistant Biotypes

Hiroyuki Shibaike[1], Akira Uchino[2], and Kazuyuki Itoh[3]

[1]National Institute for Agro-Environmental Sciences, Tsukuba, Ibaraki 305–8604, Japan
[2]National Agricultural Research Center for Tohoku Region, Ohmagari, Akita 014–0102, Japan
[3]National Agricultural Research Center for Tohoku Region, Morioka, Iwate 020–0198, Japan

To understand the occurrence of herbicide-resistant weeds, we studied the genetic profile of *Lindernia micrantha* D. Don (Scrophulariaceae), an annual paddy weed in Japan. Sixty-nine plants, including resistant and susceptible biotypes to acetolactate synthase (*ALS*), from 12 paddy fields in 3 distantly separated regions were used for DNA analysis. Comparison of the genetic relationships and patterns of amino acid substitutions of *ALS* gene between tested biotypes suggest that resistant biotypes are selected in every local. Number of *ALS* gene, as one of the factors contributing to outbreaks of weed resistance in Japan, was also discussed.

Introduction

The establishment of an agro-ecosystem within a natural ecosystem inevitably calls for the management of the wild vegetation that tries to enter. The high temperature and humidity of Asia's monsoons combined with fertile soil create ideal conditions for plant growth. Therefore, improvements in agricultural technology in Japan are aimed at finding effective weed control.

In rice cultivation, rice seedlings are initially grown in small seedling boxes, while soil is paddled and leveled to weaken competitive ability of weeds against rice before transplanting. These traditional rice cropping methods are now supplemented with the use of herbicides and agricultural machinery.

Recently, the sudden occurrence of weeds resistant to sulfonylurea (SU) herbicides, 1 of the 4 chemical classes of acetolactate synthase (*ALS*) inhibitors, has become a large problem in Japan. For example, resistant biotypes of *Lindernia dubia* var. *dubia* exhibit a high level of resistance to SU herbicides (*Figure 1*). When herbicide was applied at 16 times folds the standard dosage, a substantial number of plants of resistant biotype survived, and their growth, flowering, and fruiting were similar to those of untreated plants. In contrast, all plants of susceptible biotype died even at the recommended dosage of the herbicide.

Figure 1. Effect of Sulfonylurea herbicide to Lindernia dubia var. dubia

Genetic Relationships of Resistant and Susceptible Biotypes

Lindernia micrantha is an annual, broad-leaved paddy weed, and is widely distributed in tropical to warm-temperate regions of East Asia. Since 'one-shot' treatment by SU herbicides came into widespread use in the 1990s in Japan, *L. micrantha* had been effectively controlled or suppressed by the herbicides applied before and after emergence.

SU-resistant *L. micrantha* was first reported in Yamagata prefecture in direct-seeding paddy fields (*1*). Similar reports then published from Akita, Saitama, and Kyoto prefectures. To clarify the origin and expansion of resistant biotypes of *L. micrantha*, it is beneficial to compare genetic diversity among groups subjected to continuous use of SU herbicides, in which resistant biotypes have already appeared, and groups in which resistant biotypes have not appeared yet.

For this study, we selected 12 populations of *L. micrantha* from Kyoto, Yamagata, and Akita prefectures (*Figure 2*). Three to 8 plants were randomly collected from each population, giving a total of 69 plants examined.

Internal simple sequence repeat (ISSR) markers were generated from single-primer PCR amplifications in which the primer was designed from di- or trinucleotide repeat motifs with a 5' or 3' anchoring sequence of 1 to 3 nucleotides (*2*). The ISSR phenotype of each plant was transformed into a binary character matrix in which each locus was represented by an individual character, band present or band absent. In statistical analysis, a dendrogram was constructed to show the genetic relationships among the 69 plants (*Figure 3*).

The 9 primers used yielded a total of 23 polymorphic loci (*3*). The combination of these 9 primers recognized 40 ISSR phenotypes out of 69 plants. The mean number of ISSR phenotypes varied among the populations. In the populations of resistant biotypes, the mean number of ISSR phenotypes was close to 1. This result indicates that the populations of resistant biotypes were genetically very similar. In contrast, in the populations of susceptible biotypes, the mean number of ISSR phenotypes varied from 1 to 2.3, indicating that the genetic variability of susceptible biotypes was relatively high.

In the dendrogram (*Figure 3*), the 69 plants collected from 12 populations in 3 regions separate into 2 clusters of populations, the Kyoto cluster and the Yamagata and Akita cluster. There is a genetically large difference between the 2 clusters. Considering the geographical separation of the 2 regions, these results appear reasonable. According to the results of genetic variation within populations, the resistant biotypes segregate with short or no branch lengths (see plants indicated by the asterisks in *Figure 3*), whereas the susceptible biotypes segregate with relatively long branch lengths. The results also showed that the populations of resistant biotypes seemed to be derived from the populations of

247

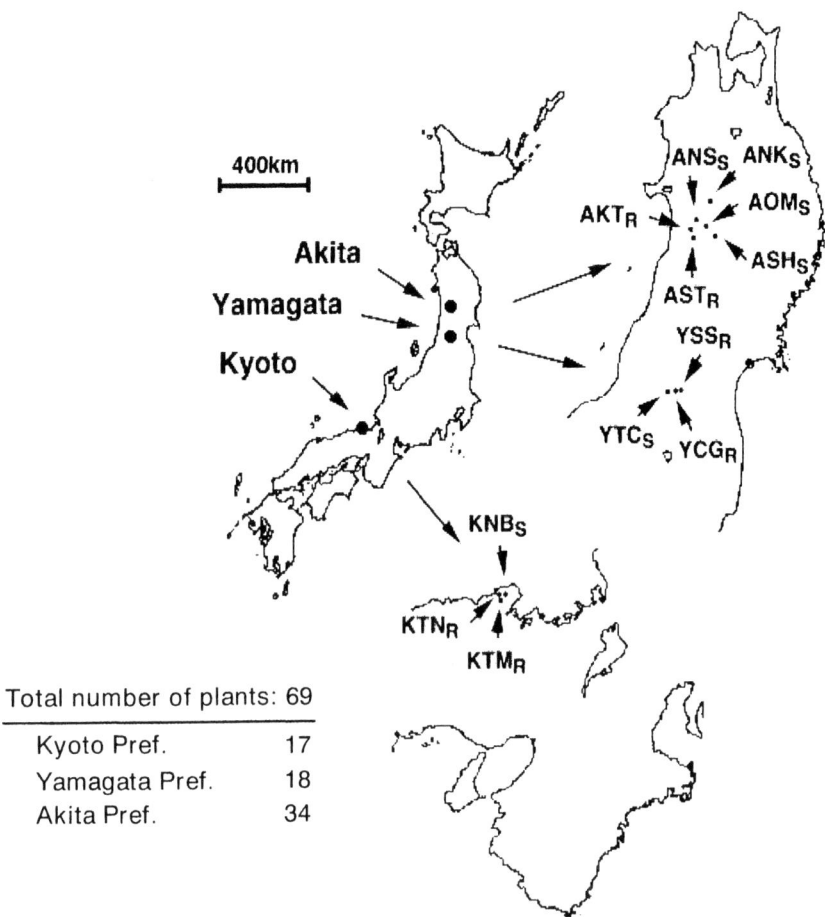

Total number of plants: 69	
Kyoto Pref.	17
Yamagata Pref.	18
Akita Pref.	34

Figure 2. Distributions of sampling sites in Japan

In the legends, the first letters of population names indicate the prefectures

(A: Akita; K: Kyoto; Y: Yamagata), and the subscripts indicate

the resistant or susceptible populations ($_R$: resistant; $_S$: susceptible).

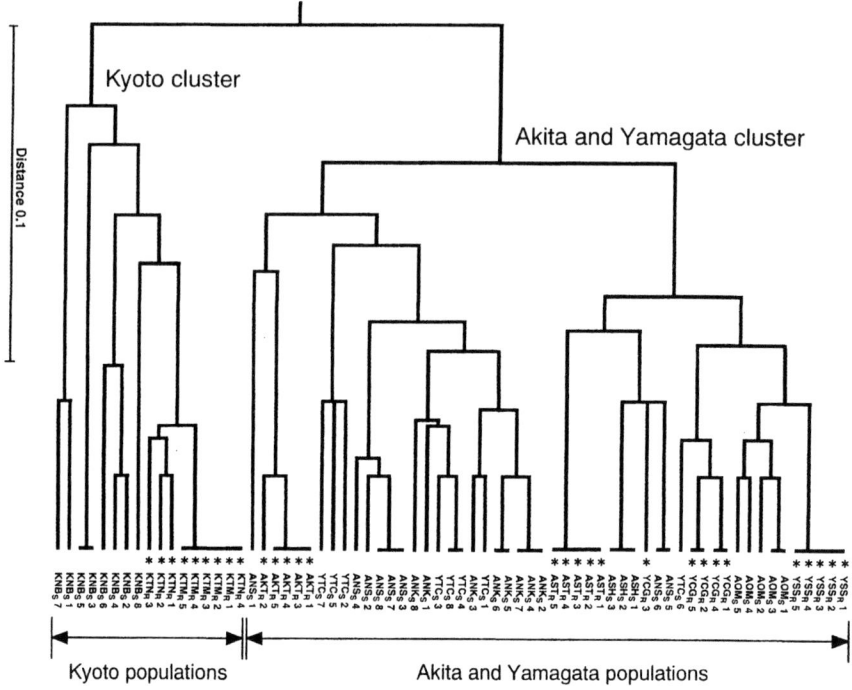

Figure 3. Genetic relationships among 69 plants of Lindernia micrantha

*: Asterisks indicate the SU-resistant biotypes.

susceptible biotypes, suggesting that there is a common genetic basis to the resistant and susceptible biotypes (see Kyoto populations in *Figure 3*).

Patterns of Amino Acid Substitutions of The *ALS* Gene

ALS is a key enzyme in the biosynthetic pathway of branched-chain amino acids, and is assumed to be the target site of ALS inhibitors such as SU herbicides (*4*). Mutations conferring resistance are located in 5 conserved amino acid sequences on the ALS enzyme (*Figure 4*). Patterns of cross-resistance to 4 classes of ALS inhibitors are different depending on the regions in which the mutations occur. Among naturally occurred resistant plants, point mutations at the proline codon in the second amino acid sequence (2) of Region A are common. Based on previous studies (*5*), we tried to examine DNA sequences of polymorphisms of the *ALS* gene conferring the resistance to SU herbicides in *L. micrantha*.

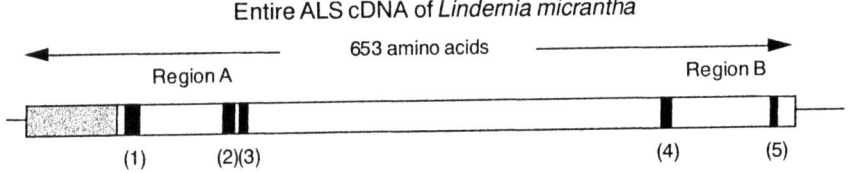

Figure 4. *Five conserved amino acid sequences (1-3 in Region A, and 4 and 5 in Region B) in* Lindernia micrantha

First-strand cDNA was synthesized from total RNA that was extracted from leaves of a single resistant plant from AKT_R (cf. *Figure 2*). This cDNA was used as template for PCR amplification of partial ALS coding regions. Newly synthesized primers in this amplified region were further used to determine the entire ALS cDNA by using 3' RACE and TAIL-PCR techniques (*6, 7*). On the basis of sequence information of *L. micrantha*, pairs of primers were synthesized to amplify a fragment containing the entire sequence of the *ALS* gene.

We first examined the nucleotide sequences of 5 resistant and 7 susceptible plants (*8*). The length of the amino acid sequence deduced from the ALS cDNA is 653 (*Figure 4*), and its homology compared with 5 other plant species (*Amaranthus* sp., *Arabidopsis thaliana, Brassica napus, B. scoparia, Nicotiana tabacum*) is about 77%. Most of the nucleotide substitutions were found only once, and the remaining 3 were polymorphic. In the 5 conserved amino acid sequences on the ALS enzyme, 2 nucleotide substitutions were detected. One was found in only 1 plant, and another was found to be polymorphic among the samples.

Table I shows the replacement of amino acid residues by nonsynonymous nucleotide substitutions in the 5 conserved amino acid sequences on the ALS enzyme. The alteration of tyrosine (Y) to histidine (H) in the first amino acid sequence (1) of Region A was found in 1 resistant biotype. But it seems that this replacement does not confer resistance, because the other resistant biotypes did not have such a replacement. All 5 resistant biotypes had various point mutations (A, S, L) in the codon for a proline residue (P) in the second amino acid sequence (2) of Region A, whereas all 7 susceptible biotypes had the proline residue (P). Because this proline codon (P) is common with other susceptible plants sequenced so far, it is possible that its replacement confers resistance to SU herbicides in *L. micrantha*. Finally, all of the resistant plants had an alteration from proline (P) to serine (A), alanine (S), or lysine (L) compared with the amino acid sequence of susceptible plants reported.

Table I. Deduced amino acid comparison for resistant and susceptible biotypes of *Lindernia micrantha*

		Region A			Region B	
Five conserved amino acid sequences reported in susceptible plants		(1) VFAYPGGASMEIHQALTRS	(2) AITGQVPRRMIGT	(3) AFQETP	(4) QWED	(5) IPSGG
Resistant biotypes	KTN$_R$3	H	A			A
	YCG$_R$1	Y	S			A
	YCG$_R$2	Y	S			A
	AKT$_R$3	Y	S			A
	AST$_R$1	Y	L			A
Susceptible biotypes	KNB$_S$1	Y	P			A
	KNB$_S$3	Y	P			A
	YTC$_S$1	Y	P			A
	YTC$_S$2	Y	P			A
	ANS$_S$1	Y	P			A
	ANS$_S$2	Y	P			A
	ASH$_S$1	Y	P			A

To prove the variation of amino acid substitutions at the proline residue, we analyzed the sequences of all 69 plants for the target region, and found that all 40 susceptible plants had proline (P), whereas the remaining 29 resistant plants had an amino acid substitution other than proline (P) without any exceptions.

Origination and Expansion of Resistant Biotypes

An UPGMA dendrogram for the 12 populations was constructed, and the results of amino acid substitutions at the proline residue in the second amino acid sequence (2) of Region A (*Table I*) were overlaid (*8*). The dendrogram consisted of 2 major clusters (*Figure 5*). The first cluster contained Kyoto populations, and the second one contained Yamagata and Akita populations. The second cluster also consisted of 2 major subclusters.

Two resistant populations in Kyoto (KTM$_R$ and KTN$_R$) showed a high level of genetic similarity, and had the same alteration from proline (Pro) to alanine (Ala). In Yamagata, 2 resistant populations (YCG$_R$ and YSS$_R$) were genetically similar, but had different amino acid substitutions (Ser and Gln). One of the resistant populations in Yamagata (YCG$_R$) had the same alteration from Proline (Pro) to serine (Ser) as a resistant population in Akita (AKT$_R$), but they showed a relatively low level of genetic similarity. Two resistant populations in Akita (AKT$_R$ and AST$_R$) were not genetically similar, and had different amino acid substitutions (Ser and Lys).

The continuous use of a particular class of herbicide leads to a marked decrease in genetic diversity within populations. At the same time, it selects plants possessing specific genetic traits conferring herbicide resistance. In addition, substitutions of amino acid residues confer resistance to SU herbicides. In this case, various point mutations occurred in the proline codon and encoded several possible amino acid substitutions. When we reviewed the data on genetic relationships and amino acid substitutions of the 69 plants, the emerged picture showed that resistant biotypes were independently selected from these areas because of at least five different amino acid substitutions (i.e., 1: KTM$_R$+KTN$_R$; 2: YCG$_R$; 3: YSS$_R$; 4: AKT$_R$; 5: AST$_R$).

What Factors Contribute to The Occurrence of Resistant Weeds?

Many factors have been shown to be important in determining the evolution of herbicide resistance in weed populations (*10*). Some of them have been included as parameters in population genetic studies, because accurate measurements of many factors that influence the evolution of resistance are generally difficult to obtain experimentally. Here, we focus on the number of *ALS* genes, and discuss whether this factor can contribute to outbreaks of weed resistance in Japan.

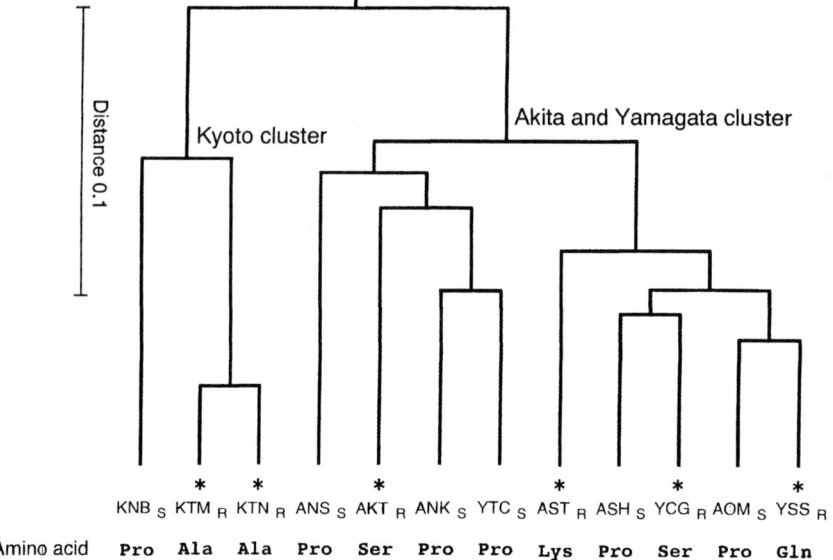

Figure 5. *Genetic relationship among 12 populations and their amino acid substitutions in ALS of* Lindernia micrantha

*: Asterisk indicates the SU-resistant populations.

Some species, such as *Arabidopsis* and sugar beet, have a single *ALS* gene, while other species, such as corn, soybean, tobacco, and brassica, have multiple *ALS* genes. In Japan, Uchino and Watanabe (*11*) reported clones encoding two different copies of partial *ALS* genes in the 4 *Lindernia* species in which resistant biotypes already occurred.

Why do paddy weeds that are resistant to *ALS* inhibitors have multiple copies of *ALS* genes? If all of the resistant weeds reported in Japan have more copies of *ALS* genes than the susceptible weeds, resistant biotypes may arise easily in these species (*Figure 6*). Because the inheritance of resistance to *ALS* inhibitors is largely dominant (*12*), it is possible that each copy has an alteration at the proline residue that confer resistance to ALS inhibitors as mentioned above. To answer this question, further investigation is required to determine the number of *ALS* genes among different species, and to clarify whether they have the same functions or not.

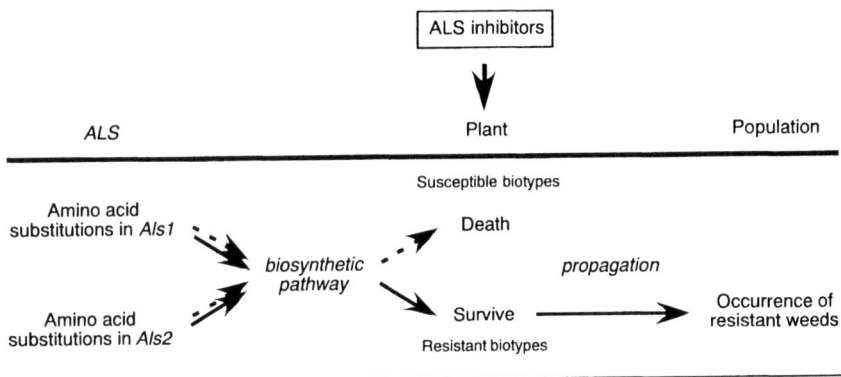

Figure 6. Genetic relationships among 69 plants of Lindernia micrantha

Straight arrows indicate resistance to ALS inhibitors, and dotted arrows do ressitance.

References

1. Itoh, K.; Wang, GX.; Ohba, S. *Weed Research* **1999**, *39*, 413-423.
2. Zeitkiewick, E.; Rafalski A.; Labuda, D. *Genomics* **1994**, *20*, 176-183.
3. Shibaike, H.; Itoh, K.; Uchino, A. *The 1999 Brighton Conference Weeds* **1999**, *Conference Proceedings Volume 1*; 1999; pp 197-202.

4. Devine, MD.; Eberlein, CV. In *Herbicide Activity: Toxicology, Biochemistry and Molecular Biology*; Roe, RM., *et al.*, Eds.; IOS Press, 1997, 159-185.
5. Wright, TR.; Bascomb, NF,; Sturner, SF.; Penner, D. *Weed Scinece* **1998**, *46*, 13-23.
6. Frohman, MA.; Duch, MK.; Martin, GR. *Proceddings of National Academy of Sciences of the United States of America* **1988**, *85*, 8998-9002.
7. Terauchi, R.; Kahl, G. *Molecular and General Genetics* **2000**, *263*, 554-560.
8. Shibaike, H.; Uchino, A.; Itoh, K. *unpublished Plant Species Biology* **2004**
9. Tranel, PJ.; Wright, TR. *Weed Science* **2002**, *50*, 700-712.
10. Cousens, R.; Mortimer, M. *Dynamics of Weed Populations*; Cambridge University Press: New York, 1995.
11. Uchino, A.; Watanabe, H. *Weed Biology and Management* **2002**, *2*, 104-109.
12. Jasieniuk, M.; Brûlé-Babel, AL.; Morrison, IN. *Weed Science* **1996**, *44*, 176-193.

Chapter 22

Molecular Characterization of Acetolactate Synthase in Resistant Weeds and Crops

Tsutomu Shimizu[1], Koichiro Kaku[1], Kiyoshi Kawai[1], Takeshige Miyazawa[1], and Yoshiyuki Tanaka[2]

[1]Kumiai Chemical Industry Company, Ltd., Kakegawa, Shizuoka 436–0011, Japan
[2]National Institute of Agrobiological Sciences, Tsukuba, Ibaragi 305–0856, Japan

The goal of this paper is to outline the studies on the molecular characteristics of acetolactate synthase (ALS) in herbicide resistant weeds and crops. For this purpose, papers and patents concerning this field were reviewed, and our studies involving novel mutated ALS genes from rice cells, synthesized rice ALS genes, performance of the gene as a selectable marker for the genetic transformation, generation of transgenic plants with the gene, and the use of the recombinant ALS's for the herbicide resistance management were briefly summarized.

Introduction

Acetolactate synthase, ALS, also referred to as acetohydroxy acid synthase, AHAS, is the first common enzyme in the biosynthetic pathway to the branched-chain amino acids; valine, leucine and isoleucine (Fig. 1).
This pathway exists in plants and microorganisms such as bacteria, fungi and algae. ALS is the primary target site of action for at least four structurally distinct classes of herbicides including the sulfonylureas (SU), the imidazolinones (IM), the triazolopyrimidine sulfonamides (TP) and our pyrimidinyl carboxy herbicides (PC)(Fig. 2).

© 2005 American Chemical Society

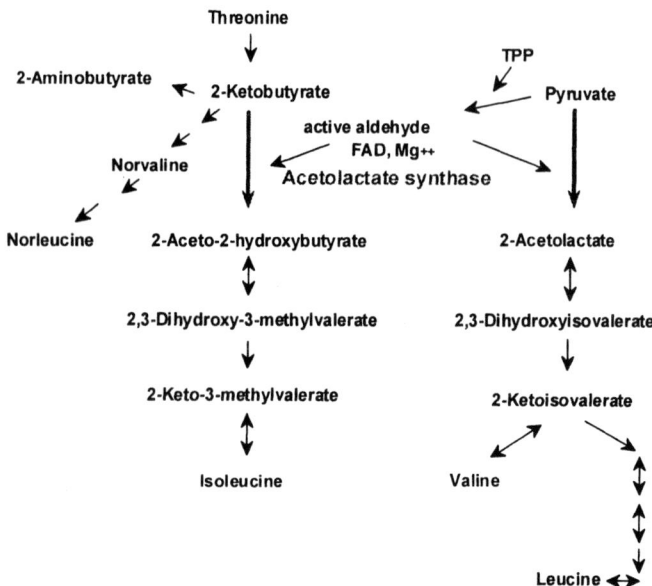

Fig. 1. Biosynthetic pathway of branched-chain amino acids

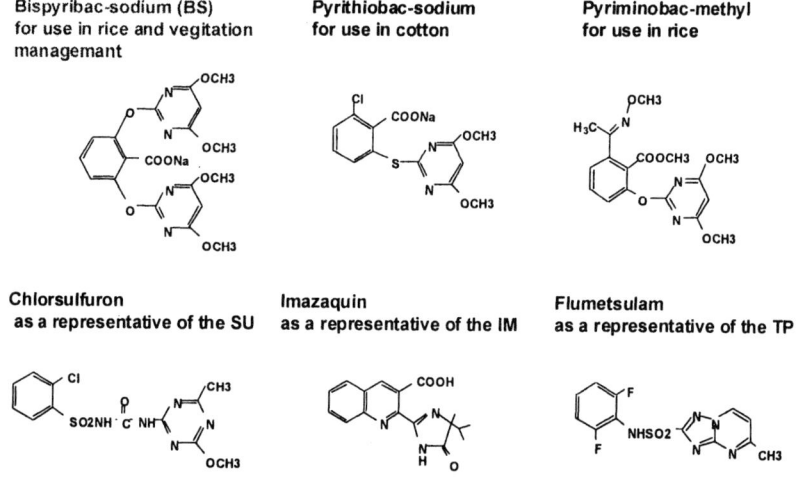

Fig. 2. ALS-inhibiting herbicides

The biological activities of the SU, the TP and the PC are extremely high with field application rates of approximately 10 g to 100 g per hectare, while the IM is approximately 10-fold less potent with field application rates of approximately 100 g to 1000 g per hectare. The SU, the TP, and the PC inhibit plant ALS at concentrations in the nanomolar range, whereas the IM does it in the micromolar range (1, 2).

Mutations in ALS's of weeds

The mutations in ALS's of weeds conferring resistance to the ALS-inhibiting herbicides are shown in Table 1.

Table 1. Mutations in ALS's of weeds conferring resistance to ALS-inhibiting herbicides

Plant species	Position	Mutation	Herbicide
Amaranthus sp.	Pro	Leu	
	Trp569	Leu	
	Ser	Asn, Thr	IM
Ambrosia sp.	Trp574	Leu	
Brassica tournefortii	Pro	Ala	SU
Lactuca serriola	Pro197	His	SU
Kochia scoparia	Pro189	Thr, Ser, Arg, Leu, Gln, Ala	SU
	Trp570	Leu	
Sisymbrium orientale	Pro	Ile	SU
	Trp	Leu	(SU)
Xanthium strumarium	Ala100	Thr	IM
	Ala183	Val	(IM)
	Trp552	Leu	IM
Lindernia dubia	Pro	Ala	SU
Lindernia dubia subsp. major	Pro	Ser	SU
Lindernia micrantha	Pro179	Ala, Gln, Ser, Lys, Gln	SU
Lindernia procumbens	Pro	Gln, Ser	SU
Scirpus juncoides	Pro	Leu	SU

The SU-resistant weeds were first found in *Kochia scoparia* and *Lactuca serriola* in the field with the repeated use of the SU, chlorsulfuron, in the US. Since then many weed species have developed resistance to the SUs and the IMs during recent years. In Japan, twelve weed species including *Lindernia* sp. have developed resistance to the SU. The mutated ALS gene conferring resistance to the SU was first shown in *K. scoparia*. The weeds possessing the mutated ALS at the proline position in the upstream region were found in the fields with the repeated use of the SU, whereas that of the alanine position was found in the fields with the use of the IM. In addition, the mutated ALS at the tryptophan position in the down stream region was found in the weeds through selection by both the IM and the SU, and a weed possessing the mutated ALS at the serine position in the down stream region was found in the fields with the use of the IM (3, 4).

Fig. 3 describes the rearrangement of the mutation sites of the ALS's in these weeds using the rice ALS numbering system.

```
1         10        20        30        40        50        60        70        80
MATTAAAAAAALSAAATAKTGRKNHQRHHVLPARGRVGAAAVRCSAVSPVTPPSPAPPATPLRPWGPAEPRKGADILVEA

          90        100       110       120       130       140       150       160
LERCGVSDVFAYPGGASMEIHQALTRSPVITNHLFRHEQGEAFAASGYARASGRVGVCVATSGPGATNLVSALADALLDS
                  *
          170       180       190       200       210       220       230       240
VPMVAITGQVPRRMIGTDAFQETPIVEVTRSITKHNYLVLDVEDIPRVIQEAFFLASSGRPGPVLVDIPKDIQQQMAVPV
              *         *
          250       260       270       280       290       300       310       320
WDTSMNLPGYIARLPKPPATELLEQVLRLVGESRRPILYVGGGCSASGDELRWFVELTGIPVTTTLMGLGNFPSDDPLSL

          330       340       350       360       370       380       390       400
RMLGMHGTVYANYAVDKADLLLAFGVRFDDRVTGKIEAFASRAKIVHIDIDPAEIGKNKQPHVSICADVKLALQGLNALL

          410       420       430       440       450       460       470       480
QQSTTKTSSDFSAWHNELDQQKREFPLGYKTFGEEIPPQYAIQVLDELTKGEAIIATGVGQHQMWAAQYYTYKRPRQWLS

          490       500       510       520       530       540       550       560
SAGLGAMGFGLPAAAGASVANPGVTVVDIDGDGSFLMNIQELALIRIENLPVKVMVLNNQHLGMVVQWEDRFYKANRAHT
                                                                 *
          570       580       590       600       610       620       630       640
YLGNPECESEIYPDFVTIAKGFNIPAVRVTKKSEVRAAIKKMLETPGPYLLDIIVPHQEHVLPMIPSGGAFKDMILDGDG
                                                                              *
644       *, Stars indicate the mutation positions.
RTVY
```

Fig. 3. Depiction of mutation sites in rice ALS gene

Five mutation positions are important for the development of resistance to the herbicides. These five mutations involve the residues of alanine at position 96, proline at position 171 and alanine at position 179 in the upstream region, and tryptophan at position 548 and serine at position 627 in the down stream region.

The ALS of *Xantium strumarium* possessing a mutation of alanine at position 96 to threonine has been shown to be resistant to the IM, but generally not to the SU. The PCs inhibited ALS of this biotype as potently as the wild-type enzyme (5). Thus this mutation is considered to confer resistance especially to the IM (Table 2).

Table 2. Mutation sites conferring resistance to ALS-inhibiting herbicides

Mutation Site[a]	SU	IM	PC
Alanine 96	S-r[b]	R[c]	S
Proline 171	R	S-R	S-R
Alanine 179	r	r	r
Tryptophan 548	R	R	R
Serine 627	S	r-R	r-R

a, rice ALS numbering system
b: little or no resistance (S); moderate resistance (r)
c, high resistance (R)

On the other hand, the ALS of *Lactuca serriola* possessing a mutation of proline at 171 to histidine has been shown to be resistant to the SU and to exhibit cross-resistance to the IM but not to the PC. The ALS of *K. scoparia* possessing a mutation of proline at position 171 to serine was inhibited by the PC herbicide, bispyribac-sodium (BS) as potently as the wild-type enzyme but expressed a cross-resistance to the PC herbicide, pyrithiobac-sodium (Fig. 4) (*5*).

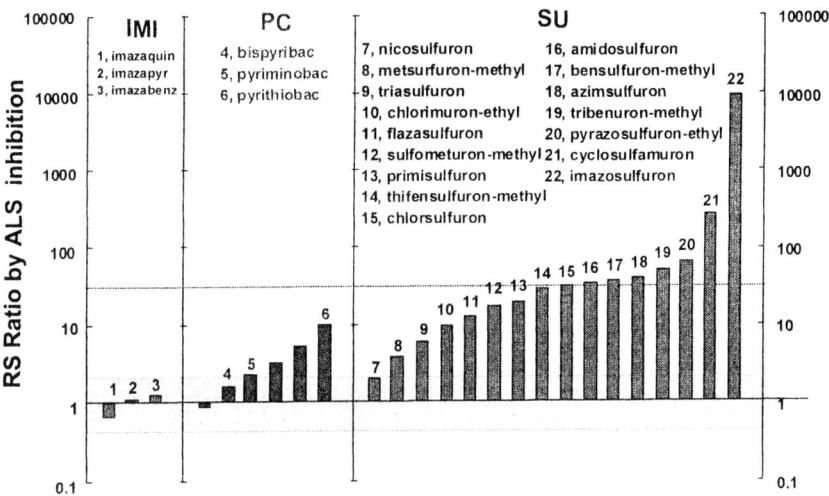

Fig. 4. Sensitivities of ALS from SU-resistant *Kochia scoparia* to ALS-inhibiting herbicides

Therefore, we should consider that the mutations at the proline 171 position confer resistance to the SU and that the cross-resistance patterns to other ALS-inhibiting herbicides varies depending on the changes in the amino acids resulting from mutation and on the tested herbicides. The mutation of tryptopahan at position 548 confers resistance to all types of ALS-inhibiting herbicides. The mutation of serine at position 627 was found in the IM-resistant *Amaranthus* sp. The mutations of this position to aspargine and histidine confer resistance to the IM and the PC but not to the SU. Fig. 4 shows the

Resistance/Sensitive ratio (RS ratio) of various herbicides for the inhibition of ALS that was prepared from chlorsulfuron-resistant *K. scoparia*. This biotype has a mutation of proline at position 189 (171 by the rice numbering system) to serine in the ALS. The RS ratios of most SUs were over 1 and those of the IM were nearly equal to 1, indicating that this mutated ALS expressed resistance to the SU but not to the IMs. The cross-resistance pattern of the PC-herbicides was moderate between those of the SU and the IM.

ALS genes of the SU-resistant *Lindernia* sp. including *Lindernia dubia* subs. Major have been studied in Japan. Uchino and Watanabe have elucidated that proline mutation in the upstream region exists in one of two ALS genes of these species, namely ALS1, which has an intron (Fig. 5)(6).

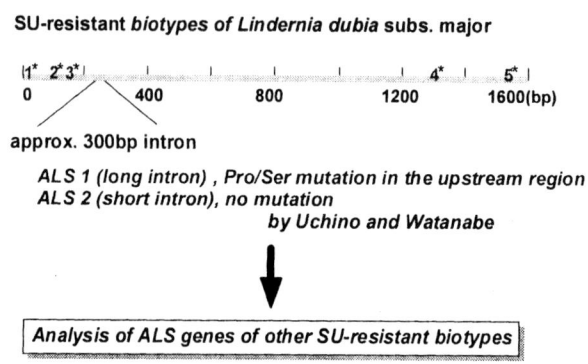

Fig. 5. ALS gene analysis of *Lindernia dubia* subs. Major (1)

This study was the first demonstration of introns in the ALS genes of higher plants as well as the detection of proline mutation in the upstream region in Japan. We examined mutations in the ALS genes of other SU-resistant biotypes of *Lindernia dubia* subs. Major by PCR with degenerate primers and by primer walking. DNA fragments with introns that included the five mutation positions were amplified using specific primers shown by SP-4 (5'-ATGGAGATCCACCAGGCGCTCA-3') and SP-5 (5'-CCACCACCTGCAGGTATCATAGG-3')(Fig. 6).

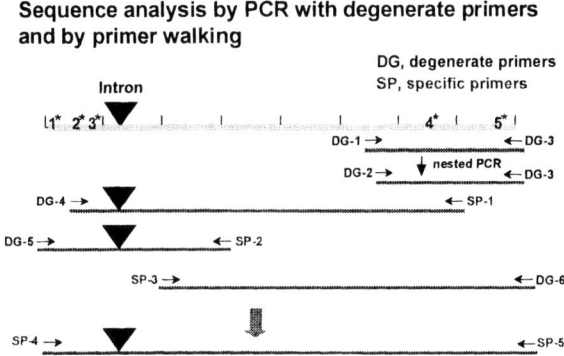

Fig. 6. ALS gene analysis of *Lindernia dubia* subs. Major (2)

The fragments were directly sequenced and cloned into the pT7 Blue T-vector. The sub-cloned fragments were then also sequenced and compared with that by direct sequencing. Consequently, it was revealed that not only the proline mutation in the upstream region but also the tryptophan mutation in the down stream region existed in the other ALS gene, namely ALS2, of these biotypes (Fig. 7).

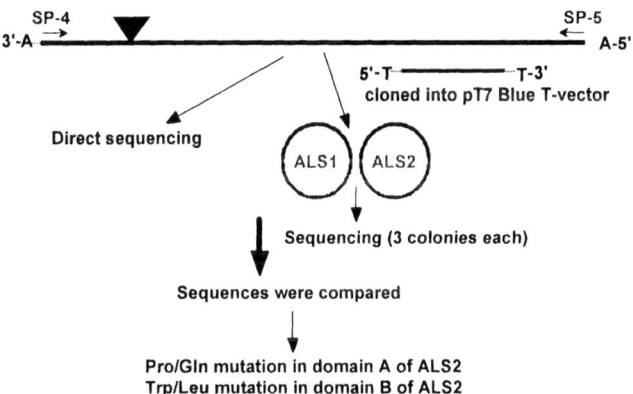

Fig. 7. ALS gene analysis of *Lindernia dubia* subs. Major (3)

This is the first detection of the tryptophan mutation in Japan and indicated that both of the two ALS genes of *Lindernia dubia* subs. Major were functionally transcribed.

Mutations in ALS's in crops

Mutations of ALS's in crops conferring resistance to the ALS-inhibiting herbicides are shown in Table 3.

Table 3. Mutations in ALS's of crops conferring resistance to ALS-inhibiting herbicides

Plant species	Position	Mutation[a]		
		CMB	SCM	SDM
Nicotiana tabacum	Ala121			Thr
	Pro196		Gln, Ala, Ser	
	Trp537		Leu	Phe
	Ser652			Asn, Thr
Beta vulgaris	Ala113		Thr	
	Pro188			Ser
Arabidopsis thaliana	Ala122			Val
	Met124			Glu, Ile, His
	Pro197	Ser		deletion
	Arg199			Ala, Glu
	Phe206			Arg
	Trp574			deletion, Leu, Ser
	Ser653		Asn	deletion, Thr, Phe
Brassica napus	Pro173			Ser
	Trp557		Leu	
Gossypium hirsutum	Trp563		Ser, Cys	
Zea mays	Ala90	Thr		
	Trp552	Leu		
	Ser621	Asp	Asn	

[a] CMB, SCM, and SDM mean the conventional cell selection, the somatic cell mutation, and the site-directed mutagenesis, respectively.

A number of plants and cultured plant cells resistant to ALS-inhibiting herbicides have been generated using the conventional mutation breeding, the somatic cell selection, and the site-directed mutagenesis. ALS genes encoding for the catalytic subunits have been cloned from some of these plants, and their sequences are shown to differ from their wild-types. For example, the mutations of proline at position 196 to glutamine, alanine and serine in the tobacco ALS have been found through somatic cells selection. The most commonly encountered mutations involve the residues of alanine and proline in the upstream region, and tryptophan and serine in the down stream region. These mutation patterns are very similar to those of the herbicide-resistant weeds described above (3).

It was expected that novel mutated ALS genes that had different mutations from those reported in papers and patents were obtained through the selection of plant cells under the pressure of BS. Fig. 8 shows the method for the generation of PC-resistant rice cells and isolation of the ALS cDNAs from phage libraries.

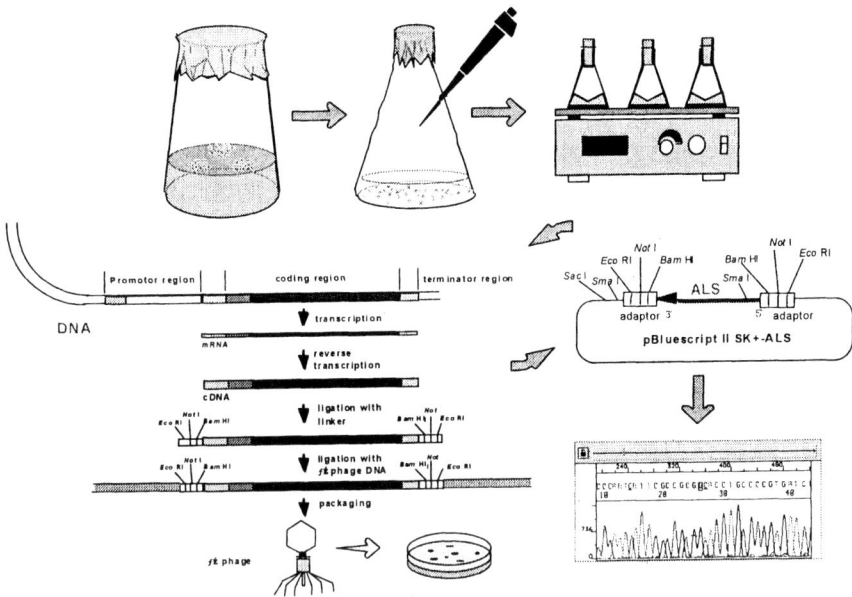

Fig. 8. Isolation of ALS genes from BS-resistant rice cells

First, the callus from rice seeds was induced. The calli were then cultured with 1 micromolar of BS for about 2 months so that the BS-resistant cells were generated. The cells were next cultured with higher concentrations of BS. And finally, several kinds of spontaneous BS-resistant cells that could grow under the pressure of 100 micromolar of BS were obtained. A wild-type ALS gene and a mutated ALS gene have been cloned from the BS-resistant cells using the partial cDNA that is an expressed sequence tag obtained from the MAFF DNA bank of Japan as a homologous hybridization probe. Fig. 9 shows a comparison of the deduced amino acid sequences between the wild-type ALS and the mutated ALS.

The first amino acid shows the sequences between position 361 and the C-terminal position 644 in the mutated ALS, and second amino acid sequence does that in the wild-type ALS. The mutations involved the residues of tryptophan 548 to leucine and serine 627 to isoleucine. This double mutation on rice is a new combination of spontaneous mutations with the novel substitution at the serine position (7). One-point mutated ALS genes were then prepared to compare the sensitivities of their recombinant ALS's to the ALS-inhibiting herbicides with that of the two-point mutant (Fig. 10).

1st Amino Acid Sequence; mutant
2nd Amino Acid Sequence; wild-type

```
361' SRAKIVHIDIDPAEIGKNKQPHVSICADVKLALQGLNALLQQSTTKTSSDFSAWHNELDQ
     ************************************************************
361" SRAKIVHIDIDPAEIGKNKQPHVSICADVKLALQGLNALLQQSTTKTSSDFSAWHNELDQ

421' QKREFPLGYKTFGEEIPPQYAIQVLDELTKGEAIIATGVGQHQMWAAQYYTYKRPRQWLS
     ************************************************************
421" QKREFPLGYKTFGEEIPPQYAIQVLDELTKGEAIIATGVGQHQMWAAQYYTYKRPRQWLS

481' SAGLGAMGFGLPAAAGASVANPGVTVVDIDGDGSFLMNIQELALIRIENLPVKVMVLNNQ
     ************************************************************
481" SAGLGAMGFGLPAAAGASVANPGVTVVDIDGDGSFLMNIQELALIRIENLPVKVMVLNNQ

541 'HLGMVVQLEDRFYKANRAHTYLGNPECESEIYPDFVTIAKGFNIPAVRVTKKSEVRAAIK
     ******  ****************************************************
541" HLGMVVQWEDRFYKANRAHTYLGNPECESEIYPDFVTIAKGFNIPAVRVTKKSEVRAAIK

601' KMLETPGPYLLDIIVPHQEHVLPMIPIGGAFKDMILDGDGRTVY
     *****************************  ****************  548; tryptophan (W)⟶leucine (L)
601" KMLETPGPYLLDIIVPHQEHVLPMIPSGGAFKDMILDGDGRTVY     627; serine (S)⟶isoleucine (I)
```

Fig. 9. Comparison of amino acid sequences between ALS's from the mutant and the wild-type

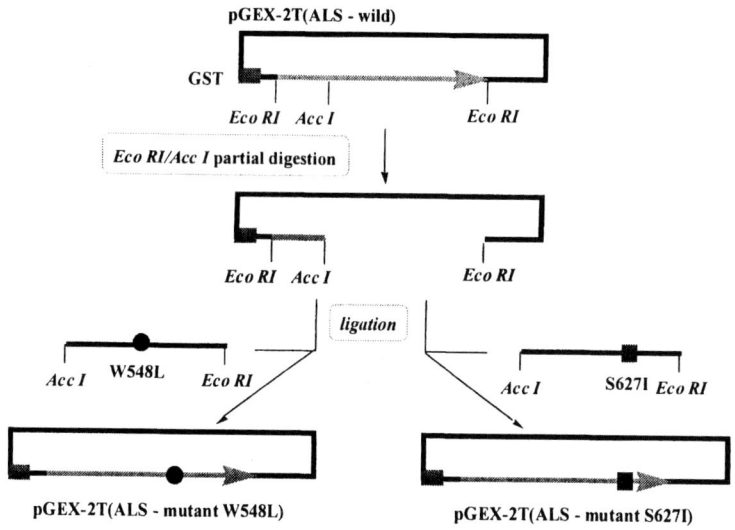

Fig. 10. Preparation of one-point mutated ALSs by PCR and self-polymerase reaction

Each one-point mutant was prepared from the two-point mutant by PCR and the self-polymerase reaction. Recombinant ALS's from these ALS genes were expressed in *Eschericia coli* as GST-fused proteins (Fig. 11) and the proteins were examined for their sensitivities to herbicides.

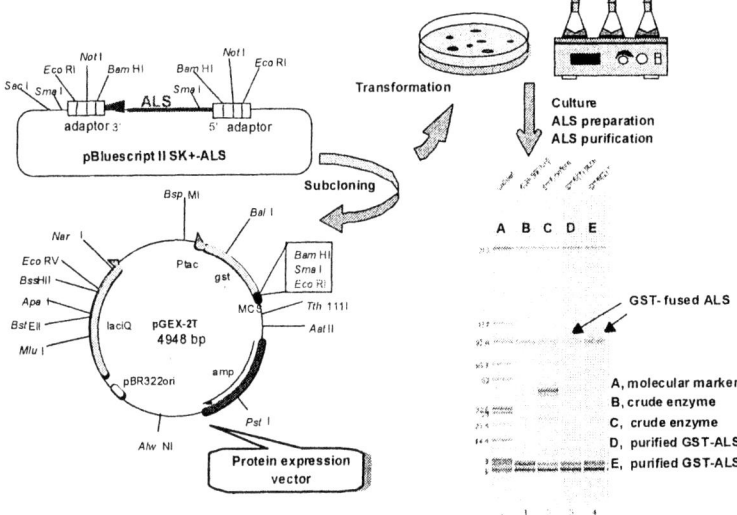

Fig. 11. Expression of recombinant ALS's as GST-fused proteins

The ALS expressed from the wild-type gene showed a similar sensitivity to BS and chlorsulfuron compared with that prepared from the natural source (Fig. 12).

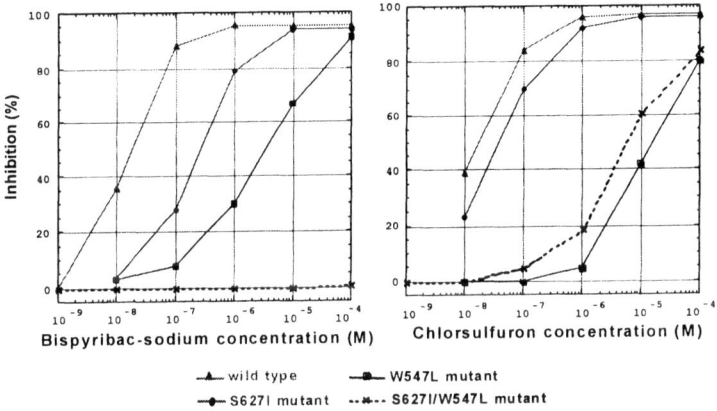

Fig. 12. Sensitivities of GST-fused ALS's to BS and chlorsulfuron

On the contrary, the ALS expressed from the two-point mutated ALS gene showed quite different sensitivities to the herbicides. This ALS showed a stronger resistance to BS than to chlorsulfuron. BS had no effect on the enzyme even at 100 micromolar, which is an approximately 10,000-fold higher concentration than the I_{50} value for the wild-type enzyme. It was notable that the two-point mutated gene imparted synergistic resistance to ALS against BS that is stronger than the additive effect predicted from the degree of each resistance of the one-point mutated ALS.

Use of the mutated ALS genes for the genetic transformation and the herbicide resistance management

As shown above, the novel mutated ALS gene from rice exhibited a high resistance to the PC herbicide. Thus we studied the use of this gene as a selectable marker for the genetic transformation of plants. Promoters and terminators derived from rice were used and a new binary vector was constructed. The two-point mutated ALS gene was driven with a rice callus specific promoter, and the GFP gene was driven with a constitutive promoter. Rice seeds were transformed with this vector by the Agrobacterium method and the transformed cells were selected by the pressure of BS. As a result, fluorescence from GFP was detected only in selected cells (Fig. 13), indicating that the two-point mutated ALS gene was an effective selection marker for rice transformation.

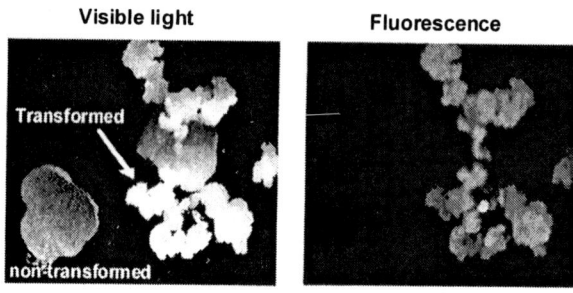

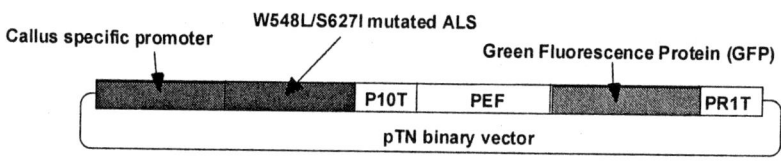

P10T, prolamine terminater of rice
PEF, elongation factor promoter of rice
PR1T, PR1 protein terminater of rice

Fig. 13. Use of the mutated ALS gene as a selectable marker for rice transformation

Transgenic rice plants were then generated to examine whether this gene works normally in the plant or not. The two-point mutated ALS gene was driven with a constitutive 35S promoter cassette with enhanced expression activity. Rice seeds were transformed with this vector and a transgenic rice plant was generated. This transgenic rice plant exhibited resistance to BS and grew normally (Fig. 14) so that it was fertile.

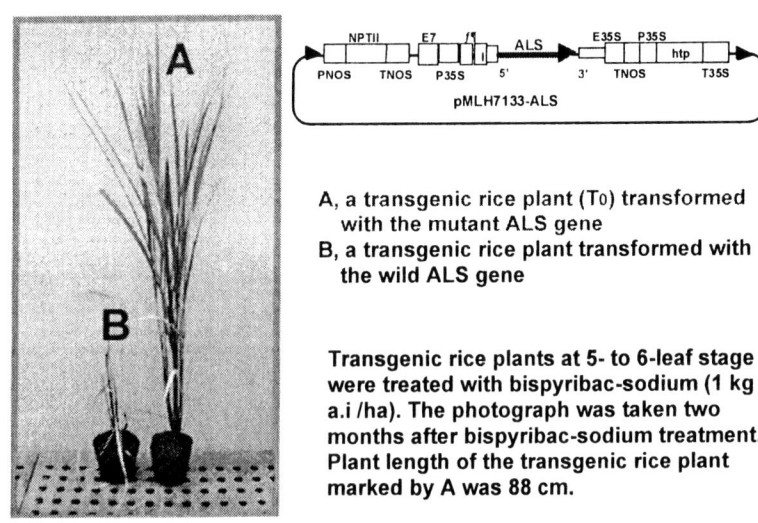

A, a transgenic rice plant (T0) transformed with the mutant ALS gene
B, a transgenic rice plant transformed with the wild ALS gene

Transgenic rice plants at 5- to 6-leaf stage were treated with bispyribac-sodium (1 kg a.i /ha). The photograph was taken two months after bispyribac-sodium treatment. Plant length of the transgenic rice plant marked by A was 88 cm.

Fig. 14. Sensitivity of the transgenic rice (T_0) to BS

T_1 seeds were collected and the BS-resistant phenotype of T_1 plants were examined. The result showed that the phenotype was segregated by approximately 3 : 1 according to Mender's law. The plants that exhibited resistance to BS were cultivated on a large scale and several kinds of T_2 seeds were collected. Consequently, homozygotes for the resistant trait were found in these T_2 seeds through examination of their sensitivities to BS. Fig. 15 shows the sensitivities of the homozygote to BS. The right figure is an enlargement of the left figure.

A, wild-type with 1ƒêM BS; B, wild-type without BS;
C, homozygote with 1ƒêM BS; D, homozygote without BS

Fig. 15. Sensitivity of the transgenic rice (T_2) to BS

This homozygote grew normally without BS and exhibited resistance to BS. These results suggested that the two-point mutated rice ALS gene functionally worked in rice and had no bad effects on rice. The marker system is needed not only for the general recombinant technology but also for the gene targeting such as the homologous recombination and the mismatch repair. The mutated rice ALS gene can be used for such gene technologies. (Fig. 16).

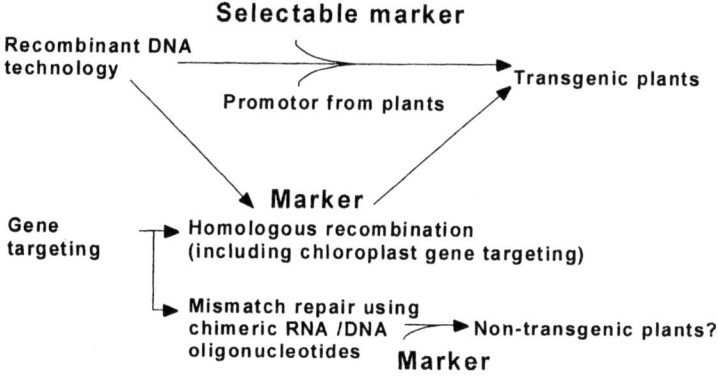

Fig. 16. Use of rice mutated ALS genes as markers for gene technologies

Using the accumulated knowledge concerning rice ALS genes, rice mutated ALS genes were artificially prepared. (Fig. 17).

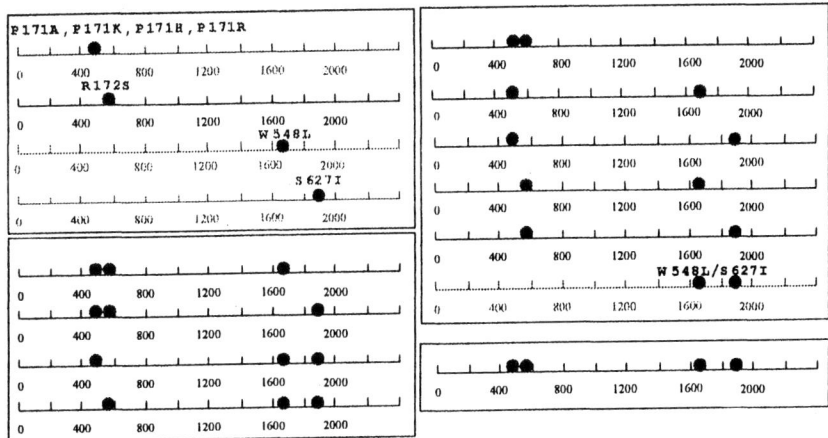

These mutated ALS genes except for the W548L/S627I double mutant were prepared by PCR and SPR using mutations detected in BS-resistance rice cells, or site-directed mutagenesis.

Fig. 17. Mutated ALS genes prepared artificially

Circles in the figure are the mutated positions. Each recombinant ALS was prepared and the sensitivity of each protein to the ALS-inhibiting herbicides was examined. Some of the results are shown in Fig. 18.

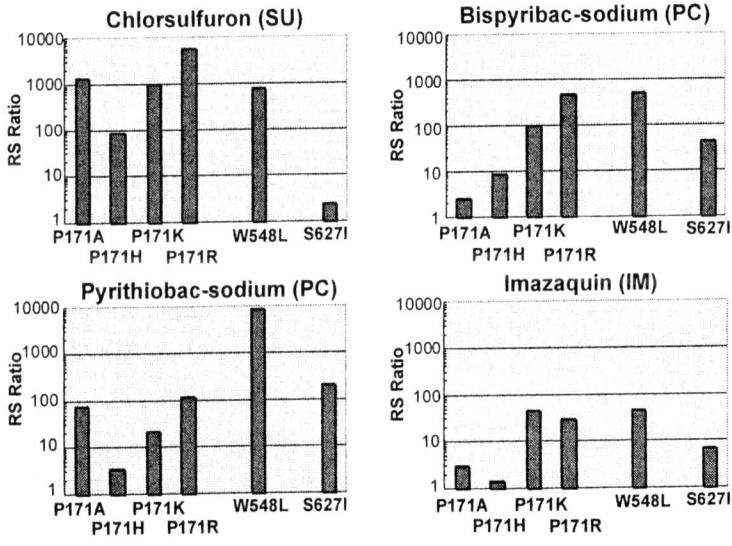

Fig. 18. Sensitivities of one-point mutated ALS's to ALS-inhibiting herbicides

The one-point mutated ALS except for the mutation of serine at position 627 to isoleucine exhibited a high resistance to the SU herbicide, chlorsulfuron. On the contrary, the resistance level of these mutated ALS's to the IM herbicide, imazaquin, was lower than that of chlorsulfuron. The mutation of proline at position 171 to alanine and histidine did not confer resistance to imazaquin. On the other hand, the resistance level of the proline mutated ALS's to the PC herbicides was moderate between those of chlorsulfuron and imazaquin. These results were correlated to the cross-resistance pattern of the proline-mutated ALS of *K. scoparia* as already described. From these results, it is considered that rice mutated recombinant ALS's, especially the proline mutants, are useful as resistant enzyme models for the herbicide resistance management at newly developed or developing ALS-inhibiting herbicides. We are now preparing other kinds of proline mutants to confirm this idea.

Summary

The first key point in this paper is that five mutation sites are important for the natural development of resistance to herbicides. The second key point is that novel rice mutated ALS genes conferring resistance to the ALS-inhibiting herbicides were isolated and synthesized. The third key point is that these genes can be used as markers for gene technologies. And finally, rice recombinant ALS's can be used as resistant enzymes for the herbicide resistance management. We hope this paper will be of use to those who are interested in the ALS-inhibiting herbicides and the resistance mechanisms to the herbicides.

Acknowledgement

We thank Dr. Masahiro Ohshima and his colleagues at the National Agricultural Research Center of Japan for their selectable marker study in this research. Parts of this research were supported by grants from the Ministry of Agriculture, Forestry, and Fisheries of Japan (Integrated Research Program for the Development of Novel Weed Control Technology by Applying Metabolic Genes in Plants and Technical Development Program for making agribusiness in the form of utilizing the concentrated know-how from the private sector).

References

1. Shimizu, T.; Nakayama, I.; Wada, N.; Nakao, T.; Abe, H.: *J. Pestic. Sci.* **1994**, 19, 257-266.
2. Shimizu, T.: *J. Pestic. Sci.* **1997**, 22, 245-256.
3. Shimizu, T.; Nakayama, I.; Nagayama, K.; Miyazawa, T.; Nezu, Y.: Herbicide Classes in Development edited by Böger, P.; Wakabayashi, K.; Hirai, K. Springer-Verlag, Berlin, Heiderberg, **2002**, 1-41.

4. Tranel, P. J.; Wright, T. R.: *Weed Sci.* **2002**, 50, 700-712.
5. Shimizu, T.; Kaku, K.; Takahashi, S.; Nagayama, K.: *J. Weed Sci. Technol.* **2001**, 46, S32-33.
6. Uchino, A.; Watanabe, H.: *Weed Biol. Management* **2002**, 2, 104-109
7. Shimizu, T.; Nakayama I.; Nagayama, K.; Hukuda, A.; Tanaka, Y.; Kaku, K. (Kumiai Chemical Industry Co., Ltd., and National Institute of Agrobiological Sciences): PCT Int. Appl. WO 0244385, 2002.

Chapter 23

Glyphosate-Resistant Horseweed (*Conyza canadensis* L. Cronq.) in Tennessee

Thomas C. Mueller[1], Joseph H. Massey[2], Robert M. Hayes[1], and Chris L. Main[1]

[1]Department of Plant Sciences, University of Tennessee, 2431 Joe Johnson Drive, Knoxville, TN 37996
[2]Department of Plant and Soil Sciences, Mississippi State University, Mississippi State, MS 39762

Horseweed is a weed native to North America, that is especially of concern in reduced tillage management. Glyphosate application of either 0.84 kg ae/ha (the standard application rate) and 3.8 kg ae/ha to susceptible plants caused complete plant death, but the same applications to putative resistant populations caused less than 15 percent growth reduction. Shikimate concentrations in all untreated horseweed plants were less than 100 µg/g, which was less than those treated with 0.84 kg ae/ha of glyphosate. Shikimate accumulated (>1000 µg/g) in both resistant populations and in the susceptible population. Differences in shikimate accumulation patterns between resistant and susceptible biotypes indicated concentrations in resistant populations declined about 40 percent from 2 to 4 DAT, while concentrations in the susceptible horseweed plants increased about 35 percent from 2 to 4 DAT. The confirmed resistance of a widespread weed implies that alternative control strategies for glyphosate-resistant horseweed may be needed.

© 2005 American Chemical Society

A common perspective in the late 1990s was that weed resistance to the herbicide glyphosate was not probable (1). This was believed because of the complex manipulations required for the development of glyphosate-resistant crops were not expected to be duplicated in nature to evolve glyphosate-resistant weeds. This assessment is no longer true. Horseweed (*Conyza canadensis* (L.) Cronq.) (also referred to as Canada fleabane or mare's-tail) is an annual plant, native to North America (2). Horseweed has been demonstrated to be a substantial problem in conservation tillage production systems in cotton in Alabama (3), in grain sorghum in Georgia (4), in corn in Wisconsin (5), and soybean and corn in Iowa and Minnesota (6), in fallow periods in the southern Great Plains (7), and in the production of container-grown ornamentals (8). Horseweed is present throughout the North American continent. Large numbers of small, wind-dispersed seeds, ranging to over 200,000 seeds per plant, are produced in late summer (2).

The first reported occurrence of glyphosate-resistant horseweed in North America was in Delaware in 2000 (9). No till corn and soybean production has been widely adopted in the mid-Atlantic region, which has favored the establishment of horseweed. Within three years of using only glyphosate for weed control in continuous cropping of glyphosate resistant soybeans, glyphosate failed to control horseweed in some fields. Seedlings originating from seed of one horseweed population in Delaware were grown in the greenhouse and exhibited greater than ten-fold resistance to glyphosate compared with a susceptible population. There were no reported differences between different salts of glyphosate. Historically, glyphosate provided essentially complete control of horseweed (10, 11, 12), so this decreased control is markedly different. This is not the same as a weed shift, where different species that were never controlled or were poorly controlled by glyphosate increase in relative abundance in that environmental setting. This glyphosate resistance represents a change at the physiological level with profound agronomic implications. Glyphosate is a potent herbicide (13). It works by competitive inhibition of the enzyme 5-enol-pyruvyl-shikimate-3-phosphate synthase (EPSPS), which catalyzes an essential step in the aromatic amino acid biosynthetic pathway. The measurement of shikimic acid accumulation in response to glyphosate inhibition of EPSPS is a rapid and accurate assay to quantify glyphosate-induced damage in sensitive plants (14). Pline et al. (14) examined the accumulation of shikimic acid in cotton varieties that were either resistant or susceptible to glyphosate. All tissues of susceptible cotton plants accumulated shikimic acid in response to glyphosate treatment, while glyphosate-resistant plants accumulated much less shimikic acid. The active site of the enzyme EPSPS has been probed using site-directed mutagenesis and inhibitor binding techniques (15). The studies suggest a high degree of structural conservation from bacteria compared to plant EPSPS enzymes.

Previous research has indicated the propensity of horseweed to develop resistance to herbicides. Populations of horseweed resistant to the herbicide paraquat were found in Ontario, Canada (16). These paraquat-resistant populations required doses > 25 times higher than susceptible populations for equivalent control. Horseweed resistant to paraquat (17) and triazines (18) was also documented from collections in Hungary.

Greenhouse Study. This research conclusively confirmed that the suspected glyphosate-resistant horseweed is resistant to glyphosate (Table I). Visual evaluations at 14 d indicated less than 15% control in resistant populations, while the susceptible plants showed 99% control.

Table I. Growth Reduction and Control of Horseweed Biotypes Treated with Glyphosate

Horseweed biotype	Glyphosate dosage	Control @ 14 d	Fresh Weight @ 17 d	Fresh Weight % of untreated
	kg ae/ha	%	G	%
Resistant-East	0	0	14.91	100
Resistant-East	0.84	4	9.10	61
Resistant-East	3.8	6	7.91	53
Resistant-West	0	0	5.79	100
Resistant-West	0.84	6	7.03	120
Resistant-West	3.8	14	5.54	96
Susceptible	0	0	11.9	100
Susceptible	0.84	99	1.94	16
Susceptible	3.8	99	1.53	13
LSD		4.4	2.1	

Source: Reproduced from ref. 21

However, all resistant plants showed some slight effect from glyphosate application. The glyphosate-resistant horseweed plant shoot apices turned light green to yellow, the plants were slightly stunted, then resumed normal growth. There were some differences between the two resistant populations. Resistant-

East plants were larger at the time of glyphosate application, and they had approximately 45 % growth reduction on fresh weight basis compared with the untreated plants. Resistant-West plants increased in size at the low glyphosate application rate, about 20 percent, or stayed the same size at the higher glyphosate application rate. The resistant populations were contrasted by the susceptible populations that had greater than 80 percent decline in plant fresh weight. This small amount of fresh weight plant material was essentially a dead stem that remained from the original plant. These results are in agreement with those of VanGessel, which first reported glyphosate resistant horseweed in Delaware (9).

Laboratory Study. Pilot studies indicated that shikimate recovery from horseweed tissue that had been finely ground in liquid N_2 and extracted for 24-h in 1 M HCl were acceptable (Table II).

Table II. Recovery of Freshly-Fortified and Endogenous Shikimate from Horseweed Tissue Using 1 M HCl as a Function of Extraction Time[a]

Treatment	Extraction Time (h)	Average Recovery	N
50 ppmw[b]	24	108.7 ± 21.5 %	2
Freshly-Fortified	48	98.9 ± 14.7 %	2
Shikimate	72	86.6 ± 15.5 %	2
500 ppmw[b]	24	95.0 ± 1.5 %	3
Freshly-Fortified	48	94.9 ± 3.0 %	3
Shikimate	72	84.1 ± 5.4 %	3
2000 ppmw[b]	24	82.5 ± 9.6 %	3
Freshly-Fortified	48	71.5 ± 3.1 %	3
Shikimate	72	71.2 ± 2.5 %	3
Endogenous[c]	24	5807 ± 129 :g/g	3
Shikimate	48	5964 ± 348 :g/g	3
	72	5854 ± 562 :g/g	3

Source: Reproduced from ref. 21
[a]Extraction time on orbital shaker using 5 mL 1 M HCl per g tissue.
[b]Applied to untreated, field-grown tissue finely ground in liquid N_2; recovery results are corrected for background shikimate concentrations which ranged from 21 to 24 ppmw.
[c]Endogenous levels of accumulated shikimate in horseweed 3 DAT with 1.9 kg ae/ha glyphosate applied as Roundup Ultramax herbicide.

Shikimate recovery from the freshly-fortified control samples was corrected using the appropriate untreated control concentrations. The average background level of shikimate in untreated horseweed was 69 ± 55 ppmw (n=16) for all untreated horseweed populations and study times. The average recoveries of shikimate from freshly-fortified horseweed tissue were 99.3 ± 19.5% (n=5) at the 50 ppmw level and 85.8 ± 4.8% (n=5) at the 500 ppmw level of fortification.

Shikimate Accumulation in Glyphosate-Resistant and Glyphosate-Susceptible Horseweed. Shikimate accumulated in concentrations significantly greater than background levels after glyphosate treatment in all horseweed populations (Table III). There were no significant differences ($\alpha = 0.05$) in shikimate levels among the glyphosate-resistant (i.e., East and West) and glyphosate-susceptible populations 2 and 4 DAT (Table III). One difference between both the two types of horseweed plants is the trend in shikimate concentration is decreasing from 2 to 4 DAT in the Resistant plants, but is increasing from 2 to 4 DAT in the Susceptible plants.

Table III. Accumulation of Endogenous Shikimate in Two (denoted East and West) Glyphosate-Resistant and Glyphosate-Susceptible (S) Horseweed Populations

Treatment R= Resistant S=Susceptible	DAT^a	Average Shikimate Concentration (:g/g)	Mean Grouping^c
R-East Treated	2	2609 ± 379^b	A
S- Treated	4	1960 ± 597	AB
R-West Treated	2	1892 ± 491	AB
R-East Treated	4	1630 ± 812	BC
S-Treated	2	1433 ± 1007	BC
R-West Treated	4	1042 ± 207	C
S- Untreated	2	99 ± 134	D
R-West Untreated	2	79 ± 54	D
R-West Untreated	4	74 ± 36	D
R-East Untreated	4	56 ± 16	D
R-East Untreated	2	53 ± 15	D
S- Untreated	4	44 ± 1	D
LSD(0.05)		808	

Source: Reproduced from ref. 21.
^{a}Days after treatment with glyphosate applied as Roundup Ultramax Herbicide.
^{b}Values are the arithmetic means with corresponding standard deviations.
^{c}Means followed by the same letter are not significantly different

Taken together with the whole plant bioassays, the shikimate accumulation data indicate that the mechanism of glyphosate resistance in horseweed is not due solely to a single, glyphosate-insensitive EPSP synthase, as we would not expect to see significant increases in shikimate in resistant plants whose EPSP synthase could not be inhibited. While the mechanism of glyphosate resistance in horseweed is not known, we hypothesize that multiple EPSP synthase genes encoding various EPSP isoforms are present that are responsible for varying levels of inhibition by glyphosate herbicide.

Glyphosate-resistant horseweed from Delaware has previously been examined to elucidate the resistance mechanism (20). Initial indications are that glyphosate uptake into the plant and subsequent translocation to the active site were not responsible for the observed resistance. However, enhanced glyphosate metabolism was also not implicated in this preliminary report. A hypothesis of this research group (20) was that an altered form of the EPSPS enzyme was present in glyphosate-resistant horseweed, although the plants retained some susceptible isoforms of the same enzyme.

Conclusions

The present study confirms glyphosate-resistance in different horseweed populations than previously reported (9), which is an important result. However, it also brings a series of interesting findings, such as shikimate accumulation in both glyphosate-resistant and glyphosate-susceptible horseweed plants. It appears as if the horseweed shikimate pathyway may be aberrant compared with what is known in other plants. Future research efforts include further studies to determine the mechanism for the observed glyphosate resistance. Additional research conducted under field conditions is currently underway to determine best management practices to control glyphosate-resistant horseweed while maintaining no tillage production practices.

Literature Cited

1. Bradshaw, L. D.; Padgette, S. R.; Kimball, S. L.; Wells, B. H. *Weed Technol.* **1997**, *11*, 189-198.
2. Weaver, S. E. *Canadian J. of Plant Sci.* **2001**. *81*, 867-875.
3. Brown, C. M.; Whitwell, T. *Weed Technol.* **1988**, *2*, 269-270.
4. Vencill, W. K.; Banks, P. A. *Weed Sci.* **1994**, *42*, 541-547.
5. Buhler, D. D. *Weed Sci.* **1992**, *40*, 241-248.
6. Buhler, D. D.; Owen, M. D. K. *Weed Sci.* **1997**, *45*, 98-101.
7. Wiese, A. F.; Salisbury, C. D.; Bean, B. W. *Weed Technol.* **1995**, *9*, 249-254.
8. Gallitano, L. B.; Skroch, W. A. *Weed Technol.* **1993**, *7*, 103-111.

9. VanGessel, M. J. *Weed Sci.* **2001**, *49*, 703-705.
10. Scott, R.; Shaw, D. R.; Barrentine, W. L. *Weed Technol.* **1998**, *12*, 23-26.
11. Bruce, J. A.; Kells, J. J. *Weed Technol.* **1990**, *4*, 642-647.
12. VanGessel, M. J.; Ayeni, A. O.; Majek, B. A. *Weed Technol.* **2001**, *15*, 703-713.
13. Daniell, H.; Datta, R.; Varma, S.; Gray, S.; Lee, S. B. *Nature Biotechnology,* **1998**, *16*, 345-348.
14. Pline, W. A.; Wilcut, J. W.; Duke, S. O.; Edmisten, K. L.; Wells, R. *J. Agric. Food Chem.* **2002**, *50*, 506-512.
15. Padgette, S. R.; Re, D. B.; Gasser, C. S.; Eichholtz, D. A.; Frazier, R. B.; Hironaka, C. M.; Levine, E. B.; Shah, D. M.; Fraleyu, R. T.; Kishore, G. M. *J. Biological Chem.* **1991**, *266*, 22365-22369.
16. Smisek, A.; Doucet, C.; Jones, M.; Weaver, S. *Weed Sci.* **1998**, *46*, 200-204.
17. Lehoczki, E.; Laskay, G.; Gaal, I.; Szigeti, Z. *Plant Cell and Environment,* **1992**, *15*, 531-539.
18. Lehoczki E.; Laskay, G.; Polos, E.; Mikulas, J. *Weed Sci.* **1984**, *32*, 669-674.
19. Singh, B.K.; Shaner, D.L. *Weed Technol.* **1998**, *12*, 527-530.
20. Bourque, J.; Chen, Y. S.; Heck, G.; Hubmeier, C.; Reynolds, T. Tran, M.; Ratliff, P. G.; Sammons, D. *Abs Weed Sci. Soc. Amer.* **2002**, *42*, 65.
21. Mueller, T. C.; Massey, J. E.; Hayes, R. M.; Main, C. L.; Stewart, C. N., Jr. *J. Food Agric. Chem.* **2003**, *51*, 680-684.

Chapter 24

Resistance Management Strategies for Fungicides

Hideo Ishii

National Institute for Agro-Environmental Sciences, Kannondai 3-1-3, Tsukuba, Ibaraki 305-8604, Japan

Introduction

Control of crop diseases largely depends on chemically synthesized fungicides despite that alternative methods of control, e.g. biofungicides, have been developed recently. Urech (1) mentioned that the evolution of crop protection technology in this decade would most likely see chemical control remaining the backbone of crop protection. At the same time, however, the growth of the fungicide market in the world has decreased over the last three decades and the average annual growth was 1.6 % from 1991 to 2000 (Table I, ref. 2). Interestingly a recovery of growth is now predicted but it is mainly based on the expectation that QoI fungicides (inhibitors of mitochondrial respiration at Qo site of cytochrome $bc1$ enzyme complex) would be developed further. It tells us that the increase of resistance problems to QoI fungicides may seriously influence the growth of the fungicide market in near future.

Although we have accumulated a lot of experience since the practical problem of fungicide resistance first occurred in the early 1970's, the recent outbreaks of resistance to QoIs and MBI-Ds (melanin biosynthesis inhibitors, which target scytalone dehydratase) demonstrated how resistance management is still difficult to achieve.

Table I. Growth of Fungicide Market (Ref. 2)

Period	Average annual growth (%)
1971 to 1980	7.1
1981 to 1990	3.8
1991 to 2000	1.6
2001 to 2006	2.2 (predicted)*

*Increase of QoIs predicted.

Resistance to QoIs and MBI-Ds

QoIs are the most commonly used fungicides these days and share about 14 % of the whole fungicide market in the world (2). Several fungicides in the same group have already been registered and many others are under development aiming for future commercialization (Table II). However, soon after their introduction, resistance development to this class of fungicides has occurred in various pathogens in Europe, Asia, and the U.S.A. (Table III).

Table II. Inhibitors of Mitochondrial Respiration at Qo Site (QoIs) of Cytochrome $bc1$ Enzyme Complex

Kresoxim-methyl*	Picoxystrobin
Azoxystrobin	Pyraclostrobin
Metominostrobin*	Orysastrobin
Famoxadone*	Fluoxastrobin
Trifloxystrobin*	Dimoxystrobin
Fenamidone	

*Registered in Japan.

QoI resistance was first reported in wheat powdery mildew in northern Germany in 1998 (3). In Japan, control failure by these fungicides has been reported in five diseases so far, i.e. cucurbit powdery mildew, cucumber downy mildew, eggplant leaf mold (4), cucumber Corynespora leaf spot (5), and wheat powdery mildew. In most cases, the level of QoI resistance was high enough to cause complete loss of fungicide efficacy at the fungicide concentration recommended for practical use.

Very recently, furthermore, the new problem of resistance against melanin biosynthesis inhibitors, which target scyta lone dehydratase (MBI-D), has also occurred in rice blast fungus in Japan (6).

QoIs such as azoxystrobin have been regarded as fungicides with low risk to the environment due to their rapid degradation in soil and are widely used for the control of a broad range of pathogens in various crops.

Table III. Occurrence of Strobilurin Resistant Isolates in the Field

Disease name	Pathogen	Ref.
Wheat powdery mildew*	*Erysiphe* (*Blumeria*) *graminis* f.sp. *tritici*	3
Wheat speckled leaf blotch	*Mycosphaerella graminicola*	19
Barley powdery mildew	*E. graminis* f.sp. *hordei*	8
Potato early blight	*Alternaria solani*	20
Cucurbit powdery mildew*	*Podosphaera* (*Sphaerotheca*) *fusca*	10,11
Cucumber downy mildew*	*Pseudoperonospora cubensis*	10,11
Cucumber Corynespora leaf spot*	*Corynespora cassiicola*	5
Cucurbit gummy stem blight	*Didymella bryoniae*	21
Eggplant leaf mold*	*Mycovellosiella nattrassii*	4
Banana black Sigatoka	*Mycosphaerella fijiensis*	22
Grapevine downy mildew	*Plasmopara viticola*	8
Grapevine powdery mildew	*Uncinula necator*	23
Apple scab	*Venturia inaequalis*	24
Pistachio Alternaria late blight	*Alternaria alternata* etc.	25
Chrysanthemum white rust	*Puccinia horiana*	26
Turf grass anthracnose*†	*Colletotrichum graminicola*	27
Turf grass leaf spot	*Pyricularia grisea*	28
Turf grass blight	*Pythium aphanidermatum*	29

*Detected in Japan. †Strobilurins not registered in Japan.

Nursery box treatment with MBI-D fungicides, e.g. carpropamid (7), is a common cultural practice in many rice growing areas in Japan as it exhibits long-lasting control efficacy against blast (pathogen: *Magnaporthe grisea*), is labor-cost effective, and contributes to diminishing the number of fungicide applications in paddy fields. Properties of these fungicides fit well to the conditions necessary for developing an integrated disease control system. Therefore, resistance to QoIs and MBI-Ds will become a limiting factor for the establishment of eco-friendly crop protection systems.

Are General Strategies Effective against Combating Resistance?

As in the case of QoI (8) and MBI-D resistance, fungal isolates usually show cross-resistance to chemicals belonging to the same fungicide group. The population of grape downy mildew pathogen, which showed resistant to QoIs was highly cross-resistant to all of the QoI-type fungicides (Table IV, ref. 8).

To combat fungicide resistance, use of the at-risk fungicide in a mixture or a rotation with a fungicide of a different type has been recommended (9).

Table IV. Cross-resistance between QoIs in Grape Downy Mildew (Ref. 8)

Fungicide	EC_{50} ppm a.i.	
	Baseline isolate	Resistant field isolate
Azoxystrobin	0.054	>10
Trifloxystrobin	0.51	>10
Famoxadone	0.11	>10
Fenamidone	0.13	>10
Cymoxanil	0.4	0.5

However, such general strategies only delay resistance development and are not necessarily effective. For example, Japanese growers received guidelines from manufacturers for both kresoxim-methyl and azoxystrobin to avoid resistance. These guidelines were based on the recommendation from the Fungicide Resistance Action Committee (FRAC) and followed by most growers. But unfortunately, a rapid decline in fungicide efficacy against cucurbit powdery mildew was reported (10,11). In this way, it is still quite hard to accurately predict the risk of new fungicides for resistance development under laboratory and field conditions before registration.

Therefore, it is less likely that no problems of resistance will occur as long as ordinary type of fungicides are used suggesting that integrated disease management will be ultimately the best method for counteracting fungicide resistance. It is important to try to reduce disease pressure and fungicide application times as low as possible so that selection pressure towards fungicide-resistant strains can also be reduced. Alternatively, the use of nonfungitoxic disease-resistance inducers, which enhance the plant's innate defense response, will be a good choice as probenazole, a major agent for rice blast control, has never encountered resistance, inspite that this product has been used for about three decades in Japan.

Can we Expect a Fitness Penalty in Fungicide-resistant Strains?

When control efficacy of a fungicide is lost due to resistance, the growers must stop using the fungicide in question immediately. But resistant isolates do not necessarily possess a fitness penalty as compared with sensitive ones, indicating that it might be difficult to expect resistant isolates to be rapidly eliminated from the agricultural environment after withdrawal of selection pressure.

We have been monitoring the distribution of QoI-resistant isolates of cucumber downy mildew and powdery mildew (12). In the commercial greenhouse, resistant isolates of downy mildew are still predominant despite that QoI fungicides were withdrawn for three years. Thus it is less likely that QoI fungicides can be reused easily when resistant fungal strains are widely detected suggesting that how it is important to keep the ratio of resistant populations under low levels.

Strategies for Newly-developed Fungicides before and after Use in Practice

When we introduce new fungicides, it is necessary to evaluate how quickly resistant strains can appear and increase with the use of fungicides. It is generally useful to elucidate the mode of action of newly developed products when predicting the inherent risk for resistance development, as most fungicides that have encountered resistance so far were site-specific inhibitors such as benzimidazoles, QoIs and MBI-Ds.

Development of optimal methods for testing fungicide sensitivity in individual combinations of a pathogen with a fungicide and collection of baseline sensitivity data using wild-type isolates is essential. It would also be ideal if large-scale field trials can be arranged before fungicide registration in order to know whether the risk for resistance development is high or low under different fungicide application programs.

The risk for resistance in *Botrytis cinerea* against anilinopyrimidine and phenylpyrrole fungicides was assessed by establishing baseline sensitivities and by monitoring sensitivity of the pathogen at various trial sites in France and Switzerland over a number of years *(13,14)*. Isolates with reduced sensitivities to anilinopyrimidines appeared and led to a reduced performance of anilinopyrimidines applied alone. However, *Botrytis* control by a mixture of anilinopyrimidines and phenylpyrroles remained excellent and the manufacturer proposed the use of this strategy for these fungicides based on the results of studies.

The routine monitoring conducted in commercial vineyards in Champagne, France, revealed the presence of resistance to benzimidazoles, dicarboximides and anilinopyrimidines in grapevine grey mold (*15*). However, the limitation in the use of these three families of botryticides prevents the build-up of resistant populations.

Use of Molecular-based Techniques for Resistance Monitoring

Molecular technology developed in the last decade has created useful methods for rapidly and precisely monitoring fungicide resistant isolates in the field (*16*). DNA-based methods mainly using PCR (polymerase chain reaction) are already in use in practice, e.g. QoI resistance (*12*).

Mutational changes of the cytochrome *b* target site seem to be the principal mechanism of QoI resistance (*11,17,18*). So we developed a molecular method for identifying resistance. Figure 1 shows the partial nucleotide sequences of cytochrome *b* gene in resistant and -sensitive isolates of cucumber downy mildew. A single point mutation at codon 143 was found in resistant isolates and the mutated nucleotide sequence could be recognized by the restriction enzyme *Ita*I. Then leaf discs were cut from pathogen-infected cucumber plants, total DNA directly purified and a fragment of cytochrome *b* gene amplified by PCR. The PCR products purified from resistant isolates were clearly cut with this enzyme but products from sensitive isolates were not digested with the same enzyme.

Using this DNA-based method, the changes of mutated cytochrome *b* gene in pathogen populations were monitored in cucumber greenhouses where resistant isolates had been widely distributed. However, when the use of QoI fungicides was stopped for over three years, the proportion of mutated DNA, resistance-type DNA in another word, dramatically decreased in individual isolates and most DNAs remained as a wild type. Inspite that all the isolates tested were still highly resistant to QoI fungicides, *Ita*I digestion pattern showed that recovery of the wild type DNA proceeded in the pathogen isolates. In this way, the shift-back of mutated DNA to the wild type was observed after withdrawal of the selection pressure with QoI fungicides for more than three years.

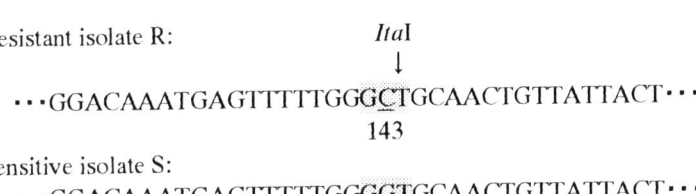

Figure 1. Partial Nucleotide Sequences of Cytochrome *b* Gene in Strobilurin-Resistant and -Sensitive Isolates of Cucumber Downy Mildew and Restriction Site of the Enzyme *Ita*I (Ref. *12*)

To explain this interesting phenomenon, the following hypothesis on heteroplasmy of mitochondrial DNA was proposed. Multi-copy mitochondrial DNA is known to encode cytochrome *bc1* complex, the target protein of QoI fungicides. When the use of QoI fungicides was stopped, intracellular selection of the normal wild-type DNA might have occurred in resistant isolates resulting in a decrease in the proportion of mutated DNA. In order to detect the low proportion of the mutated cytochrome *b* gene in mitochondrial cells, we used a rhodamine-labeled reverse primer in PCR experiments. Only PCR products, which carry the mutated DNA, were digested with the restriction enzyme *Ita*I and the ratio of mutated DNA measured on a polyacrylamide gel using a fluorescence image analyzer. With this method, we could detect the mutated DNA at the level of 1 %. However, expensive equipment or materials are still needed, so more cost-effective methods must be developed in the near future. This is demanded in particular when diagnosis of resistance is carried out in public sectors like regional agricultural experiment stations or plant protection stations.

How can we Utilize Resistance Monitoring Data?

It would not always be difficult to routinely monitor fungicide resistance in the field if we have enough manpower. But one simple question is what extent the monitoring data itself makes sense. It is common that fungicide-resistant populations coexist with sensitive populations in the same field. In a model experiment, QoI-resistant populations of cucumber powdery mildew were spray-inoculated with sensitive populations in various proportion s and control efficacy of the fungicide azoxystrobin was examined. A decrease in efficacy was observed even when resistant populations were included at 10 to 1 % levels in the spore suspensions used for inoculation.

However, even when the proportion of resistant populations was same in spore suspensions inoculated, the control efficacy of fungicide varied depending on the absolute amounts of the pathogen. Therefore, one thing, which we must clarify more, is the relationship of size and ratio of resistant populations in the field with expected efficacy of the fungicide. When this relationship becomes clearer and is combined with more sophisticated disease prediction system, one could expect to make a decision whether the fungicide in question should be applied or not and manage resistant populations below acceptable level s.

Necessity and Role of Regional Activities

When QoI resistance became a practical problem for the first time in Japan, the QoI-manufacturing companies urgently distributed a pamphlet to related organizations so that the latter could be made more aware of the risk for resistance development. However, the manufacturers still proposed a 'European-style' guideline based on the recommendation s from FRAC. The total application times of strobilurins, which were allowed to use, were too many in our understanding. So it may be necessary to design regional strategies for resistance management as cultural practice and disease pressure vary from place to place and recommendations for fungicide use made in one country are not always applicable to the others.

The Research Committee on Fungicide Resistance has been established in 1991 and now belongs to the Phytopathological Society of Japan. The committee aims to contibute to the solution of resistance problems through the activities including regularly organized symposium, establishment of fungicide-sensitivity testing methods, monitoring of fungicide sensitivity etc. The committee edited a book entitled 'Manuals for Testing Fungicide-sensitivity in Plant Pathogens' (published by the Japan Plant Protection Association in 1998) and this book also covers a list of literature relating with fungicide resistance, which had been reported since 1971 within Japan. The activities of this committee and other information will be introduced soon on the homepage of the Phytopathological Society of Japan as follows: http://ppsj.ac.affrc.go.jp/PPSJ_J.html.

References

1. Urech, P. A. *Plant Pathol.*, **1999**, *48*, 689-692.
2. Margot, P. *Abstr. 8th Intr. Cong. Plant Pathol.*, 2003, *1*, 71.
3. Anonymous. *AGROW*, 1998, *318*, 11.
4. Yano, K.; Kawada, Y. *Jpn. J. Phytopathol.* 2003, *69*, 220-223.
5. Date, H. *Abstr. 13th Symp. Fungic. Resis.* (Phytopath. Soc. Jpn.), 2003, 1-8.
6. Yamaguchi, J.; Kuchiki, F.; Hirayae, K.; So, K. *Jpn. J. Phytopathol.*, **2002**, *68*, 261.
7. Kurahashi, Y.; Kurogochi, S.; Matsumoto, N.; Kagabu, S. *J. Pestic. Sci.*, **1999**, *24*, 204-216.
8. Heaney, S. P.; Hall, A. A.; Davies, S. A.; Olaya, G. *Proc. BCPC Conf. - Pests & Dis.*, 2000, 755-762.
9. Hewitt, H. G. *Fungicides in Crop Protection*, CAB International, Wallingford, 1998, pp 1-221.
10. Fuji, M.; Takeda, T.; Uchida, K.; Amano, T. *Proc. BCPC Conf. -Pests & Dis.*, 2000, 421-426.
11. Ishii, H.; Fraaije, B. A.; Sugiyama, T.; Noguchi, K.; Nishimura, K.; Takeda, T.; Amano, T.; Hollomon, D. W. *Phytopathology*, **2001**, *91*, 1166-1171.
12. Ishii, H.; Sugiyama, T.; Nishimura, K.; I shikawa, Y. *Modern Fungicides and Antifungal Compounds*, Dehne, H.-W.; Gisi, U.; Kuch, K. H.; Russell, P. E.; Lyr, H. Eds.; *III*, 2001, AgroConcept, Bonn, pp. 149-159.
13. Forster, B.; Staub, T. *Crop Protect.*, **1996**, *15*, 529-537.

14. Baroffio, C. A.; Siegfried, W.; Hilber, U. W. *Plant Dis.*, **2003**, *87*, 662-666.
15. Leroux, P.; Fournier, E.; Brygoo, Y.; Panon, M. -L. *Phytoma*, **2002**, *554*, 38-42.
16. Ishii, H. *Agrochemical Resistance: Extent, Mechanism, and Detection*, Clark, J. M.; Yamaguchi, I. Eds., 2002, Amer. Chem. Soc., Washington, DC, pp. 242-259.
17. Gisi, U.; Sierotzki, H.; Cook, A.; McCaffery, A. *Pest Manag. Sci.*, **2002**, *58*, 859-867.
18. Fraaije, B. A.; Butters, J. A.; Coelho, J. M.; Jones, D. R.; Hollomon, D. W. *Plant Pathol.*, **2002**, *51*, 45-54.
19. Anonymous. *AGROW*, 2003, *422*, 13.
20. Pasche, J. S.; Wharam, C. M.; Gudmestad, N. C. *Proc. BCPC Conf. - Pests & Dis.*, 2002, 841-846.
21. Olaya, G.; Holm, A. *Phytopathology*, **2001**, *91* (No. 6, Suppl.), S67-68.
22. Chin, K. M.; Wirz, M.; Laird, D. *Plant Dis.*, **2001**, *85*, 1264-1270.
23. Wilcox, W. F.; Burr, J. A.; Riegel, D. G.; Wong, F. P. *Phytopathology*, **2003**, *93* (No. 6, Suppl.), S90.
24. Steinfeld, U.; Sierotzki, H.; Parisi, S.; Poirey, S.; Gisi, U. *Pest Manag. Sci.*, **2001**, *57*, 787-796.
25. Ma, Z.; Felts, D.; Micaailides, T. J. *Pestic. Biochem. Physiol.*, **2003**, *77*, 66-74.
26. Cook, R. T. A. *Plant Pathol.*, **2001**, *50*, 792.
27. Syngenta (Personal Commu., 2001).
28. Vincelli, P.; Dixon, E. *Plant Dis.*, **2002**, *86*, 235-240.
29. Olaya, G.; Cleere, S.; Stanger, C.; Burbidge, J.; Hall, A.; Windass, J. *Phytopathology*, **2003**, *93* (No. 6, Suppl.), S67.

Advances in Formulation and Application Technology

Chapter 25

Effects of Surfactant Concentration on Stability of Dispersion

Tatsuo Sato

Monsanto, Midorigaoka, Chofu, Tokyo 182-0001, Japan

The effect of concentration of a cationic surfactant (polyoxyethylene diethylenetriamine dialkylamide) on the stability of aqueous titanium dioxide (TiO_2) dispersion was studied by measuring adsorption and zeta potential. It was found that as the surfactant concentration increases, stability decreases initially and then increases, showing a minimum. As the concentration increases further, stability decreases, showing a maximum. The zeta potential changes from negative to positive as the adsorption increases and reaches a maximum when the adsorption reaches a plateau value. The dispersion flocculated when the absolute value of the zeta potential dropped below 10 mV, indicating that the destabilization by the surfactant in the low surfactant concentration is due to reduction of the electrical repulsion. Destabilization at the higher surfactant concentration is likely caused by the depletion effect since there is no change in adsorption and zeta potential in the surfactant concentration range, and the surfactant molecules are not sufficiently large to form bridges between particles (1).

Introduction

Aqueous dispersion formulations such as concentrated suspensions (SC) and concentrated emulsions (EW) are becoming more and more important as regulation of the safety of pesticides becomes more and more strict. The most important physical property required for these formulations is dispersion stability. Dispersion stability is important not only for storage stability but also for biological activity. In general, the biological activity decreases remarkably when the active compound particles flocculate (2,3). Therefore, biologically active compounds must be dispersed as fine particles for stability and bioactivity. Dispersion is usually stabilized by the addition of surfactants. Therefore, it is essential for formulation chemists to know how to use surfactants efficiently in developing dispersion formulations. Particularly, selection of appropriate type of surfactants and optimization of the concentration are very important.

In this work, the effect of surfactant concentration on the stability of aqueous suspensions was studied by measuring surfactant adsorption and zeta-potential of dispersed particles (1).

Aqueous TiO_2 suspensions were used as a model of SC and a cationic surfactant was used as an example of surfactants.

Experiments

Materials

Surfactant
The surfactant used in this work was a cationic surfactant of the following structure supplied by Takemoto Oil and Fat Co. in Japan.
Polyoxyethylene(6moles)diethylenetriaminedialkylamide
($C_{12}OCNHCH_2CH_2NHCH_2CH_2NHCOC_{12}(CH_2H_4O)_6$),
Average Mw: 744

Titanium Dioxide
Titanium dioxide used in this work was rutile-type Tipaque R-680 supplied by Ishihara Industrial Company in Japan.
Surface area: 8 m^2/g; Average diameter: 0.21 microns; Specific gravity: 4.2.

Procedure

The stability of dispersion was determined by a sedimentation method. The suspensions were prepared by dispersing titanium dioxide pigments by a paint shaker with glass beads in the surfactant solutions. The solution was then poured into a sedimentation test tube.

Stability was determined by measuring the settling rate and the sedimentation volume. As the stability increases, the settling rate decreases and the sedimentation volume formed in long-term storage (about one month) decreases. The stability was graded 0 (very unstable) to 5 (very stable) by an overall evaluation.

Results and discussion

Change of dispersion stability with surfactant concentration

Fig. 1 shows the change in the stability of the TiO_2 suspension and the zeta potential with surfactant concentration. The stability is initially very good in low surfactant concentration up to 0.05%. As the surfactant concentration increases from 0.05 %, the stability decreases and reaches a minimum (very poor) at about 0.2 %. As the concentration increases further, the stability increases rapidly and then decreases, showing a maximum at about 1.0 %.

The zeta-potential of TiO_2 is initially negative. As the surfactant concentration increases, the potential changes from negative to positive, passing the zero point of charge (zpc). The stability decreases as the absolute value of the zeta-potential decreases and reaches a minimum at the zpc.

These results indicate that the decrease of the stability with the increase of the surfactant concentration in the low surfactant concentration range (0 – 0.2 %) is caused by the reduction of electrical repulsion between negatively charged particles and the increase of the stability in the higher surfactant concentration range (0.2 – 1.0 %) is caused by the increase of the electrical repulsion between positively charged particles.

The adsorption isotherm is shown in Fig. 2. Fig. 2 shows that as the surfactant concentration increases, the adsorption increases and reaches a plateau at about 1.0 %, indicating that the adsorption follows the Langmuir isotherm.

It can be seen by comparing Fig. 1 with Fig. 2 that the dispersion stability reaches a maximum at the concentration where the adsorption reaches a plateau.

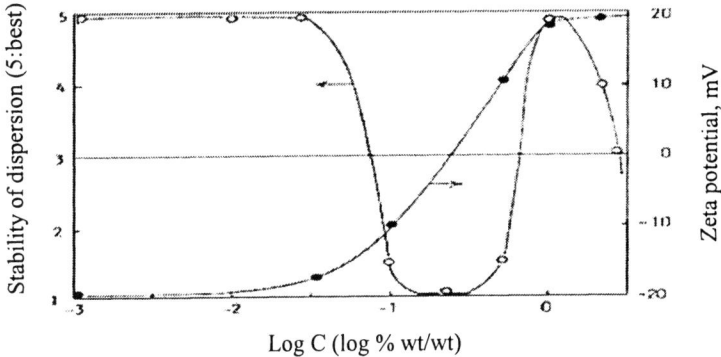

Fig. 1. *Change in dispersion stability of TiO_2 and ζ potential with surfactant concentration at TiO_2 20%(w/w)*

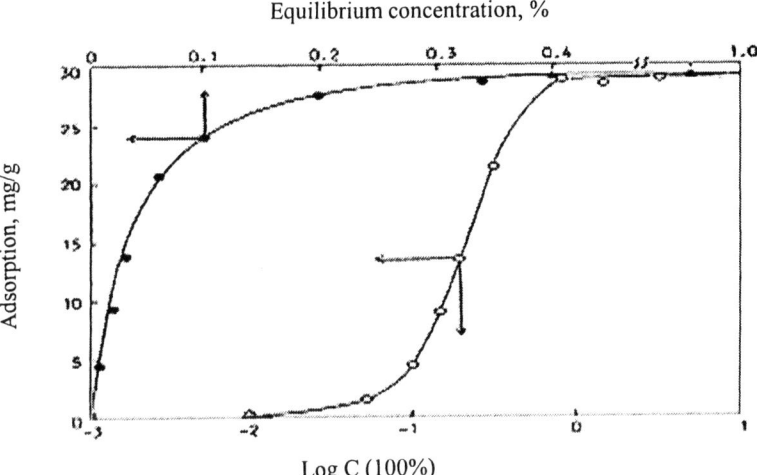

Fig. 2. *Adsorption isotherm of POE-DETADAA on TiO_2 at 20 % (w/w)*

Computation of total potential energy

According to the DLVO theory, total potential energy between two particles is expressed by the sum of van der Waals attractive potential energy and electrical repulsive potential energy and the dispersion is stable when the total potential energy is higher than 15 kT (4).

The electrical repulsive potential energy due to the electrical charge of the particles in aqueous solution can be approximated by;

$E_R^{el} = (\varepsilon r \zeta^2 / 2) \ln\{1 + \exp(-\kappa H)\}$ [1]

where ε is the dielectric constant of medium, r is the radius of particles, ζ is zeta potential, κ is the Debye-Huechel reciprocal length parameter, and H is the shortest distance between two particles.

The van der Waals attractive potential energy between particles at small distance can be approximated by;

$E_A = - Ar/12H$ [2]

where A is the Hamaker constant.

The potential energy between TiO_2 particles calculated from Eq. [1] and Eq. [2] is shown in Fig. 3.

Curve 3 is the potential energy when the dispersion is stable and Curve 4 is the potential energy when the dispersion is not stable.

The maximum height of Curve 3 and Curve 4 are 67 kT (higher than 15 kT) and 8 kT (lower than 15 kT), respectively as shown in Fig. 3.

The results are well accounted for by the DLVO theory.

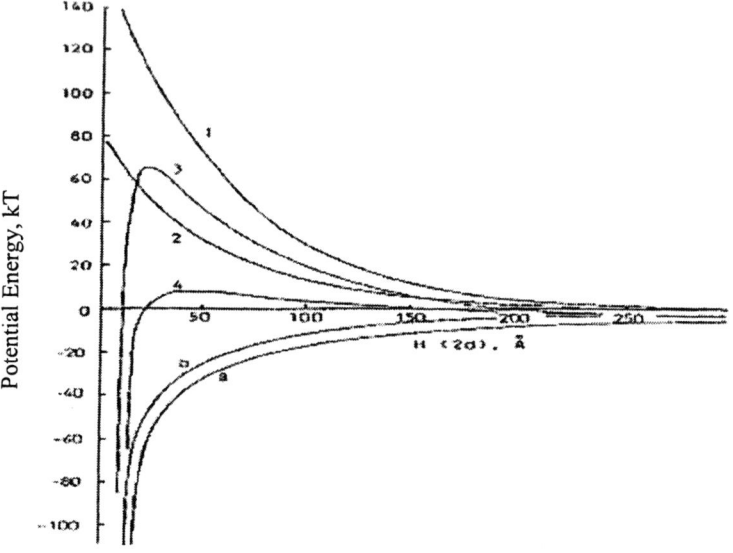

Fig. 3. Potential energies between TiO_2 particles. (a)E_A with no adsorbed layer (b) E_A with adsorbed layer of thickness 50 A (1) $E_R{}^{el}$ at 15 mV (2) $E_R{}^{el}$ at 10 mV. (3) Total potential energy at 15 mV (1 + b) (4) Total potential energy at 10 mV (2 + b)

Destabilization in High-Surfactant Concentration Solutions.

Fig.1 shows that the stability decreases when the surfactant concentration becomes higher than the concentration where the adsorption reaches a maximum, although there is no change in the zeta potential in this surfactant concentration range. This flocculation can not be explained by the DLVO theory (4 - 6) which states that the dispersion stability depends on electrical repulsion.

It was found fairly recently that free (non-adsorbing) polymer can affect colloid stability (4 – 11). Flocculation caused by free polymers is called "depletion flocculation" (5,6). The first theory for the depletion flocculation was the theory proposed by Asakura and Oosawa (10, 11) of Nagoya University in Japan. According to them, when two particles approach each other in a polymer solution to a distance of separation that is less than the diameter of polymer molecules, polymer may be extruded from the inter-particle space. This leads to a polymer-depleted-free zone between two particles. An osmotic force is then exerted from the polymer solution outside the particles and this results in flocculation.

It is likely that the flocculation of TiO_2 dispersion observed in the high surfactant concentration is caused by the depletion effect of free surfactant molecules in this solution. A number of theories derived to explain the mechanism of depletion flocculation were reviewed by Napper in his textbook (6).

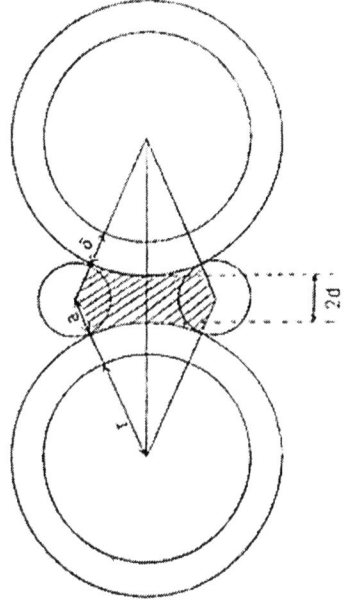

Fig. 4. Model for calculation of osmotic attraction between two particles containing adsorbed layer in a solution of spherical surfactant molecules.

Conclusions

The following conclusions were obtained in this work.

1. Destabilization of TiO_2 in low-surfactant-concentration solutions (0.05 - 0.2 %) is due to the reduction of electrical repulsion between the particles caused by the reduction of negative charge on the particles.

2. Stabilization in medium–surfactant-concentration solutions (0.2 – 1.0 %) is due to the increase in electrical repulsion caused by the increase in positive charge on the particles.
The results are well accounted for by the DLVO theory.

3. Destabilization in high-surfactant-concentration solutions (1.0 – 3.0 %) is likely due to the depletion effect caused by free (non-adsorbing) surfactant molecules.

References

1. Sato, T; Kohnosu, S.; *J. Colloid Interface Sci.,* **1991**, 143, 434.
2. Sato, T.; Application in Agriculture in *"Electrical Phenomena at Interface"* Ed. Ohshima, H.; Furusawa, K.; Marcel Dekker, N. Y. **1998**.
3. Sato, T.: *News Letter* (Chemical Soc. Japan) **1996**, 21(4), 8.
4. Sato, T.; Ruch, R.; *"Stabilization of Colloidal Dispersions by Polymer Adsorption",* Marcel Dekker, N. Y., **1980**.
5. Sato, T.; *J. Coatings Technology,* **1993**, 65, 113.
6. Napper, D. H.; *"Polymeric Stabilization of Colloidal Dispersions",* Academic Press, London, **1983**.
7. Sato, T.; *J. Applied Polymer Sc.,* **1971**, 15, 1053.
8. Sato, T.; *J. Applied Polymer Sci.,* **1979**, 23, 1693.
9. Sato, T.; Sieglaff, C. F.; *J. Applied Polymer Sci.,* **1980**, 25, 1781.
10. Oosawa, F.; Asakura, S.; *J. Chemical Physics;* **1954**, 22, 1255.
11. Asakura, S.; Oosawa, F.; *J. Polymer Sci.,* **1958**, 33, 183.

Chapter 26

Pesticide Formulation Technology Innovation

Curtis Elsik[1], Scott Tann[2], and Andrew Kirby[3]

[1]Austin Research Laboratory, Huntsman LLC,
Austin, TX 78752
[2]Surface Sciences Headquarters, Huntsman LLC,
Houston, TX 77056
[3]Asia Pacific Technical Center, Huntsman Corporation
Australia, Ascot Vale, Melbourne, Australia

Advances in pesticide formulation technology are required to keep up with a multitude of demands. Advances in application technologies, changes in active ingredient mode of action or rate, and regulatory pressures are just some of the many driving forces that require novel formulations to succeed. The classical formulation types include granule (GR), aqueous solutions (SL), tank mix adjuvants (TMA), emulsifiable concentrates (EC), concentrated emulsions (EW, EO), suspension concentrates (SC), suspoemulsions (SE), wettable powders (WP), water dispersible granules (WG, WDG) and capsulated suspensions (CS).

Introduction

Once an active ingredient has been discovered, there must be some way to deliver it to the target pest. The agrochemical formulation is used to provide both a vehicle to sell a product to the consumer as well as to allow for a stable method of its application. The physical form of the active compound, its solubility and its mode of action all combine to form the basis for how the compound can be formulated.

If the active ingredient can be adsorbed onto or absorbed into a carrier granule and broadcast in this coarse granular form, then a granule (GR) formulation can be commercialized. If the active is soluble in water or can be made water soluble by forming a salt, then an aqueous (SL) formulation can be developed. If the technical is a liquid or can be dissolved in an oil phase, then an emulsifiable concentrate (EC) formulation is appropriate. If the constituent is a solid that is not soluble in oil or water, then either a suspension concentrate (SC) or a water dispersible granule (WG) can be evaluated. If the product needs to contain two active ingredients and one is water soluble and the other is not, then either a concentrated emulsion (EW) or suspension concentrate (SC) can be developed based on individual component requirements. For systems requiring two water insoluble active ingredients where one is a liquid or soluble in a solvent, then a suspoemulsion (SE) formulation is possible.

This leaves two basic formulation types to consider. The first is the capsulated suspension (CS) formulation. This formulation is used to deliver a version of the active ingredient that has a reduced toxicity, reduced volatility, or a reduced release rate. The agrochemicals application area is one of the key areas when encapsulation technology has yielded commercial products. The second category is tank mix additive (TMA). The TMA is not really a formulation type on its own, but is rather any material that is added to a tank mix to aid or modify the action of the agrichemical, or the physical characteristics of the mixture(1). It can be in almost any form, and can function in one or more of a multitude of mechanisms.

Granules

If the active ingredient is a liquid or can be dissolved in a solvent to form a homogeneous liquid, then this liquid can be adsorbed onto or absorbed into a carrier granule. This coarse granule is then broadcast as is to a location where contact can be made with the target species. Granule (GR) formulations are common for insecticide formulations, where the pest movement results in contact with the active ingredient.

Granule formulations have used many substrate materials, with one of the most common being clay. The substrate must have a high porosity or high surface area to provide a sufficient loading of technical material. Advances in granule formulations in the past include using waste material like peanut hulls. Another was using the carrier as a bait for the pest. One particular example of this technology is using ground corn cobs as the carrier for insecticide.

Current granule development has centered on the use of recycled material as the carrier. Some developmental products have been made out of waste materials generated in the recycled paper industry.

Aqueous Solutions

If the active is soluble in water, or if it can be made water soluble by forming a salt, then an aqueous (SL) formulation can be developed. The concentrated formulation must be physically and chemically stable. The dilution into the spray tank must also be physically and chemically stable long enough to apply the product successfully.

The aqueous solution can be one of the easiest formulations to develop, as it typically requires only the addition of a wetter or spreader for spray efficacy. If required, a preservative, buffer, antifoam or antifreeze can also be present.

Current research into glyphosate formulations has driven the technology to high loading, low viscosity, low irritancy and inexpensive formulations. With the proliferation of herbicide resistant crops, the formulation must also be non-phytotoxic to the crop it is applied to.

Future trends with glyphosate formulations are package mixing with other actives to address resistant weed species. This can not always be done with a technical material that is also water soluble, so this trend results in more complex formulations.

Emulsifiable Concentrates

If the technical is an oil miscible liquid or can be dissolved in an oil phase, then it can be formulated as an emulsifiable concentrate (EC). Surfactants that function as emulsifiers are typically incorporated into the formulation at around 5-10 w/w%. An emulsion is formed spontaneously when the concentrate is diluted into the spray tank water.

Commercial production of emulsifiable concentrates often uses a matched pair emulsifier system. The matched pair uses two surfactant blends that are mixed at the appropriate ratio to maximize the kinetic stability of the emulsion that is formed. One surfactant blend has a relatively low HLB, and the other surfactant blend has a relatively high HLB. These two blends are mixed at various ratios until the optimal HLB for the desired system is found.

A current trend is the development of single blend emulsifiers, typically using experimental design techniques(2). With the proper optimization, the single blend can be robust enough to handle major inconsistencies in technical purity and supplier variability. In some cases the single blend is sufficient for muliple loading levels of active ingredient.

A second trend is the reformulation to greener solvents such as the alkyl lactates or alkylene carbonates(3). Propylene and butylene carbonate have high boiling points and high flash points. They have low odor and low evaporation rates. They have low toxicity and are safer to handle. They are readily biodegradable and have good solvency.

A major shift in the anionic component of the agrochemical EC emulsifier systems is finally taking place in the U.S. as linear alkylbenzene sulfonate is replacing hard, or branched alkylbenzene sulfonate. The linear version is more easily biodegraded, and has replaced the branched version in almost all application areas in the U.S. outside of agrochemicals already. In Asia and Europe, the switch to linear has already taken place even in ag applications.

Regulatory pressures are driving a trend towards lower application rate of pesticides. There is a corresponding demand being placed on formulations to deliver the lower loadings successfully. In the case of emulsifiable concentrates this demand can be satisfied by developing a microemulsifiable concentrate. This is a nonpolar solution of active ingredient with emulsifiers that dilutes to a microemulsion in the spray tank instead of macroemulsion. The benefits of microemulsion formulations have been well documented elsewhere(4).

Concentrated Emulsions

If the product needs to contain two active ingredients and one is water soluble and the other is oil soluble, then either a concentrated emulsion or a suspension concentrate (SC) can be developed based on individual component requirements. The concentrated emulsion can be either oil in water (EW) or water in oil (EO, OD). If one of the actives is a solid then the system can be formulated as a suspoemulsion, which is a combination of an emulsion and a suspension.

One advantage of a concentrated emulsion is that it can be used to deliver two actives with different solubility characteristics in one package. The system can usually be formulated with reduced solvent. There are two main disadvantages of a concentrated emulsion. The first is that high shear production equipment is required. The second is the difficulty in developing a physically stable formulation for a commercial product.

A concentrated emulsion is kinetically stable and not thermodynamically stable. It is only a matter of time before the system becomes physically unacceptable. The emulsion can either cream or sediment depending on the relative densities of the continuous and discontinuous phase. Flocculation can accelerate this separation. Ostwald ripening can lead to larger drop size. Coalescence can occur where the emulsion drops come together and physically become one larger drop. This coalescence is referred to as oiling. In extreme cases there can be a phase inversion where the continuous phase becomes the discontinuous phase.

Current research into developing concentrated emulsions includes the study of forming nano or mini-emulsions. These are systems that would normally exist as a macroemulsion. Processing techniques have been used to produce a particle size about an order of magnitude less that the typical 1-10 microns that exist in concentrated emulsions. These systems can be made using either extremely high shear or high pressure equipment, or by using the phase inversion temperature (PIT) technique. This is where the system is heated to above the PIT, and then the process temperature is reduced. The interfacial tension is at a minimum at the PIT, and smaller drops can be formed with relatively lower shear equipment than normal. Care must be taken to not confuse these so-called mini or nano-emulsions with thermodynamically stable microemulsions.

Future development will likely have microemulsions replacing concentrated emulsions as the preferred formulation type. The microemulsion is thermodynamically stable and is not prone to the same physical instabilities as a concentrated macroemulsion. Microemulsions are optically isotropic, have a low viscosity, and an ultrafine dispersed particle size of 10-200 nm. This ultrasmall particle size provides a much larger number of drops containing active ingredient in the spray solution. This should come the closest of any formulation type other than a true solution to providing a uniform distribution of the pesticide during application.

The microemulsion should not require any special high shear or high energy processing equipment. A well formulated product will show no separation over a finite temperature range. If a phase change temperature is encountered during transport or storage, the microemulsion will usually reform spontaneously or with only mild agitation. Suspomicroemulsions could replace Suspoemulsions due to their enhanced physical stability.

Suspension Concentrates

If the pesticide constituent is a solid that is not soluble in oil or water, then either a suspension concentrate (SC) or a water dispersible granule (WG, WDG) can be evaluated. A suspension concentrate is a dispersion of active ingredient in a continuous aqueous phase. The concentrated suspension is then diluted in the spray tank prior to application. If the solid has to be dispersed in an oil phase, then the oil flowable (OD) has to spontaneously emulsify if it is to be diluted into an aqueous spray tank.

A suspension concentrate contains the solid active ingredient, a surfactant for wetting, a surfactant for dispersing, and a thickener as a suspension aid. The result is a very complex system that can be very difficult to optimize. The solubility of the active has to be very low to prevent Ostwald ripening. There are limits to the particle size and loading levels that can be achieved with suspension concentrates, and these are two areas of current research.

The suspension concentrate can also be prone to hard settling with very poor resuspensibility. The process to make a suspension concentrate is very energy intensive and some type of wet milling is usually required. The multiple formulation obstacles and limitations in performance of the suspension concentrate has led to one future research area that completely replaces the suspension concentrate with a structured surfactant formulation alternative(5).

Structured Surfactant Formulations

Structured Surfactant Formulations (SSF) are close-packed three-dimensional matrices of a liquid crystalline phase that suspend insoluble pesticide materials. The active ingredient can be either solid or liquid. Additional actives and/or adjuvants can be dissolved or suspended in the formulation if desired. The main advantage of an SSF over standard suspension concentrates is that no thickening or suspending agents are required. The SSF formulation can be solvent-free if desired, or include an oil adjuvant built in for improved efficacy.

When the surfactant loading in aqueous solutions reaches high levels the surfactants begin to associate in liquid crystalline phases. Liquid crystalline phases can also be formed at low surfactant concentration by a technique known as salting out. This phase behavior can be exploited by suspending dispersions of pesticide active ingredients using economical levels of surfactant. A typical SSF contains from 0-60 w/w% active, surfactants for wetting and dispersing as well as structuring, and finally antifreeze and antifoam, if required. The typical properties of an SSF are excellent concentrate physical stability, relatively low viscosity (after the yield stress is overcome), good storage stability, good wetting, and the ability to form stable suspensions on dilution in the spray tank.

SSF technology can be applied to a wide range of pesticide materials, including herbicides, fungicides, and insecticides. Surfactant chemistry can be used to optimize the formulation efficacy in the development of an herbicide SSF. Combination packs that include both water-soluble and water-insoluble actives can also be developed. Formulation adjuvants such as oils or electrolytes can be incorporated into an SSF formulation to increase its biological activity.

Structured Surfactant Formulations take advantage of basic aqueous surfactant phase behavior. Figure 1 shows a schematic of general surfactant aqueous equilibrium phase behavior as a function of surfactant concentration.

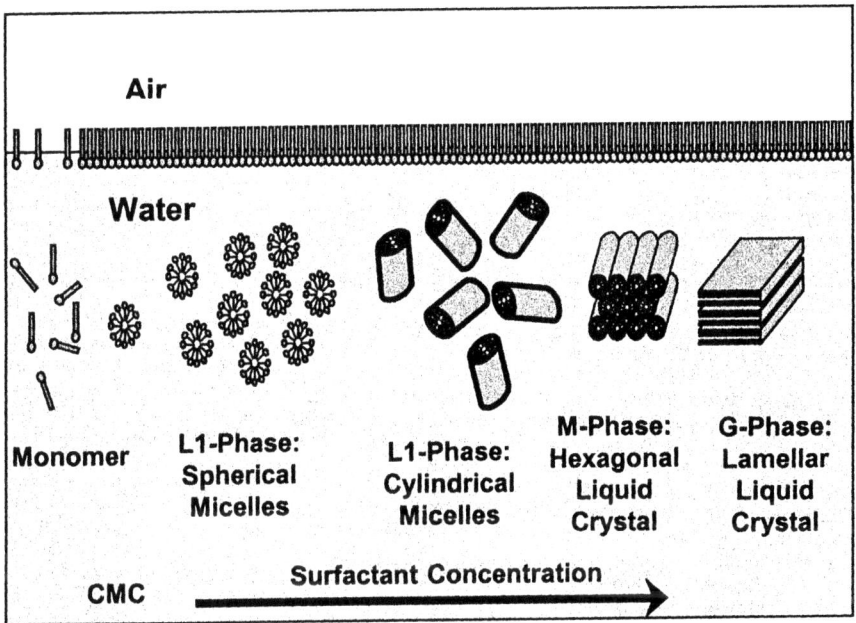

Figure 1. General Surfactant Aqueous Equilibrium Phase Behavior

At very low surfactant concentrations the surfactant exists as monomers dispersed in water, with an incomplete coverage at the surface. At the critical micelle concentration (CMC) the surface monolayer is saturated, the surface tension has reached a minimum, and the surfactant molecules associate into micelles in equilibrium with a few monomers. The so-called L_1 phase is optically isotropic and continues to be dominant as the surfactant concentration is increased.

Eventually, a surfactant concentration is reached where the spherical micelles become elongated, or slightly cylindrical in shape. This phase is optically isotropic at rest, but shows the unique property of streaming birefringence. This means that the elongated micelles orient under shear, producing an optically anisotropic system. This birefringence can be observed using crossed polarizers.

At a significant level of surfactant the system structures into a liquid crystalline phase. First the M-phase or hexagonal liquid crystal is formed consisting of stacked cylinders of very long range. Then a G-phase is formed, which is a lamellar liquid crystal formed with long-range sheets of surfactant bilayer.

The hexagonal and lamellar liquid crystal phases noted above require high surfactant concentrations. However, surfactants can be forced to form a structured phase at relatively low surfactant concentrations using a technique known as salting-out. At the required electrolyte concentration a surfactant rich phase precipitates in the form of spherulites. The spherulites contain a significant amount of water, and a stable SSF can be formulated at a relatively low surfactant concentration. A schematic of an SSF is shown in Figure 2.

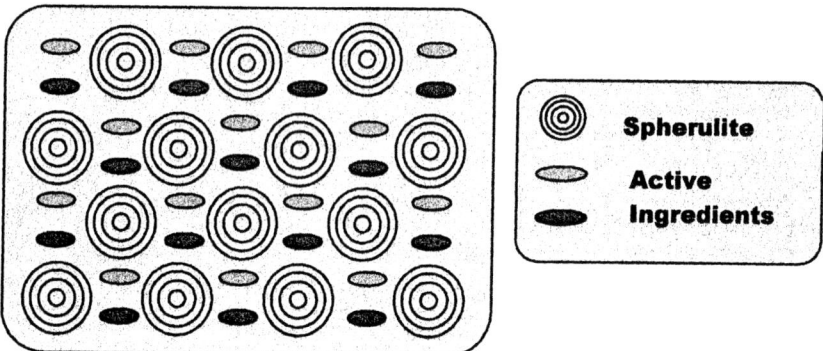

Figure 2. Schematic Representation of a Structured Surfactant Formulation

The SSF products exhibit unique pseudoplastic rheology. Figure 3 shows the general behavior of Diuron SSF system. At zero or very low shear rates the system has a very high viscosity. This high viscosity contributes to the high physical stability of SSF products with no settling of the suspended solids. After the yield stress is reached the structure of the system breaks and the viscosity quickly decreases to very low values. This low viscosity allows the SSF product to be easily pumped or dispensed. This unique rheology of an SSF is attained without the inclusion of thickening or suspending agents.

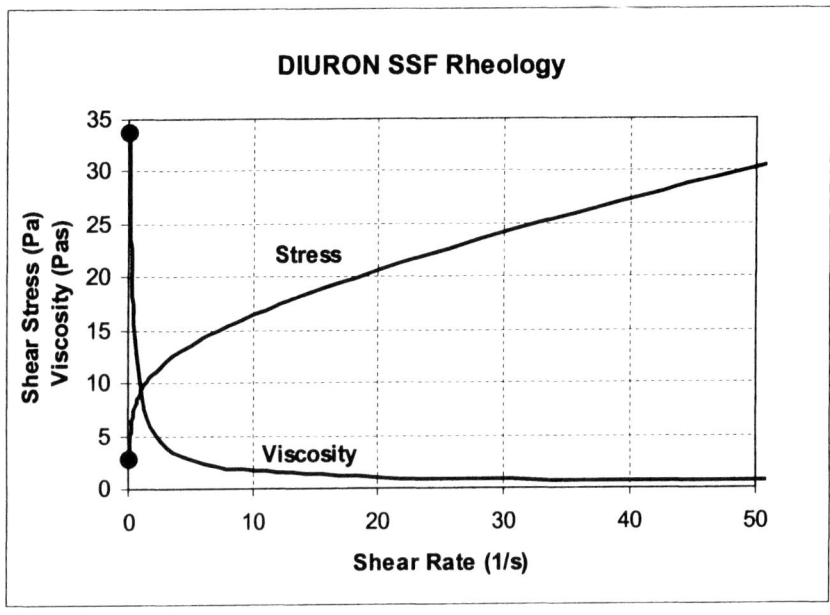

Figure 3. Diuron SSF Rheology

The SSF technology gives the agricultural formulator a new formulation type to evaluate in place of standard SC formulations. SSF technology may be used to optimize biological activity by the incorporation of oils, additional surfactants, and/or other adjuvants including electrolytes. SSF technology may also broaden the constituent size range that can be successfully formulated. It is possible to evaluate the formulation of sub-micron size solid active ingredients. This technique could lower the amount of pesticide applied if the biological efficacy is improved. With a properly designed yield strength it would be possible to evaluate SSF technology with 100 micron size actives and above.

Structured Surfactant Formulations do not require high levels of surfactant. The spherulitic phase that precipitates contains a significant amount of water that is defined by the equilibrium phase behavior. The structured system that suspends the active ingredient is therefore made up of significant amounts of water and not just surfactant. Another technique that can be used to minimize the total amount of surfactant required is to have the surfactants perform multiple functions. The structuring surfactant system can be designed to also act as the wetter and dispersant in both the original formulation production and the final dilution for product application.

When the Structured Surfactant Formulation is diluted in the spray tank, a stable, low viscosity suspension is formed. The spherulites dissolve at the reduced spray tank surfactant concentration. The dilute suspension can then be applied as any stable SC product.

Water Dispersible Granule

A water dispersible granule (WG, WDG) is a finely milled solid active ingredient that has been agglomerated to reduce dust exposure. When added to the spray tank the granule should rapidly break down to its primary particle size. The process to make WGs can be very difficult and can include pan granulation, extrusion, spray drying, and fluid bed agglomeration.

Current research is under way to apply WG formulations in fertilizer solutions. This will require that the granule rapidly disintegrate and form a stable dispersion in high electrolyte solutions.

Futuristic granules have been developed that contain components that aid in dispersing the active ingredient. These so-called dancing granules have found commercial application in rice fields.

Capsulated Suspensions

A capsulated suspension (CS) is basically a suspension concentrate where the active ingredient has been microencapsulated. The technical may have been coated by interfacial polymerization or one of many other techniques. The encapsulation is used to reduce a.i. toxicity or volatility, or in an attempt to control the active release rate.

A summary of the research activities in the CS formulation area would be a paper in itself. One of the main areas of research is to manage the process so that the release rate can actually be controlled, and not just slowed.

There is also a desire to reduce the dispersed particle size to lower levels. A typical CS may have a solid particle size of from 3-10 microns. For a truly nanoencapsulated formulation there is hope to develop stable formulations with a submicron particle size. A smaller particle size will yield a higher number of particles and should produce a more uniform application in the field.

Tank Mix Adjuvants

The final formulation category is the tank mix additive (TMA). The TMA is not really an active ingredient formulation type on its own. It is any material that is added to a tank mix to aid or modify the action of the agrichemical, or the physical characteristics of the mixture. It can be formulated in any form previously discussed. It can perform one or multiple functions to enhance efficacy.

The most common type of TMA is the aqueous solution of surfactant that functions as a wetter, spreader, sticker, etc. A second type of TMA is the crop oil concentrate, which is similar to an EC formulation.

While the TMA can take just about any form, there has been a trend of combining multiple functions into one package mix. This can result in a concentrated emulsion, suspoemulsion, or other intricate colloidal formulation. Just as has been seen in pesticide development, there is a similar trend to develop multiple function TMAs as microemulsions(6).

Conclusions

Technology, regulatory pressure, and environmental concerns are combining to drive down application rates. Colloid science principles can be used to formulate systems with reduced particle size. For example, a microemulsion can be formulated to replace a concentrated emulsion. The reduced particle size should yield a more uniform pesticide application, and this should maximize efficacy.

Combination products are becoming more important, and adjuvants can be formulated into the pesticide concentrate as well. The result can be sophisticated products like suspomicroemulsions. The structured surfactant formulation can be evaluated in lieu of the suspension concentrate. The SSF's main advantages are the absence of thickener, and technology that allows an adjuvant to be built into the pesticide formulation.

On a final note, nicosulfuron with an optimized adjuvant has efficacy at an application rate of only 1 g/ha(7). When the calculations are performed to see just where we currently stand with ultralow application rates, one finds that the nicosulfuron is active at the concentration of 1.5 molecules/10 nm^2! Ten square nanometers is a pretty small area, but it is amazing to find that only one and a half molecules are all that is required to be effective. What does the future hold: one molecule per target pest?

Acknowledgements

The authors wish to acknowledge the Huntsman LLC for permission to publish this work. The authors greatly appreciate the assistance of Joe Arzola, Howard Stridde, Al Stern, Yolanda Lopez, Mark Rollinson and David Ross.

References

1. ASTM E1519 "Standard Terminology Relating to Agricultural Tank Mix Adjuvants," *Annual Book of ASTM Standards*, Vol. 11.05, **2000**.
2. Ashrawi, S. S., Elsik, C. M., Stridde, H. M., and Smith, G. A., "Strategies for Agricultural Formulations: A Statistical Design Approach," *Pesticide Formulations and Application Systems: A New Century for Agricultural Formulations, Twenty First Volume, ASTM STP 1414*, J. C. Mueninghoff, A. K. Viets, and R. A. Downer, Eds., American Society for Testing and Materials, West Conshohocken, PA, **2002**.
3. Elsik, C. M., Stridde, H. M., and Machac, J. R., "Physical Properties of Alkylene Carbonate Solvents and Their Use in Agricultural Formulations," *Pesticide Formulations and Delivery Systems: Meeting the Challenges of the Current Crop Protection Industry, ASTM STP 1430*, R. A. Downer, J. C. Mueninghoff, and G. C. Volgas, Eds., American Society for Testing and Materials, West Conshohocken, PA, **2003**.
4. Tadros, Th. F., "Microemulsions in Agrochemicals", in Solans, C. and Kunieda, H. (Eds.), *Industrial Applications of Microemulsions*, Surfactant Science Series, Vol. 66, Marcel Dekker, New York, **1997**.
5. Elsik, C. M., Perdreau, L., Rollinson, M. and Diu, M. L., "Agricultural Applications of Structured Surfactant Formulations," *Pesticide Formulations and Application Systems: 23rd International Symposium, ASTM STP 1449*, G. Volgas, R. Downer, and H. Lopez, Eds., ASTM International, West Conshohocken, PA, **2003**.
6. Elsik, C. M., Stridde, H. M., and Ashrawi, S. S., "Microemulsion Formulation of Agricultural Adjuvants," *ISAA Proceedings*, Amsterdam, The Netherlands, **2001**.
7. Green, J. M., and Hale, T., "Effect of Adjuvant Type on the Control of Yellow Foxtail (Setaria Glauca) with Nicosulfuron," Proceedings Weed Science Society of America, 43:58, **2003**.

Chapter 27

Dinotefuran Release Properties from Controlled Release Formulations

Daisuke Kishi and Seiichi Shimono

Functional Chemicals Laboratory, Mitsui Chemicals, Inc., 1144 Togo, Mobara-shi, Chiba 297–0017, Japan

Dinotefuran is a novel neonicotinoid compound of low toxicity discovered by Mitsui Chemicals, Inc. This compound is highly systemic and controls a broad range of pests in various crops, however, release control of dinotefuran is necessary for rice nursery box application because of its high water solubility so a controlled release granule containing wax was developed. It was suggested that the ester value and crystallinity of the wax were effective to control the release rate and there was a good linear relationship between the release rate and water temperature. The release rate was controlled by modifying the composition, especially the amount of amorphous silicone dioxide.

Introduction

Dinotefuran (DT) is a novel neonicotinoid compound discovered by Mitsui Chemicals, Inc. It has a tetrahydrofuran ring in its structure instead of aromatic heterocycles such as chloropyridine, which is a characteristic for other neonicotinoids. The melting point is 107.5 °C and water solubility is comparatively high (about 4%, Figure 1).

DT has a broad spectrum of activity, being effective for not only sucking pests, Hemiptera, but also for some Diptera, Lepidoptera, Coleoptera species and so on. DT exhibits very low toxicity on birds, fish, and water fleas as well as mammals, but on the other hand, it is toxic against the honey bee (1, 2).

DT has a wide insecticidal spectrum and low toxicity, however, it is released completely within 1 day from the conventional formulation because of its high water solubility, which causes leaching, run-off and the outflow of DT to underground and neighboring areas, so bioavailability to rice plants will be reduced, consequently decreasing the efficacy on pests. So, release control of the active ingredients is necessary for formulations containing DT which are applied to the rice paddy field, which is the main focus for Japanese pesticides.

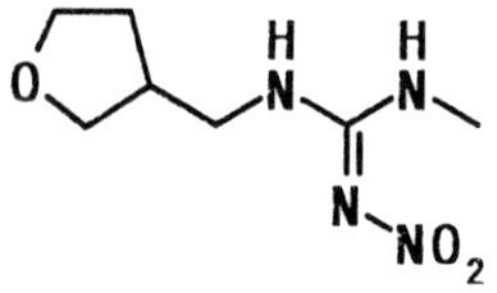

ISO name : Dinotefuran
Chemical name (IUPAC)
: (*RS*)-1-methyl-2-nitro-3-(tetrahydro-3-furylmethyl) guanidine
Melting Point : 107.5 °C
Relative density : 1.40 (20°C)
Vapor pressure : < 1.7×10^{-6} Pa (30°C)
Appearance : White Powder
Solubility in water : 39.8g/L
Log Pow : - 0.549 (25°C)

Figure 1. Chemical and physical properties of dinotefuran

Preparation of Controlled Release Granule

We have developed a controlled release granule containing wax, as a result of manufacturing trial of several types of controlled release formulations,the formula and five components of which sre shown in Table 1. To manufacture the controlled release granule, all of the components were mixed and heated at about the melting point of the wax during mixing, and then the mixture was extruded with heating at about the solidifying point, under the melting point of the wax.

So, the wax was melted but DT was not melted in these processes, because the melting point of DT was higher than that of the wax. The extruded product was cooled to room temperature. After a disintegration and sieving process, the controlled release granule (CR-Gr) was obtained.

Table I. Formula of controlled release granule (CR-Gr)

Components (Function)	(%)
DT : Dinotefuran (active ingredient)	2 – 3
Wax (binding agent and matrix former)	16 – 20
ASD : Amorphous silicone dioxide (oil-absorbent)	2 – 6
Talc and CaCO3 (lubricant and filler)	balance
Total	100

Methods for Release Rate Determination

The obtained CR-Gr was put into water. If the active ingredient is released completely, the concentration in water will be 800 ppm. After standing for a prescribed period, the concentration of DT in water was measured by HPLC.

Mt, the cumulative concentration of DT released was calculated by the following equation:

Mt (%) = amount of DT in water / amount of DT in the granule × 100

In this case, the results could be considered as a "sink condition" because the DT water solubility is 50 times as high as the possible maximum concentration in this study. All studies were examined under these conditions.

Release Properties of Controlled Release Granule
I. Effect of kind of wax

The first study of release properties was conducted in order to examine the effect of the kind of wax on the release profiles, so several types of controlled release granules were prepared containing different kinds of wax. Table II shows the characteristics of the wax tested. Each wax has a different acidic value, saponification value and ester value. Montan ester wax has the highest ester value among them, and oxidized polyethylene wax has the lowest. Figure 2 shows the DT release profiles of the granules containing these waxes, and it was clear that released DT varied depending on the kind of wax. We noticed a correlation between released DT and the ester value of the wax.

Table II. Characteristics of wax tested

Wax	AV	SV	EV (mg KOH / g)
Oxidized polyethylene wax	22	37	15
Montan acid wax	143	163	20
Montan ester wax	21	151	130
Partially saponified montan ester wax	11	118	107
Carnauba wax	9	84	75

AV, acid value : SV, saponification value : EV, ester value, which was calculated by subtracting AV from SV.

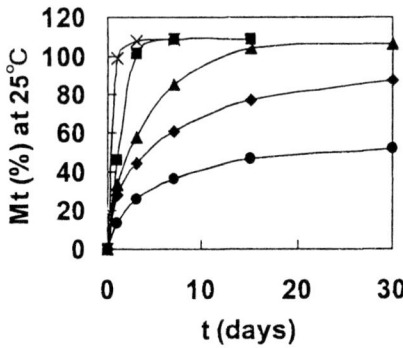

Figure 2. Effect of kind of wax on release profiles of 2% DT CR-Gr (21% of wax, 5% of ASD)

×, oxidized polyethylene wax : ■, montan acid wax : ▲, partially saponified montan ester wax : ◆, montan ester wax : ●, carnauba wax

The cumulative concentrations of DT 1 day after application which were plotted against the reciprocal of the ester value, showed that there was a positive correlation between them (Figure 3). This result indicates that the release rate tended to be slow when the ester value of the wax was high. The acidic wax might form a relatively simple matrix, because only linear or simply branched chains can be formed by a hydrogen-bond at the carbonate end. By contrast, it appeared that the ester wax might form a more complicated matrix, because of some fatty acid residues in a molecule. Accordingly, montan ester wax, which has the highest ester value, was selected for further studies.

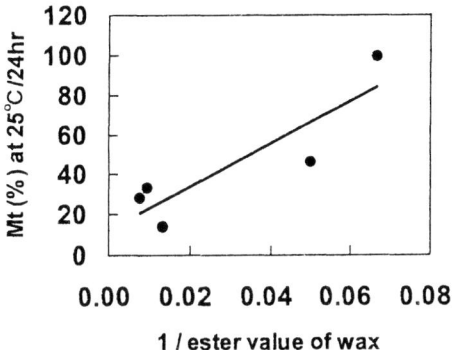

Figure 3. Correlation between released DT 1 day after application and the reciprocal of the ester value.

Release Properties of Controlled Release Granule
II. Effect of Annealing

The effect of annealing on the release profiles of the controlled release granules was examined. When the calorie of the heating wax in the mixing or extrusion process was changed, CR-Grs having different release rates were obtained even if they had the same formula. So it was thought that adding a calorie of heating was one of the important factors to control the release rate. However, when these products annealed at 50 °C for 1 day, under the melting point of the wax, released DT was restored within a certain range.

To analyze those phenomena, the properties of the waxes themselves were examined. Three types of wax conditions were prepared: non-treatment wax, cooled wax after melting completely, and annealed wax after melting and cooling. They were studied by X-ray powder diffraction analysis (XRD) and differential thermal analysis (DTA). The X-ray diffractogram showed two very intense signals ($2\theta = 21.4, 23.8°$) as shown in typical waxy materials and was separated into two regions: part of an intense signal was in the crystalline region and parts of others were in the noncrystalline region, and then the degree of crystallinity was calculated from these ratios. On the other hand, the heat of melting of the wax was obtained by integral calculus in a DTA curve.

As shown in Table III, the degree of crystallinity and heat of melting declined after the wax was melted once but were restored to the original value by annealing.

Table III. Degree of crystallinity and heat of melting in several thermal treatments of montan ester wax itself

Conditions of Thermal Treatment	Degree of Crystallinity (%)	Heat of Melting (μ Vs/mg)
Non-treatment	80.8	156
Cooled after melting completely	74.1	121
Annealed after melting and cooling	79.7	140

annealing : stored at 50°C for 3 days after melting and cooling

The release of DT and heat of melting of the CR-Gr products before and after annealing were examined and the results are shown in Table IV. The release of DT was decreased by annealing, but heat of melting was increased by it.

In a word, a high degree of crystallinity was important to stabilize the release rate within a certain range and annealing was effective to increase the crystallinity.

Table IV. Released DT and heat of melting of CR-Gr before and after annealing

Products	Released DT (%, at 25 °C × 24hr)	Heat of Melting (μ Vs/mg)
Before annealing		
CR-Gr A	27.2	117
CR-Gr B	37.9	122
CR-Gr C	45.2	117
After annealing		
CR-Gr A	23.3	133
CR-Gr B	19.9	139
CR-Gr C	22.5	139

annealing : stored at 50°C for 1 day

Release Properties of Controlled Release Granule
III. Typical release pattern and analysis by Higuchi Model

Figure 4 shows the typical release pattern of CR-Gr. This profile was separated into two steps. In the first step, less than ten percent of DT was released within 10 minutes. This burst effect might be due to the presence of the compound on the surface of CR-Gr. In the 2nd step, after 10 minutes, the rest of DT was released slowly over 30 days.

We speculated that DT release from the granule was simple diffusion by reason of its formula, so the release profile was analyzed by the Higuchi diffusion model equation (3, 4). The Higuchi diffusion model can be simplified, as in the following equation (3 - 7).

$$M_t = k t^{1/2} + C$$

M_t is the cummulative concentration of released DT at time t (hr), k is the release rate coefficient, and C is the intercept. When the results in the 2nd step after ten minutes were plotted by the equation, a very good linear relationship, in which the coefficient of correlation was almost one, was obtained (Figure 5). Accordingly, the DT release profile of the controlled release granule was considered as simple diffusion.

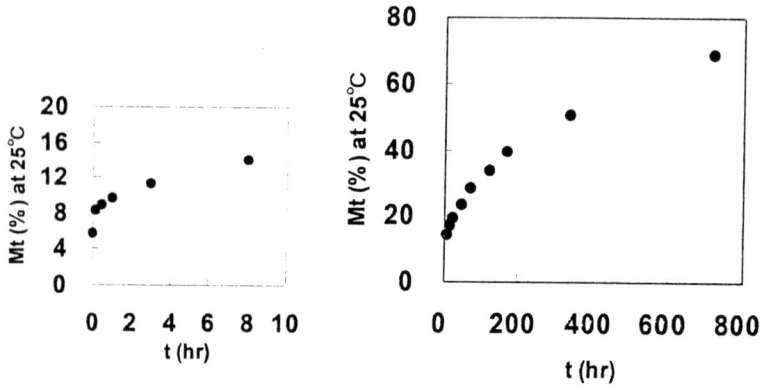

Figure 4. Typical release profiles of 2% DT CR-Gr (18% of wax, 3.75% of ASD). Right graph within 8 hr was magnified and presented in the left.

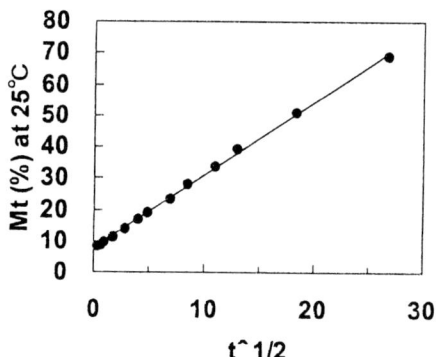

Figure 5. Analysis of release pattern by Higuchi diffusion model equation

Release Properties of Controlled Release Granule
IV. Effect of Water Temperature

The effect of the water temperature on the release rate was analyzed by the Higuchi diffusion model in the same way and the results are shown in Table V. It was revealed that the release rate was affected by the water temperature and agreed with the Higuchi diffusion model at each temperature.

Table V. Effect of water temperature on the release rate. The results were analyzed by the Higuchi diffusion model equation.

Temperature	k	C	r
55°C	32.338	1.476	0.988
45°C	11.808	5.225	0.995
35°C	4.884	5.097	0.999
25°C	2.307	7.884	0.999
15°C	1.234	4.047	0.996

The relation between the water temperature and the release rate coefficient, k, was studied further. Figure 6 shows the Arrhenious plot of the absolute temperature and the release rate coefficient. Then, a straight line was obtained with a high correlation coefficient. This result indicates that the release rate at a prescribed temperature was predictable, although it was effected by the water temperature.

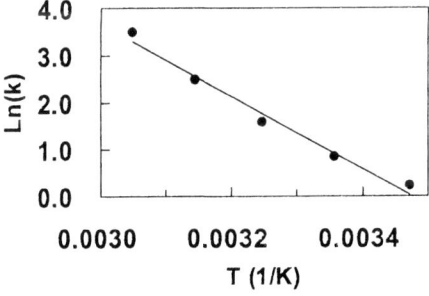

Figure 6. Effect of water temperature analyzed by arrhenious plot of the absolute temperature and the release rate coefficient.

Release Rate Control Technique

The last study was about the release rate control technique, in other words, the effect of the amount of the wax and the amorphous silicone dioxide (ASD) on the release profiles of CR-Gr was examined.

The effect of the amount of the wax or ASD was analyzed by the Higuchi diffusion model equation. The result was that the release rate varied depending on the amount of these ingredients and agreed with the Higuchi equation. Then, the release rate coefficient, k, was plotted against the amount of the wax or ASD and linear relationships were obtained as shown in Figure 7 and 8.

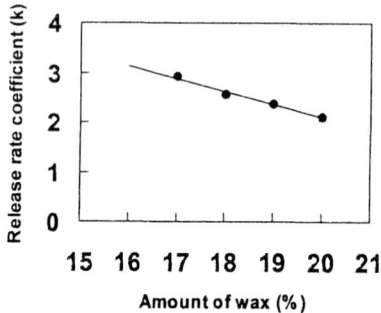

Figure 7. *Effect of the amount of the wax on the release profiles from 3% DT CR-Gr (3.75% of ASD).*

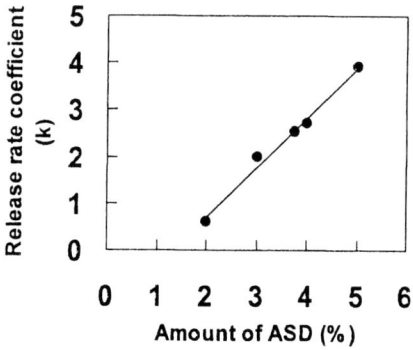

Figure 8. *Effect of amount of the ASD on the release profiles from 3% DT CR-Gr (18% of wax)*

As a result, it was shown that the amount of the wax and that of the ASD influenced the release rate, and the release rate was controlled by modifying the composition, especially the amount of the ASD. This effect might be due to two reasons: First, the matrix density varied depending on the amount of the wax, and then the release rate was changed. Second, the ASD absorbs about twice as much oil as its own weight. Melted wax can be regarded as oil in this case. The ASD is very efficient in changing the matrix density.

Conclusion

We developed a controlled release granule of DT containing a certain wax, and its release properties agreed with the Higuchi diffusion model. There were four factors which have an effect on the DT release properties.

1) The dinotefuran release rate varied depending on the kind of the wax; the release rate was slow when the ester value of the wax was high.

2) A high degree of crystallinity was important to stabilize the release rate within a certain range.

3) Annealing was effective to increase the crystallinity.

4) There was a good relationship between the release rate and the water temperature. The release rate at a prescribed temperature was predictable.

Finally, the release rate was controlled by modifying the composition, especially the amount of the ASD. In a word, granules showing appropriate release rate can be manufactured in response to various application requirements.

Acknowledgement

We would like to express our thanks to a number of colleagues at the Functional Chemicals Laboratory and Material Science Laboratory, Mitsui Chemicals Inc. and at Santou Chemicals Inc.

References

1. Kodaka, K. et al.: *Proc Brighton Crop Prot. Conf. - Pests Dis.* **1**, 21 – 26 (1998)
2. Wakita, T. et al.: *Pest Manag. Sci.* (2003), in press
3. Higuchi, T.: *J. Pharm. Sci.* **52**, 1145 – 1149 (1963)
4. Higuchi, W. I.: *J. Pharm. Sci.* **56**, 315 – 324 (1967)
5. Ritger, P. L. and Peppas, N. A.: *J. Control. Rel.* **5**, 37 – 42 (1987)
6. El-Shanawany, S.: *J. Control. Rel.* **26**, 11 – 19 (1993)
7. Cotterill, J. V. et al.: *J. Control. Rel.* **40**, 133 – 142 (1996)

Chapter 28

Advances in Pesticide Applications and Their Significance to the Agrochemical Industry Servicing Tropical Farming

William Taylor and Per Gummer Andersen

Hardi International A/S, Helgeshoj Alle 38, Taastrup, DK 2630, Denmark (www.hardi-international.com)

Pesticide application has benefited from many advances in key areas that are associated with machine design and use. Field efficiency, drift control, safety whilst loading products and in use, the cleaning of internal and external surfaces, being some examples. The spurs for these activities are diverse, but legislation that demands improved water quality, lower operator risks, reduced Plant Protection Product use, and improved operator productivity, have been vital. To make, monitor and claim the many potential gains that were needed, has been stimulated by the activities of internationally agreed Standards. These are protocols that describe what needs to be considered and how performance should be measured as well as threshold limits that need to be reached. Although many of these activities are intended for field crop sprayers, or those used with air assistance in tree and bush crops, sprayers carried by operators and are manually operated, are also included. The application of pesticides as sprayed drops in the open environment of fields and plantations has also triggered Standards that define how drift is to be assessed, the categorisation of low drift equipment as well as protocols that may fully account for the fate of all the active ingredient within and beyond the treatment zone. Communication and training are equally critical; approved courses for operators, the periodic testing of sprayers both monitor and help sustain high standards on farms and plantations as well as ensuring effective transfer of knowledge. All countries and crops are likely to benefit from these broad initiatives that encourages more responsible Plant Protection Product use. This paper will focus on these advances – using tropical crop sprayers and spraying methods as examples that will illustrate what has been achieved and where advances still await.

Introduction

Pesticide application has benefited from many advances in key areas that are associated with machine design and use. These advances can be grouped into those aspects of design that have improved the safe use of the equipment, the loading of Plant Protection Products [PPP] into the sprayer, the application of a diluted solution for optimal effects, drift containment to the treated zone and the cleaning of internal and external surfaces. Spray machinery design has also been influenced by its needs to meet the interests of Regulatory Bodies who may wish to improve safety in a more general way – such as drift control or may require a specified product to be loaded into the machine in a particular manner.

Safe use

Sprayers – in almost every country of the world – are getting larger, are fitted with wider booms, are spraying at faster speeds to meet the pressures introduced with the growth in farm size and the reduction of available labour. Tank capacities that exceed 4000 litres and booms that reach beyond widths of 38 metres and spraying at speeds in excess of 15 km.hour are no longer unusual. .These changes further increase the needs for engineered controls on these more sophisticated machines Engineering control systems that protect the operator from physical injury as well as reduce the threat of exposure to PPPs are in common use. Thus, hydraulic mechanisms to fold and raise booms both avoid the physical demands on the operator but also discourage him from making contact with contaminated surfaces on the machine and those he may have just sprayed. Tank drainage is remotely controlled without having to get under the machine to unscrew a plug whilst induction bowls avoid the need for the operator to climb up on top. Bayonet fixing nozzles reduce potential exposure since the time required to change nozzles is a fraction of that demanded from the older screw fittings. Multi-turreted nozzle bodies can be used to replace blocked nozzles instantly in the field without having to remove them for cleaning. Clean water supplies are carried on the machine and are conveniently located to where the operator works. Spillage onto, for example, gloves or nozzle-blocking debris can be quickly removed. Modern filtration systems – coupled to improved formulations - are now so effective that nozzle blockage or disturbances to their patterns are rare. The occasions that operators find themselves making running repairs or adjustments – are very rare. Most modern sprayers have their functions remotely controlled from within the protected environment of the driving cab; a control area that is ideally pressurised to stop the ingress of particles. Research has shown that cabs with windows or doors open and not sealed – readily permit particulate ingress irrespective of wind direction.

PPP loading

Perhaps the quickest and most rapid uptake of any new sprayer technology – has been the speed with which operators have recognised the need for – and used - induction bowls to load the undiluted product into the main tank. These devices are fitted to the sprayer in a position that the operator can safely reach from the ground in order to both avoid spillage and time [Fig 1]. Almost every formulation can be loaded into the sprayer with ease; liquids, WDG, sachets are sucked into a pipe that directs the product into a violently agitated zone of the main tank. Although, induction bowl use was encouraged by safety needs – operators must not be encouraged to climb up onto sprayers especially when carrying product – their popularity was gained by the speed with which they could now load the bigger sprayers and the increased quantities of product loaded each time; a trend that still continues as water volumes come down and hence, product concentration, increases. Loading speed is limited only by the time it takes the operator to open the package and release its contents.

Fig 1: Operators can load sprayers quickly and safely using facilities such as this induction bowl with its container cleaning facility, PPP store and work area

Optimising nozzle use

PPP are almost always diluted with water before use and sprayed using hydraulic [often flat fan] nozzles that break this liquid up into drops. Both the volume of spray solution and the drop size can be changed and this facet is used to optimise retention on target plant surfaces, disperse more product to preferred locations on a plant or within a leaf canopy. In some instances, the same mechanism may be used to gain crop selectivity by discouraging retention by the crop yet enhancing the same spray on weeds. These efficacy driven needs for a range of nozzles to meet each and every situation is now extended to meet further – environmentally driven – demands [Fig 2]; but the same volume rate/drop size approach is being used to reduce drift beyond the treatment zone – a move that may jeopardise product performance. Regulators may also influence the final advice on how a product should be applied. Some products are approved for use with a stated maximum concentration to protect operators from inadvertent exposure. This concentration will then define a lower volume rate for that active which may restrict the operators nozzle range. Similarly, Regulators may require that some products specify a larger drop size than that needed for its optimal efficacy to limit the risk of drift or operator inhalation. In addition to all these interests that advise on how a product should be applied – there are those the operator may also wish to exploit. In his wish to increase field timing – critical when optimising biological responses – he will exploit faster work rates by using lower water volumes [that will avoid time wasted in transporting water or going to its source] and he may use a coarse drop sizes to keep spraying when wind speeds may be restrictive. The trend now in Northern Europe is to spray at volumes of 100 to 150 l/ha using speeds of 10 to 15 km/hour rather than 200 to 400 l/ha at 6 to 8 km/hour. In reality, many compromises between conflicting needs are made but a range of nozzle sizes and types have to be considered when meeting all these demands; demands originally imposed by the product but now conditioned by other needs.

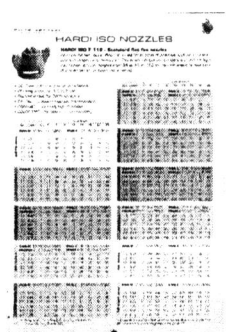

Fig 2: Nozzle selection tables need to describe not just physical performance but also likely uses and restraints

Drift control

Larger less drifting drops have been traditionally generated by increasing the size of the nozzles orifice; an option that increases liquid throughput as well as coarsen the resultant drop size. Such drops were also faster and entrained air; features that further increase drift reducing benefits. However, the higher water volumes that then have to be used are at variance with field work rates, efficacy and other demands. New nozzle designs evolved from these conventional, high output, flat fans to those with a pre-orifice and now a type that induces air to generate coarser sprays but at low liquid throughputs [Fig 3]. Sprays which are Coarse, Very Coarse and Extra Coarse can be applied at low volumes. Thus, for those products that will tolerate large widely dispersed drops on the target surfaces, there are great user benefits to be gained. Not surprisingly,

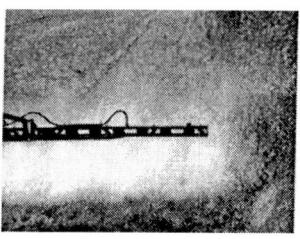  

Fine spray   Coarse spray   Coarse spray with Buffer Zone

Fig 3: The most appropriate nozzle choice now has to consider - not just the needs of the target plant and/or best but - environmental interests too; especially the need to reduce contamination - of those areas beyond the treatment zone - by drift

many foliar applied products show less than optimal biological effects from these applications. These products can be applied in low volumes of finer sprays - and with low drift hazards – by using well designed air assistance that entrains the drops immediately after the full swath across the boom has been formed.

In-field cleaning

Carry over of spray liquid from one crop type to another –risks crop damage and spurred the original interest in better sprayer hygiene. Recently, surface water contamination by pesticides has been attributed to inappropriate sprayer cleaning too. If internal residues are discharged at a point – for example – that is close to surface water or on freely draining soil, the resultant, localised but very high dose exceeds that for which it is approved - to increase this hazard. Today, dried spray deposits on the outside of the sprayer have also been associated to the same problem as well as posing exposure risks to operators and bystanders.

In-field cleaning encourages the following of a routine that ensures all the product is removed and confined to the intended treatment zone [Fig 4]. Equipment that makes this possible has to encourage the operator to do this extra task frequently; a goal that has to be reached with ease and at speed. Modern sprayers have fewer hold back points for residual liquid; better sumps, non-stick build surfaces and bleeding control valves are some advanced features. Dedicated tanks carry clean water to the field where – after spraying is finished – can be pumped through sprinklers to rinse internal main tank surfaces, flush all control pipe work and still be retained for it to be sprayed out. Field areas preferred for this discharge are those that have not been treated or exposed to less than the full approved dose. Advanced internal field cleaning is so efficient that the volume of clean water used for internal use is reduced- allowing some to be retained for external use. High pressure hoses effectively remove external deposits with lower rates of water use. Operators know where deposits are likely to be greatest and can focus the time spent at the more critical areas. Appropriate in-field cleaning techniques may do more to remove hazards of ground water contamination from pesticide use than any other single development.

Fig 4: Cleaning internal and external surfaces of the sprayer – when done in the last field of use –safeguards the safety of other crops, the environment and the machine - but does so, whilst still containing all the PPP to its approved area of use.

Regulation

Sprayers – just as with pesticides – are subjected to independent scrutiny and controls. International Standards encourage a minimal level of performance or a common identity that avoids confusion. Thus, colour coded nozzles convey to the operator that if all his nozzles are one colour then all flow rates will be the same and thus the dose across the boom will be uniform. Pictograms used to show operator functions are common to all conforming manufacturers and will convey the same message to the operator irrespective of whether he is literate or not or his native language. Standards describe protocols on how to make performance measurements and set threshold limits that must be reached. Internal residual volumes within sprayers are specified. For example, field crop sprayers require such volumes to be < . These control measures are welcomed and encouraged by leading manufacturers as they seek the means for improvement in pesticide use. Until quite recently, little supportive data was needed to back commercial claims made by manufacturers of sprayers or nozzles. This situation is changing with some Regulators demanding that low drift equipment be independently approved and listed as a prerequisite for some specified pesticide uses. Today, sprayers are tested in the EU before and during their use; tested to meet requirements that ensure the PPP is applied with the greatest care and accuracy. Condition of pipes, effectiveness of gauges and uniformity of nozzle output are just some criteria that are examined.

Record keeping for crop assurance/tracking interests

Calibration for dose, water volume and drop size is now much easier to achieve. Nozzle selection guides help select most appropriate types and tractors – or prime mover – more readily attain, hold and record – spraying speeds. Electronic monitors can record total spray liquid emission, volumes of product used as well as the area treated and when. GPS based systems are recording where the product has been applied; many developments that make PPP use fully accountable and traceable [Fig 5].

Communication and training

Sprayers and spraying is a far more sophisticated technology today then it was just a few years ago. The modern sprayer is an advanced piece of equipment that is expected to complete – efficiently, effectively and safely – a vitally important task; a task that has a high public profile. To meet that challenge, training takes place at every level – from the operator to the machinery dealer who helps to advise on the purchase. Whilst instruction books remain essential, the

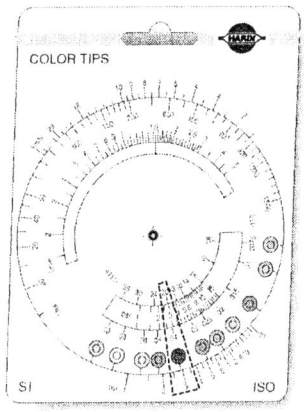 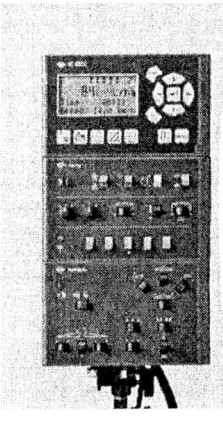

This controller will record:
Volume rates Spraying speed Tank contents Volume sprayed Area treated Area left for treatment

Fig 5: Sprayer calibration and recording is much eased with devices such as this nozzle selection device and in cab controller

operator is further supported with CDs, active flow charts, moving images and devices to help ensure accuracy of adjustment and recording [Fig 5]. Web sites such as that used by Hardi International A/S [www.hardi-international.com] make a wealth of information immediately available too. Users can interact with the site as well be updated on recent developments. Courses for all spraying needs are available; operators may wish to attend locally held events whilst specialist interests are also met. Week long Application Technology Courses are held on application technology; an activity that attracts university, regulatory, agrochemical folk from around the world.

Conclusion

The sprayer remains one of the most important farm machines. However, to meet current needs for efficiency as well as safety, it has evolved into larger more complex designs. Modern manufacturers are meeting today's challenges by introducing technological advances that both meet these demands but doing so in a manner that encourages the operator to understand and adopt the required improved practice.

Chapter 29

Tape Formulations and Their Application Technology

Masao Inoue[1], Satoshi Nakamura[2], and Toshiro Ohtsubo[2]

[1]Planning and Coordination Office, Agricultural Chemicals Sector, Sumitomo Chemical Company, Ltd., Chuo-ku, Tokyo 104–8260, Japan
[2]Agricultural Chemicals Research Laboratory, Sumitomo Chemical Company, Ltd., Takarazuka, Hyogo 665–8555, Japan

Pyriproxyfen yellow tape formulation is a novel pesticide delivery system utilizing whitefly attraction to the yellow color and pyriproxyfen's excellent unhatching activity against whitefly. The tape formulation attracts whitefly by the yellow color and such attraction causes whitefly to contact with pyriproxyfen, which results in the unhatching activity against whitefly. The formulation has an excellent efficacy against whitefly when it is applied horizontally along ridges as a preventive measure. The tape formulation has a great advantage to reduce worker's exposure and environmental impact because it offers whitefly control without spraying. This technology will be useful as a tool for integrated pest management (IPM).

Introduction

Recent demand for safer pesticide products has led the industry to seek safer pesticide formulations and application technology (Figure 1). With this goal in mind, pesticide industry developed several new formulations such as flowables, water dispersible granules, water-soluble packaging and microcapsules with the object of reducing worker's exposure to the toxicological effects of pesticides and the adverse impact to the environment. However, up until now, despite success with the development of safer pesticide formulations, the development of new application technologies have not particularly been successful.

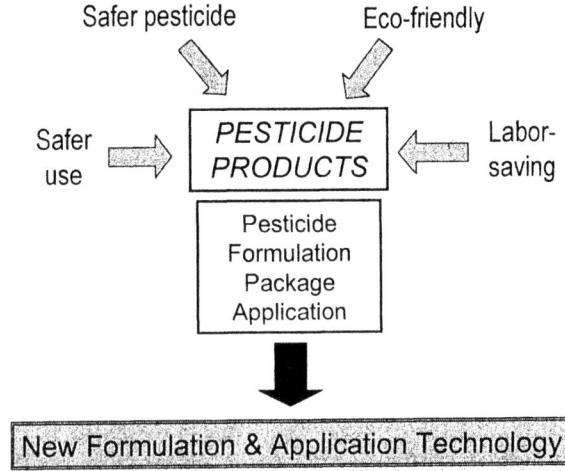

Figure 1. Recent Demand for Pesticide Formulation and Application Technology

Pyriproxyfen is a juvenile hormone mimic offering protection against whitefly. This compound belongs to an insect growth regulator and has an excellent unhatching activity against whitefly eggs. Since it has a high safety against beneficial insects, it is a useful tool in integrated pest management (IPM) programs.

Whitefly has a unique characteristic to be attracted to yellow color. We have developed a novel pesticide delivery system utilizing whitefly attraction to yellow color and pyriproxyfen's excellent unhatching activity against whitefly.

Concept of Pyriproxyfen Tape Formulation

The concept of pyriproxyfen tape formulation and application technology is described in Figure 2. Essentially, the concept of the delivery system is to utilize the whitefly's attraction to the yellow tape formulation. Such attraction causes whitefly to land on the tape formulation and thereafter come into contact with pyriproxyfen which will result in the unhatching activity against whitefly eggs. Finally the tape formulation controls the next genereation of whitefly.

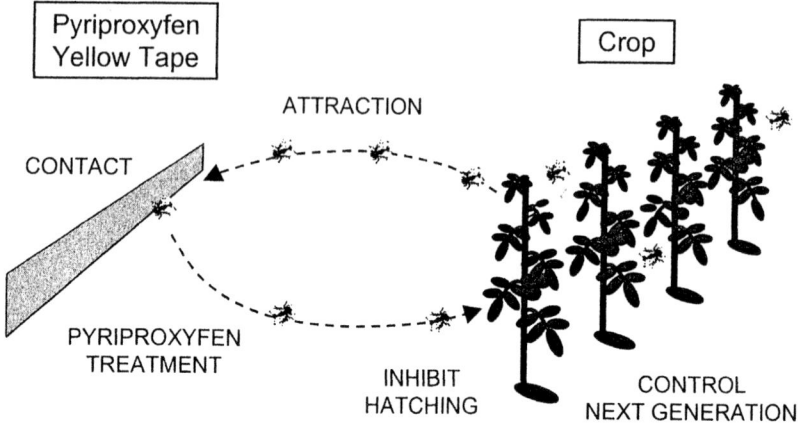

Figure 2. Concept of Pyriproxyfen Tape Formulation

Whitefly Attraction to Yellow Tape

Previous reports suggested that bright yellow attracted whitefly more efficiently than dark yellow. Our own study also demonstrated that whitefly was attracted to a normal yellow and a greenish yellow most efficiently (*1*).

In order to investigate the effect of yellow color on efficacy, biological test using yellow tape formulation and white tape formulation was conducted. As shown in Figure 3, while the white tape formulation and untreated plot did not inhibit hatching of whitefly eggs, yellow tape formulation showed high unhatching efficacy (*2*). This result inidicated that the attraction by yellow color plays an important role in biological efficacy of the tape formulation.

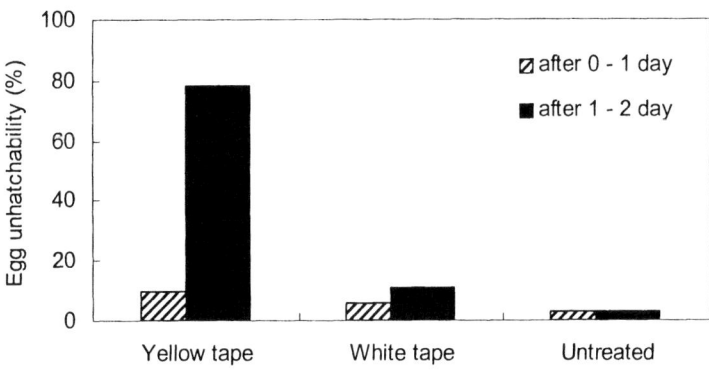

Whiteflies and dwarf bean were kept in a cage with tapes (10cm×10cm). Unhatchability of eggs laid after 0-1 day and 1-2 day were investigated.

Figure 3. Efficacy of Pyriproxyfen Yellow Tape and White Tape Formulations

Efficacy of tape formulations with same pyriproxyfen amount and different sizes was evaluated. As shown in Figure 4, tape size affected the biological efficacy. This result also suggested that attraction efficiency is related to the biological performance of the tape formulation.

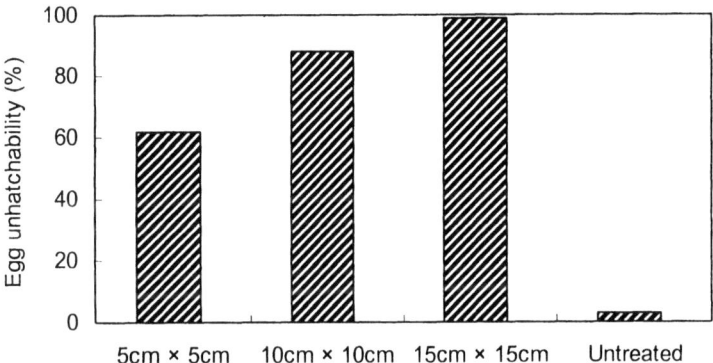

Whiteflies and dwarf bean were kept in a cage with a square tapes with 22.5mg pyriproxyfen. Unhatchability of eggs laid after 2-3 day were investigated.

Figure 4. Efficacy of Tape Formulations with Different Sizes

Sustainability of yellow color under sunlight is also an important property of the tape formulation. By using a special coloring agent, the pyriproxyfen tape formulation showed excellent color sustainability. As shown in Figure 5, reflection spectrum of the tape formulation is not changed after 240 hours radiation by the Sunshine Fademeter (*3*).

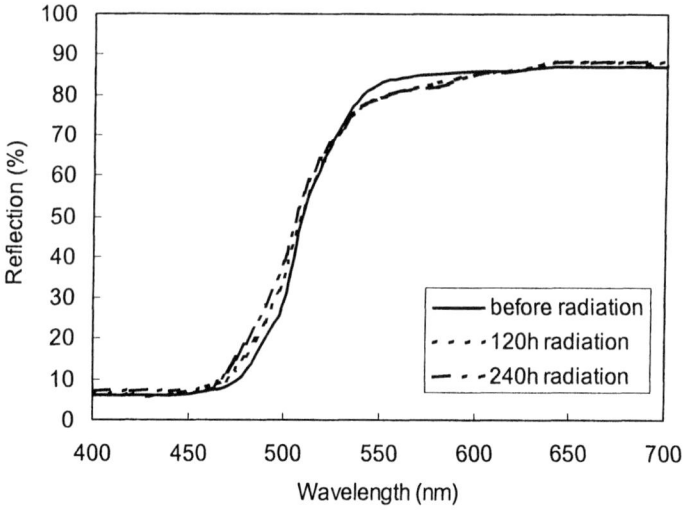

Figure 5. Sustainability of Yellow Color of Tape Formulation

Formulation Development

Formulation study was conducted to optimize the pyriproxyfen content, tape width and its physico-chemical properties. Figure 6 shows the result of efficacy test using tape formulations with three levels of pyriproxyfen. The tape formulations with 1 g/m^2 and 2 g/m^2 of pyriproxyfen have higher efficacy than the tape with 0.5 g/m^2 of pyriproxyfen.

Figure 7 shows the result of efficacy test using tape formulations with different widths. The tapes with 50 mm and 100 mm width showed better control than 25 mm tape. This result suggests that the area of yellow tape plays an important role in whitefly attraction affecting biological efficacy against whitefly. This result is consistent with the result shown in Figure 4.

333

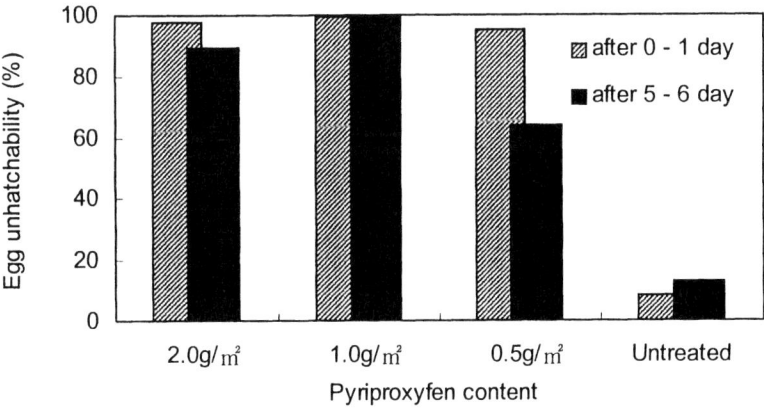

Whiteflies were treated with pyriproxyfen tapes by conpulsory contact.
After the treatment, whiteflies were kept with fresh cabbage for egg-laying.

Figure 6. Efficacy of Tape Formulations with different pyriproxyfen contents

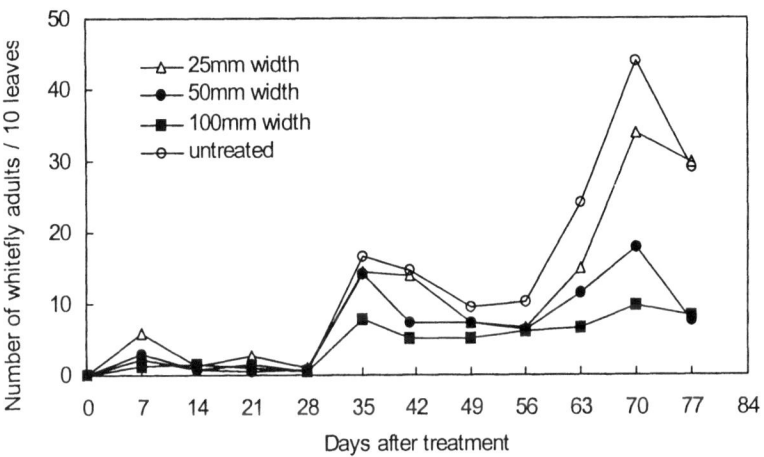

Pyriproxyfen 2g/m² tapes were set along ridges horizontally. (150cm height)

Figure 7. Efficacy of Tape Formulations with different widths

Effect of bleeding rate of pyriproxyfen on the biological and physico-chemical performance of tape formulation was investigated. Tape formulation with fast bleeding showed excellent efficacy but slight blocking in the rolled tape, while tape formulation with slow bleeding showed good efficacy and no blocking. Additionally environmental impact was reduced by controlling the bleeding rate of tape formulation.

The formulation study demonstrated that the biological efficacy and physico-chemical properties of the tape formulation are optimized by using a yellow tape containing 1 g/m^2 of pyriproxyfen with 50 mm width and slow release characteristics (4).

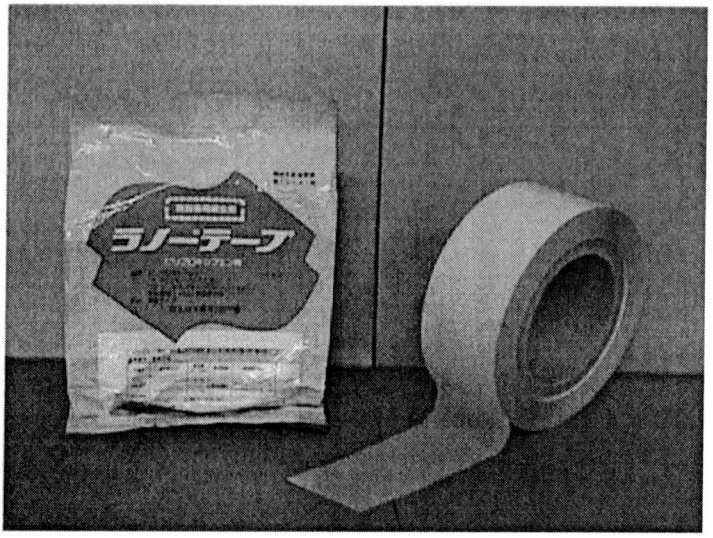

Plate 1. A Photograph of Pyriproxyfen Tape Formulation

Application Technology

Application research of tape formulation was done to determine the best arrangement, application timing and setting position.

Efficacy test in a tomato greenhouse was conducted by two typical arrangements, horizontal application and vertical application. As shown in

Figure 8, the tape formulation showed good whitefly control by both methods. However, from the viewpoint of working time needed for application, horizontal method appeared to be more practical. This test was conducted by preventive measure where the tape formulation was set at the beginning of the cultivation period.

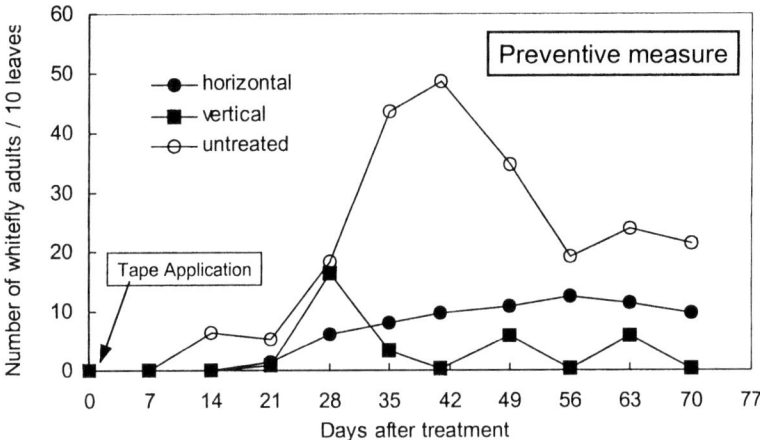

Pyriproxyfen $2g/m^2$ tapes of 50mm width were set. (150cm height, preventive application)

Figure 8. Efficacy of Tape Formulations by Different Arrangement

Figure 9 shows the result of efficacy test by curative measure. Tape formulation was applied horizontally to tomatoes where the whitefly population was significantly high. Under the conditions, tape formulation did not result in good control of whitefly.. Compared to the result shown in Figure 8, priproxyfen tape formulation shows better biological performance when it is used by a preventive measure. It is reasonable to expect that the biological efficacy by the tape formulation to be slow as pyriproxyfen, an insect growth regulator, has a slow activity on whitefly in addition to the time it takes to attract whitefly to yellow color and make contact with the tape formulation.

Regarding the setting position, tape formulation should be set over the top of crop so that tape formulation does not get hidden into the crop canopy.

Application research indicated that the tape formulation should be applied horizontally along the ridges by preventive measure at the beginning of cultivation (4). Pyriproxyfen tape formulation keeps an excellent whitefly control through the cultivation period by this application method. Additionally, working time needed for application was significantly reduced by using tape formulation in comparison with the use of spray application.

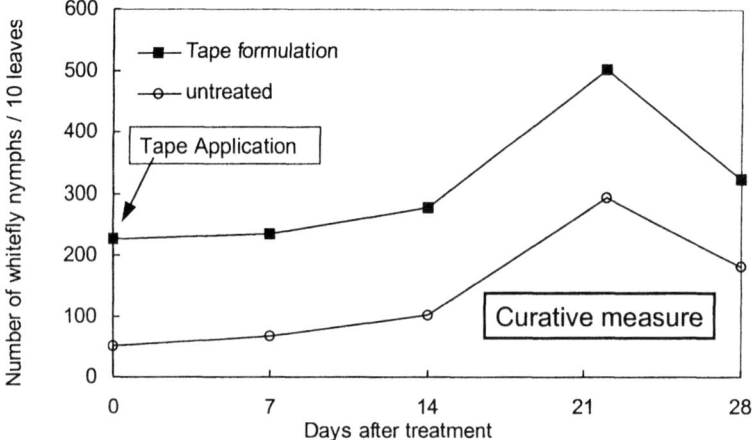

Pyriproxyfen 1g/m² tapes of 50mm width were set along ridges horizontally. (150cm height, curative application)

Figure 9. Efficacy of Tape Formulations by Curative Measure

Plate 2. A Photograph of Tape Formulations Applied in a Greenhouse

Tape formulation does not only help to reduce worker's exposure but also reduces the time needed for insect control.

Conclusion

A novel pesticide delivery system using pyriproxyfen tape formulation was successfully developed. The tape formulation and its application technology can significantly reduce the worker's exposure to pesticides and environmental adverse impacts. The technology also has an advantage to provide laborsaving pest-control method by reducing the amount of pesticide and the time of application. Furthermore, it will also serve as a useful tool in integrated pest management (IPM) programs.

Acknowledgements

We would like to express our sincere thanks to a number of colleagues at Sumitomo Chemical Co., Ltd., Agricultural Chemicals Research Laboratory, Takarazukla, Plant Protection Division, Tokyo and Plant Protection Division International, Tokyo, Japan for their continuous support and encouragement in this research.

References

1. Inoue, M.; Nakamura, S.; Ogawa, M.; Ohtsubo, T. : *14th Japan Pesticide Formulation and Application Symposium*, Abstract, **1994**
2. Nakamura, S.; Inoue, M. : *Applied Entomology and Zoology*, **1994**, 29, 454-457
3. Inoue, M.; Nakamura, S.: *Sumitomo Chemical*, **1999**, I , 16-24
4. Inoue, M.; Nakamura, S.; Ohtsubo, T. : *21st Japan Pesticide Formulation and Application Symposium*, Abstract, **2001**

Author Index

Agui, Noriaki, 243
Akamatsu, Miki, 159
Akiyama, Yumi, 38
Andersen, Per Gummer, 320
Campbell, Sonia, 62
Carr, J. A., 124
Carr, Wesley C., Jr., 138
du Preez, L. H., 124
Elsik, Curtis, 297
Endo, S., 112
Führ, F., 50
Fujisawa, Takuo, 205
Fukui, Mitsuru, 82
Gammon, Derek W., 138
Giesy, J. P., 124
Gross, T. S., 124
Hall, J. Christopher, 168
Hayatsu, Masahito, 82
Hayes, Robert M., 273
Hebert, Vincent R., 70
Hoagland, Robert E., 168
Horio, T., 112
Inoue, Masao, 328
Ishihara, S., 112
Ishii, Hideo, 280
Ishii, Y., 112
Ishizaka, M., 112
Itoh, Kazuyuki, 244
Iwata, Hitoshi, 28
Kaku, Koichiro, 255
Kasai, Shinji, 243
Katagi, Toshiyuki, 205
Kawai, Kiyoshi, 255
Kendall, R. J., 124
Kim, Jang-Eok, 92

Kim, Jong-Soo, 92
Kim, Y. J., 50
Kirby, Andrew, 297
Kishi, Daisuke, 309
Kobara, Y., 112
Komatsu, Kazuhiro, 28
Kurihara, Norio, 159
Kyung, K. S., 50
Lee, E. Y., 50
Lee, J. K., 50
Li, Qing X., 62
Main, Chris L., 273
Massey, Joseph H., 273
Mihara, Minoru, 243
Miyagawa, Hisashi, 159
Miyashita, Masahiro, 159
Miyazawa, Takeshige, 255
Molin, William T., 216
Mueller, Thomas C., 273
Nakagami, Shizuka, 159
Nakamura, Satoshi, 328
Nomura, Takakazu, 28
Oba, Shigehiro, 2
Ohtsu, K., 112
Ohtsubo, Toshiro, 328
Ohyama, K., 185
Papiernik, Sharon K., 101
Park, S. H., 50
Pfeifer, Keith F., 138
Preston, Christopher, 195
Riechers, Dean E., 216
Ruzo, Luis O., 205
Sato, Futoshi, 28
Sato, Tatsuo, 290
Seiber, James N., 14

Sekiya, Emi, 82
Shibaike, Hiroyuki, 244
Shimada, Takahiro, 159
Shimizu, Tsutomu, 255
Shimono, Seiichi, 309
Smith, E. E., 124
Solomon, Keith R., 124
Spencer, William F., 101
Tago, Kanako, 82
Takimoto, Yoshiyuki, 205
Tanaka, Yoshiyuki, 255
Tann, Scott, 297

Taylor, William, 320
Teranishi, Kiyoshi, 38
Tomigahara, Yoshitaka, 205
Tomita, Takashi, 234
Uchino, Akira, 244
Ueji, M., 112
Van Der Kraak, G. J., 124
Vaughn, Kevin C., 216
Yaguchi, Noboru, 243
Yates, Scott R., 101
Yoshioka, Naoki, 38
Zablotowicz, Robert M., 168

Subject Index

A

ABC (ATP binding cassette) transporter, 220, 221
Acetaminophen metabolism, 213–214
Acetohydroxy acid synthase (AHAS). See Acetolactate synthase (ALS)
Acetolactate synthase (ALS)
ALS genes from bispyribac-sodium (BS)-resistant rice cells, 263–264, 263f, 264f
ALS inhibiting herbicides, 245, 255, 256f
Amaranthus sp., herbicide resistance from ALS mutations, 257t, 259
Ambrosia sp., herbicide resistance from ALS mutation, 257t
amino acid substitution in resistant and susceptible *Lindernia micrantha*, 250, 250t, 257, 257t
Aribidopsis thaliana (thale cress), herbicide resistance from ALS mutations, 262t
Beta vulgaris (common beet), herbicide resistance from ALS mutations, 262t
biosynthetic pathway of branched-chain amino acids, 255, 256f
Brassica napus (rape), herbicide resistance from ALS mutations, 262t
Brassica tournefortii, herbicide resistance from ALS mutation, 257t
cDNA, 249–250, 249f
effect of multiple copies of ALS genes on resistance, 251, 253, 260
Gossypium hirsutum (upland cotton), herbicide resistance from ALS mutation, 262t
GST-fused recombinant ALSs, 264–266, 265f
inhibition by sulfonylurea (SU), 245, 255
Kochia scoparia, herbicide resistance from ALS mutations, 257, 257t, 259–260, 259f
Lactuca serriola, herbicide resistance from ALS mutation, 257, 257t, 259
Lindernia dubia, ALS gene analysis, 260–262, 260f, 261f
Lindernia dubia, herbicide resistance from ALS mutations, 257t, 260, 260f
mutated ALS genes
 artificially prepared genes, 268–270, 269f
 as selectable marker for rice transformation, 266–268, 266f, 268f
 sensitivity of transgenic rice to BS, 267–268, 267f
 sensitivity to herbicides, 269–270, 269f
mutation sites in resistant weeds (rice ALS numbering system), 258, 258f, 258t
mutations in crops, 262–266
Nicotiana tabacum (tobacco), herbicide resistance from ALS mutations, 262t
nucleotide sequences of resistant and susceptible *Lindernia micrantha*, 249–250, 250t
one-point mutated ALS genes, 263–264, 264f
Scripus juncoides, herbicide resistance from ALS mutation, 257t
Sisymbrium orientale, herbicide resistance from ALS mutations, 257t

Xanthium strumarium, herbicide resistance from ALS mutations, 257t
Zea mays (maize, corn), herbicide resistance from ALS mutations, 262t
See also *Lindernia micrantha*
Acrylamide, 20–21
Aflatoxins, 21
Agrochemical industry, 4–6
Agrochemical regulation
  in European Union, 5
  in Japan, 4f, 5, 39
  insecticides, 15–16
  sprayers and spraying, 321, 323, 326
Alachlor, 227, 228t
Algae and periphytonic algae, 113, 115–117, 118–122, 120f
Algal Growth Inhibition Test, 113
ALS. See Acetolactate synthase (ALS)
Amaranthus sp., herbicide resistance from ALS mutations, 257t, 259
Ambrosia sp., herbicide resistance from ALS mutation, 257t
Aminotriazole, 16
Amphibians
  androgen and anti-androgen mediated mechanisms, 129
  Bidder's organ and skin coloration in *Bufo marinus*, 131
  estrogen and anti-estrogen mediated mechanisms, 128–129
  gonadal development, 130–132
  hermaphroditism, 126
  induction of secondary sexual characteristics and gonadal development in *Xenopus laevis*, 126, 127f, 130–131
  lack of standardized tests and biomarkers for, 126, 133
  population-level effects, 132
  possible mechanisms of atrazine effects on sexual characteristics and gonadal development, 127–132, 128f
  sex reversal, 126
  thyroid mediated mechanisms, 129–130
Anaphes Iole wasps, 63
Anopheles mosquitoes, 3
Anthocyanin, 224–225
Aqueous solution (SL) formulations, 298, 299
Aribidopsis thaliana (thale cress), 262t
Aryl acylamidase catalyses, 197–198, 197f
Atrazine
  androgen and anti-androgen mediated mechanisms, 129
  Bidder's organ and skin coloration in *Bufo marinus*, 131
  chemical and physical properties, 227, 228t
  conjugation with glutathione catalyzed by glutathione transferases, 199f
  estrogen and anti-estrogen mediated mechanisms, 128–129
  fish and, 125
  gonadal development and, 130–132
  Koch's postulates and, 133
  lack of standardized tests and endpoints for amphibians, 126
  metabolism in *Pseudomonas*, 175–176, 176f
  population-level effects, 132
  possible mechanisms of atrazine effects on sexual characteristics and gonadal development, 127–132, 128f
  Science Advisory Panel (SAP) and, 126, 134
  and secondary sexual characteristics and gonadal development in *Xenopus laevis*, 126, 127f, 130–131
  thyroid mediated mechanisms, 129–130
  traditional toxicity studies, 126
Aventis Environmental Sciences, 9

Azimsulfuron
  degradation in lysimeter soil, 58, 59f
  degradation in rice paddies, 51
  distribution in rice plants, 57
  distribution in soil profile, 57–58
  fate in rice plant-grown lysimeter, 58, 59t, 60
  in leachates, 53, 55, 56t
  in soil and plants, 54
  lysimeter test methods, 51–53
  mineralization and volatilization, 53, 54, 55f
  partition in aqueous phase, 55–56
  partition in organic phase, 55–56
  properties and use, 51
  [pyrazole-4-^{14}C]azimsulfuron, structure and properties, 52, 53f

B

Bacillus thuringiensis (B.T., Bt) pesticides, 7, 20
Bacteria. *See* Pesticide-degrading bacteria; Prokaryote pesticide metabolism
Bayer CropScience, 9
Bensulfuronmethyl, 114–115, 116t, 121t
Bentazon and 8-hydroxybentazon
  binding to humic substances, 93
  catechol reaction with bentazon, 93, 94f, 95
  catechol reaction with 8-hydroxybentazon, 94, 95–97, 98f–99f
  products of transformation with catechol by laccase, 95–97, 95f, 97f–99f
Bentazone, 116t
Beta vulgaris (common beet), 262t
Biological Plausibility (IPCS criteria), 124, 127, 132
Biorational insecticides, 21

Body louse *(Pediculus humanus humanus)*
  base substitutions in Na$^+$ channel cDNA, 239–240, 240f
  and disease, 235
  open reading frame (ORF) in Na$^+$ channel cDNA, 239
  sensitivity to phenothrin (knockdown rates), 237f, 238t
  See also Head louse
Brassica napus (rape), 262t
Brassica tournefortii, 257t

C

Cafenstrole, 114–115, 116t, 121t
California Department of Pesticide Regulations (DPR), 149, 156
CalPuff model, 22, 23f
Capillary GC methods, 29
Capsulated suspension (CS) formulations, 298, 306
Carbaryl (1-napthyl-*N*-methylcarbamate)
  carbaryl-degrading bacteria, 83, 87, 88–90, 89f
  carbaryl-hydrolases (CehSB, CehNC, and CehAC), 87–88, 88f, 88t
Carboxylesters, 197
CCRs (Crop Protection Products), 113
Chloracetamide, 221–223, 222f
Chlorpyrifos, 77–78
Concentrated emulsion (EW) formulations, 291, 298, 300–301
Concentrated suspensions (SC), 291–295, 293f, 294f, 298, 302
Consistency (IPCS criteria), 124, 127, 132
Consumer insecticides, 9–10
Continuing Surveys of Food Intake by Individuals (CSFII), 139
Controlled release granules (CR-Gr)
  effect of amorphous silicon dioxide, 318–319, 318f

preparation, 310–311, 311t
release profiles, analysis by Higuchi model, 315, 316f, 317f, 317t, 318, 318f
release profiles, effect of annealing, 313–314, 314t, 315t
release profiles, effect of type of wax, 311–313, 312f, 312t, 313f
release profiles, effect of water temperature, 317, 317f, 317t
release profiles, typical release pattern, 315, 316f
release rate determination, 311
Cr-Gr. See Controlled release granules (CR-Gr)
Crayfish, 63
Crop Protection Products (CPPs), 113
Crop selectivity of herbicides
  aryl acylamidase catalyses and propanil selectivity in rice, 197–198, 197f
  carboxylesters and chlorfenprop-methyl selectivity in wheat, 197
  cytochrome P450 monooxygenases (CYP) detoxification of herbicides, 198, 198f
  export pumps for herbicide conjugates, 200, 200f
  glutathione transferases (GST), conjugation reactions in maize and soybeans, 199, 199f
  herbicide resistance in weeds and CYP-dependent detoxification, 198, 199t
  pathways for metabolism of herbicides in plants, 196f
  See also Herbicide detoxification; Herbicides; Plant pesticide metabolism
CropLife International, 6
CS. See Capsulated suspension (CS) formulations
Culex mosquitoes, 3

Cytochrome P450s
  cytochrome P450 monooxygenases (CYP) detoxification of herbicides, 198, 198f
  herbicide resistance in weeds and CYP-dependent detoxification, 198, 199t
  induced expression by safeners, 219
  inducers and, 201, 202f
  open reading frames (ORFs) in genomes, 198
  Phase I pesticide metabolism in plants, 171
  transformations of methoxychlor and its metabolites in vertebrates, 189, 190–191

D

DDT, 16, 235, 236
DDT-related compounds and their metabolites. See Methoxychlor
Dengue fever, 3
Depletion effect of surfactants, 295
Diamuron, 114–115, 116t
Diatoms, 118–122, 121f
Diazinon, 77–78
Diethofencarb and its metabolites
  diethofencarb structure, 207f, 209f
  distribution of metabolites in grape, 209–210, 210t
  excretion of diethofencarb and metabolites in rats, 211
  identification of metabolites, 207–209, 211
  metablite structures, 209f, 212f
  proposed metabolic pathway in grape plants, 209f
  proposed metabolic pathway in rats, 212f
  thiolactic acid conjugated metabolite formation, 206, 209f, 213–214
  thiolactic acid conjugates, 206

Dimepiperate, 114–115, 116t
Dimethametryn, 114–115, 116t, 121t
Dinotefuran (DT)
   properties, 309–310
   release profiles, effect of annealing
      in CR-Gr, 314, 315t
   release profiles, effect of type of wax
      in CR-Gr, 311–313, 312f, 313f
   release profiles, effect of water
      temperature, 317
   release profiles, typical release
      pattern of CR-Gr, 315, 316f
   structure, 310f
Dioxins, 19
Dispersion stability
   adsorption and, 292, 293f, 294–295
   background, 291
   concentrated emulsions (EW), 291
   concentrated suspensions (SC),
      291
   depletion flocculation, 295
   effects of surfactant concentration,
      292, 293f, 294–296
   electrical repulsive potential energy,
      293–294
   total potential energy computations,
      293–294
   van der Waals attractive potential
      energy, 294
   zeta potential, 292, 293f, 295
Diuron SSF system, 304–305
DPR (California Department of
   Pesticide Regulations), 149, 156
DT. *See* Dinotefuran (DT)

Endocrine disruption and endocrine
   disruptors
   binding to estrogen receptor (ER),
      159–160
   chemicals listed by Food Quality
      Protection Act, 21
   IPCS assessment guidelines, 124,
      127, 132–133
   See also Atrazine
Environmental health business, 7–9, 9t
Environmental health chemicals, 4
Enzyme-linked immunosorbent assay
   (ELISA), 19
EPA (U.S. Environmental Protection
   Agency), 16, 126, 134, 139
ER (estrogen receptor). *See* Estrogen
   receptor (ER)
Esprocarb, 114–115, 116t
Estrogen receptor (ER)
   binding mode of methoxychlor
      metabolites, 164–165, 165f
   enantiomerselective ER binding of
      mono-demethylated methoxychlor
      metabolites, 162–163, 163f, 164t,
      166
   ER binding activity of DDT-related
      compounds and metabolites, 161–
      162, 162f
   estradiol interactions with ER, 164–
      165, 165f
EW. *See* Concentrated emulsion (EW)
   formulations
Export pumps for herbicide
   conjugates, 200, 200f

E

EC. *See* Emulsifiable concentrate (EC)
   formulations
Electron capture (EC)-GC, 19
ELISA (enzyme-linked
   immunosorbent assay), 19
Emulsifiable concentrate (EC)
   formulations, 298, 299–300

F

Fenitrothion (*O,O*-dimethyl-*O*-4-nitro-
   m-tolyl phosphorothioate)
   cometabolism by *Sinoorhizobium*, 84
   fenitrothion-hydrolyzing gene *(fed)*,
      84, 85
   fenitrothion oxygenase genes *(mhqA*
      and *mhqB)*, 84, 86

isolation of fenitrothion-degrading
bacteria, 83–84
metabolism by *Burkhoderia* strains,
84–86, 85*f*
16S rDNA analysis, 83–84
synergistic utilization by
Sphingomonas and Burkhoderia,
86–87, 86*f*
Fipronil
desulfinyl fipronil (metabolite C), 63,
64–67
fipronil sulfide (metabolite A), 63,
64–67
fipronil sulfone (metabolite B), 63,
64–67
gas chromatography/mass
spectrometry (GC/MS), 65, 66*f*
low mammalian toxicity, 63
pressurized fluid extraction (PFE),
63–65, 66*t*
quantification of soil and gauze
samples, 66, 67*t*
recovery from fortified and aged soil,
65, 66*t*
thermal lability, 64, 65*f*
toxicity to *Anaphes Iole* wasps, 63
toxicity to crayfish, 63
toxicity to tarnished plant bug *(Lygus Lineolaris),* 63
use as insecticide, 62–63
FMC Corporation, 9
Food and Agriculture Organization
(FAO) Code of Conduct on the
Distribution and Use of Pesticide, 6
Food Quality Protection Act of 1996
(FQPA)
aggregate risk assessment
requirement, 139
cumulative risk assessment
requirement, 139, 156
endocrine effects, 156–157
pre-/post natal sensitivity, 156
risk assessment and tolerance setting
procedures, 155
Food safety, improvements in, 19–21

Formulations
aqueous solution (SL) formulations,
298, 299
background and general concerns,
297–298, 307
capsulated suspension (CS)
formulations, 298, 306
concentrated emulsion (EW)
formulations, 291, 298, 300–301
emulsifiable concentrate (EC)
formulations, 298, 299–300
general surfactant aqueous
equilibrium phase behavior, 303–304, 303*f*
granule (GR) formulations, 298–299
safer formulations and technology
development, 321
structured surfactant (SSF)
formulations, 302–306, 304*f*
suspension concentrate (SC)
formulations, 291, 298, 302
suspoemulsion (SE) formulations,
298
tank mix additives (TMA), 298, 307
water dispersible granule (WG)
formulations, 298, 306
water in oil emulsion (EO, OD)
formulations, 300
yellow tape formulation
(pyriproxyfen), 330–337
See also Controlled release granules
(CR-Gr)
Fungicide Resistance Action
Committee (FRAC), 283
Fungicides
cross-resistance of fungal isolates,
282, 283*t*
detection of resistant mutants with
*Ita*I restriction enzyme, 285–286
fitness of fungicide-resistant strains,
283–284
general strategies for combating
resistance, 282–283
growth of fungicide market, 280,
281*t*

Japan Plant Protection Association, 287
MBI-Ds (melanin biosynthesis inhibitors), 280, 281–282
QoI fungicides (inhibitors of mitochondrial respiration at Qo site), 280–281, 281*t*, 282*t*, 285, 285*f*
regional strategies for control of resistance, 286–287
Research Committee on Fungicide Resistance, 287
strategies for newly developed fungicide use, 284

G

Gas chromatography (GC), 19
Gas chromatography/mass spectrometry (GC/MS)
multiresidue methods (MRMs) for pesticides, 39–44
recoveries and degradation rations of pesticides, 44
selective-ion monitoring (SIM), 40, 42*t*–43*t*
total ion chromatograms (TICs) of pesticides, 40*f*, 41*f*
General surfactant aqueous equilibrium phase behavior, 303–304, 303*f*
Global warming and diseases, 3
Glutathione *S*-transferases (GST)
atrazine conjugation with glutathione, 199, 199*f*
conjugation reactions in maize and soybeans, 199, 199*f*, 217–218
correlation with anthocyanin and phytochrome distribution, 224–225
gene expression regulation by safeners, 219–220
open reading frames (ORFs) in genomes, 199

Phase II pesticide metabolism, 171, 172, 174–175, 217, 226–227
plant response to stress, 199, 218
resistance in weeds, 200, 217
subcellular location, 223
tau class GST proteins (*Tt*GSTU1, *Tt*GSTU2), 221
tissue distribution, 223–224, 225–227
tissue-specific induction, 225–227
transport of glutathione-herbicide conjugates to vacuole, 220–221
Glyphosate
inhibition of 5-enol-pyruvyl-shikimate-3-phosphate synthase (EPSPS), 274
replacement of older insecticides, 20
resistance in horseweed *(Conyza canadensis)*, 273–278, 275*t*
shikimic acid accumulation in sensitive plants, 274, 277–278, 277*t*
Gossypium hirsutum (upland cotton), 262*t*
Granule (GR) formulations, 298–299
GST. *See* Glutathione *S*-transferases (GST)

H

Head louse *(pediculus humanus capitis)*
incidence of infestations, 236–237, 236*t*
reduced Na^+ channel sensitivity in pyrethroid-resistant head lice, 236
resistance-associated Na^+ channel mutations, 239–241, 240*f*, 241*f*
resistance-associated Na^+ channel mutations in lice, 239–241, 240*f*, 241*f*

sensitivity to phenothrin (knockdown rates), 237–238, 237f, 238t, 239f
Heptachlor, 16
Herbicide detoxification
  cytochrome P450 monooxygenases (CYP) detoxification of herbicides, 198, 198f
  enzymes in herbicide detoxification, 196–200, 196f
  herbicide resistance in weeds and CYP-dependent detoxification, 198, 199t
  regulation of enzymes, 201, 202f
  triazine and acetamide detoxification in maize and soybeans by GSTs, 199, 217–218
  See also Crop selectivity of herbicides; Plant pesticide metabolism; Safeners
Herbicide resistance in weeds
  acetolactate synthase (ALS) mutations in resistant weeds, 257–262, 257t, 258t, 259f, 260f, 261f
  aryl acylamidase catalyses and propanil resistance, 197–198
  CYP-dependent detoxification and, 198, 199t
  effect of multiple copies of ALS genes, 251, 253
  glutathione S-transferase (GST) resistance, 200, 217
  inducers, 201, 202f
  multiple herbicide resistance, 201
Herbicide selectivity. See Crop selectivity of herbicides
Herbicides
  chemical and physical properties, 227, 228t
  molinate (Ordram), 23–24
  use in rice paddies, 113
  See also Crop selectivity of herbicides; Herbicide detoxification; Herbicide resistance in weeds; Plant pesticide metabolism; specific types
High performance liquid chromatography (HPLC), 19
Horseweed *(Conyza canadensis)*
  glyphosate resistance, 273–278, 275t
  in conservation tillage systems, 274
  paraquat resistance, 275
  shikimate accumulation, 274, 277–278, 277t
  shikimate recovery from horseweed, 276–277, 276t
  triazine resistance, 275
Human welfare mission of agrochemical industry, 4–5

I

Imazosulfuron, 114–115, 116t, 121t
Imidazolinones (IM), 255, 256f, 257
Industrial Source Complex Short Term model, 22, 23t
Insecticides
  biorational, 21
  consumer, 9–10
  distribution, 9
  first generation, 15
  public health, 10
  regulation, 15–16
  second generation, 15
  See also specific types
Integrated pest management (IPM), 6–7, 329, 337
Intellectual property strategies, 6
Internal simple sequence repeat (ISSR) markers and phenotypes, 246
International Program on Chemical Safety (IPCS) assessment criteria of causative agents of disease, 124, 127, 132–133
IPCS criteria. See International Program on Chemical Safety

(IPCS) assessment criteria of causative agents of disease

J

Japan Plant Protection Association, 287
Japanese Food Sanitation Law (FLS)
MRL violation rates for pesticides in Japan, 45–48, 49*t*
number of pesticides and agrochemicals regulated, 4*f*, 5, 39
official methods, 39*f*

K

Kasumigaura, Lake, 114, 114*f*, 117–118, 119*f*
See also Rice herbicides
Kochia scoparia, 257, 257*t*, 259–260, 259*f*
Koch's postulates, 133

L

Lactuca serriola, 257, 257*t*, 259
Lindernia dubia, 257*t*, 260–262, 260*f*, 261*f*
Lindernia micrantha
amino acid substitution in resistant and susceptible *L. micrantha*, 250, 250*t*, 257, 257*t*
distribution of sampling sites, 246, 247*f*
effect of multiple copies of ALS genes on resistance, 251, 253
genetic relationships and amino acid substitutions in *ALS*, 251, 252*f*
genetic relationships between plants, 246, 248*f*, 249

nucleotide sequences of resistant and susceptible *L. micrantha*, 249–250, 250*t*
origination and expansion of resistant types, 251
sulfonylurea (SU) resistance in, 245–246, 245*f*
See also Acetolactate synthase *(ALS)*
Liquid chromatography/mass spectroscopy (LC/MS)
background, 28–29
instruments and conditions, 30*t*–31*t*, 31, 33
ion chromatograms for blank extracts, 34*f*
pesticide elution on GPC and ENVI-Carb/NH$_2$, 32*t*–33*t*, 34
pesticide polarities and molecular weights, 37*f*
recoveries of pesticides, 35*t*–36*t*, 36
sample preparation method, 29, 31
LOAEL (Lowest Observed Adverse Effect Level), 23*t*
LOEL (lowest observed effect level), 141
Lures, in pest control, 21
Lysimeters, 51, 52, 52*t*

M

Malaria, 3
Mass spectrometry (MS), 19
Maximum residue limit (MRL)
Japanese Food Sanitation Law (FLS), 39, 140
methamidophos, 140
MRL violation rates for pesticides in Japan, 45–48, 49*t*
risk assessment, 39, 140
MBI-Ds (melanin biosynthesis inhibitors), 280, 281–282
Mefenacet, 114–115, 116*t*
Methamidophos

acepate conversion to methamidophos, 139, 156
acetylcholinesterase (AChe) inhibition, 139, 141, 142t, 151–152
agricultural use, 138–139
AIC (Akaike Information Criteria), 141
BMD (Benchmark Dose), 141
BMDL (lower confidence limit of BMD), 141
chronic toxicity, 141–142, 152
crop residue data, 140, 140t
deterministic (point estimate) model, 139, 146, 149
developmental neurotoxicity, 144–146, 145t, 151, 152–153
developmental toxicity, 143–146, 145t, 147t, 151
dietary exposure, 146, 148t, 149, 153–155
hazard identification, 151–153
IRED (Interim Reregistration Eligibility Document), 139
LOEL (lowest observed effect level), 141
maternal toxicity, 144, 145t, 151
MOS (margin of safety), 146, 148t, 149–151, 150t
NOEL (no observed effect level), 139, 141–142
P-glycoprotein transporters (PGTs) and, 153
probabilistic (Monte Carlo) model, 140, 146
properties, 140
reproductive toxicity, 143
Risk Management Decision Document, 139
TASS and DEEM computer models, 139–140, 146, 148t, 149, 154
tolerance assessments, 140, 149–151, 157
uncertainty factor, 141

USEPA Index Chemical for relative potency factor (RPF), 139
Methoxychlor and its metabolites
binding mode of methoxychlor metabolites, 164–165, 165f
cytochrome P450s transformations, 189, 190–191
demethylation, 186–189, 189f, 190, 190f
enantiomerselective ER binding of mono-demethylated methoxychlor metabolites, 162–163, 163f, 164t, 166
estrogen receptor binding activity, 161–162, 162f
estrogenic effects, 160, 185, 192
in vitro metabolism by precision-cut rat, mouse, Japanese quail, and rainbow trout livers, 186–188
metabolic pathways, 191–192, 192f
bis-OH-MXC [2,2-bis(p-hydroxyphenyl)-1,1,1-trichloroethane]
estrogenic activity, 160
production by rat and trout liver enzymes, 186–188, 187f, 188f
structure, 160f
mono-OH-MXC [2-(p-hydroxyphenyl)-2-(p-methoxyphenyl)-1,1,1-trichloroethane]
enantiomers, 162–163, 163f, 189–190, 189f
estrogenic activity, 160
production by mouse, Japanese quail, and trout liver enzymes, 186–188, 187f, 188f
stereoselective formation in rat, mouse, Japanese quail, and rainbow trout, 189–190, 190f, 191
structure, 160f
radio-HPLC chromatograms of metabolites, 187f

similarity to DDT, 159, 160, 185
structures, 160f, 161f
toxicity in vertebrates, 185, 192
Methyl bromide, 22, 23t
Methylisothiocyanate (MITC), 22
Metofluthrin, 9–10
Ministry of Agriculture, Forestry and Fisheries (MAFF) Japan, 113
Molinate, 114–115, 116t
Molinate (Ordram), 23–24
MOS (margin of safety), 146, 148t, 149–151, 150t
MRL. *See* Maximum residue limit (MRL)
Multiresidue methods (MRMs)
capillary GC methods, 29
gas chromatography/mass spectrometry (GC/MS) methods, 39–44
See also Liquid chromatography/mass spectroscopy (LC/MS)

N

Nicotiana tabacum (tobacco), 262t
NOEL (no observed effect level), 139, 141–142

O

Olyset, 10
Orchard spraying and dermal/respiratory exposure for applicators, 21–22, 22t
Organochlorine (OC) pesticides, 20
See also specific types
Organophosphate (OP) insecticides. *See* Chlorpyrifos; Diazinon; Methamidophos
Orius strigicollis (Oristar-A), 7

P

Pediculicides, 235
See also Body louse; Head louse
Pentachlorophenol, 177–178, 177f
Pentoxazone, 114–115
Pesticide application
communication and training, 326
drift control, 324, 324f
in-field cleaning, 325, 325f
loading of pesticides into sprayer, 322, 322f
optimizing nozzle use, 323f
record keeping, 326, 327f
regulation and standards, 321, 323, 326
safe use of equipment, 321
Pesticide-degrading bacteria
carbaryl degradation by *Arthrobacter*, 88–90, 89f
carbaryl-hydrolases (CehSB, CehNC, and CehAC), 87–88, 88f, 88t
carbofuran, results of repeated application, 83
enhanced or accelerated degradation of pesticides, 83
fenitrothion cometabolism by *Sinoorhizobium*, 84
fenitrothion-hydrolyzing gene *(fed)*, 84, 85
fenitrothion metabolism by *Burkhoderia*, 84–86, 85f
fenitrothion oxygenase genes *(mhqA and mhqB)*, 84, 86
isolation, 83, 87
isolation from soil, 83–84
16S rDNA sequencing, 83
significance, 82–83
Sphingomonas enzymes and substrates, 87, 177–178
synergistic carbaryl utilization by *Mesorhizobium* and *Pseudomonas*, 87

synergistic fenitrothion utilization by
Sphingomonas and *Burkhoderia*,
86–87, 86*f*
See also Carbaryl; Fenitrothion;
Prokaryote pesticide metabolism
Pesticide formulation technology
aqueous solution (SL) formulations,
298, 299
background and general concerns,
297–298, 307
capsulated suspension (CS)
formulations, 298, 306
concentrated emulsion (EW)
formulations, 291, 298, 300–301
emulsifiable concentrate (EC)
formulations, 298, 299–300
general surfactant aqueous
equilibrium phase behavior, 303–
304, 303*f*
granule (GR) formulations, 298–
299
structured surfactant (SSF)
formulations, 302–306, 304*f*
suspension concentrate (SC)
formulations, 291, 298, 302
suspoemulsion (SE) formulations,
298
tank mix additives (TMA), 298, 307
water dispersible granule (WG)
formulations, 298, 306
water in oil emulsion (EO, OD)
formulations, 300
Pesticide photochemical oxidation
reactions
elevated temperature systems, 74,
75*f*, 77
hydroxyl-mediated reactions, 73, 76–
78
laboratory chemical oxidation
evaluations, 76–78
laboratory photochemical
evaluations, 73–76
laboratory photochemical pathways,
72–73
relative rate evaluations, 76–77

sunlight-induced tropospheric
reactions, 71
use of solid phase microextraction
(SPME) fibers, 74
wall interference, 71, 74
Pesticide residue chemistry. *See*
Residue analysis
Pesticide resistance in weeds
ALS mutation sites in resistant weeds
(rice ALS numbering system),
258, 258*f*, 258*t*
Amaranthus sp., 257*t*, 259
Ambrosia sp., 257*t*
Brassica tournefortii, 257*t*
Kochia scoparia, 257, 257*t*, 259–
260, 259*f*
Lactuca serriola, 257, 257*t*, 259
Lindernia dubia, 257*t*, 260–262,
260*f*, 261*f*
Scripus juncoides, 257*t*
Sisymbrium orientale, 257*t*
Xanthium strumarium, 257*t*
Pesticide volatilization. *See*
Volatilization
Pesticides
annual U.S. application rates,
70
multi-pesticide residues in food, 45–
48
pesticides detected in agricultural
products in Japan, 45–49
phytotoxicity, 113
reduced use and lower residues in
foods, 19–20
tropospheric fate, 71–72
tropospheric transport, 70–71
See also Insecticides; Pesticide
photochemical oxidation reactions;
Plant pesticide metabolism;
Prokaryote pesticide metabolism
Phase I pesticide metabolism
cytochrome P450s transformations in
plants, 171
cytochrome P450s transformations in
vertebrates, 189, 190–191

methoxychlor demethylation, 189f, 190, 190f
oxidation, reduction, and hydrolysis in plants, 171
Phase II pesticide metabolism conjugative processes in plants, 171–173
glucose conjugation in plants, 171
glutathione (GSH) conjugation in plants, 172–173, 174–175, 217, 226–227
glutathione (GSH) conjugation in prokaryotes, 174–175
glutathione S-transferases (GSTs) in plants, 171, 172, 217
Photochemical oxidation reactions. See Pesticide photochemical oxidation reactions
Photochemical oxidation reactions of pesticides. See Pesticide photochemical oxidation reactions
Phytochrome, 224–225
Phytoplankton. See Algae
Plant pesticide metabolism
bound xenobiotic residues, 173
carboxylesters and chlorfenprop-methyl selectivity in wheat, 197
conjugation of pesticides as co-metabolic process, 171
cytochrome P450s transformations, 171
2,4-D (2,4-dichlorophenoxyacetic acid) metabolism, 171
enzymes in herbicide detoxification, 196–200, 196f
fenchlorazole-ethyl (FCE), 173
fenoxaprop-ethyl (FE) metabolism, 172–173, 174f, 217
glucose conjugation, 171
glutathione (GSH) conjugation, 172–173, 174–175
glutathione S-transferases (GSTs), 171, 172
herbicide resistance in weeds, 196, 197

herbicide selectivity, 171
herbicide tolerance, 171, 172, 179–180
pathways for metabolism of herbicides in plants, 196f
Phase I oxidation, reduction, and hydrolysis, 171
Phase II conjugative processes, 171–173, 217
Phase III secondary conjugate formation, 171, 173, 217, 218
Phase IV secondary degradation in vacuole, 218–219
prokaryote genes for creating herbicide-resistant crops, 179–180, 180t
transport of glutathione-herbicide conjugates to vacuole, 220–221
See also Crop selectivity of herbicides; Herbicide detoxification; Safeners
Plant Protection Product application. See Pesticide application
Plants, 169–171, 170t
Precision-cut liver slice technique, 185–186, 187f, 190
Predator insects, 7
Pressurized fluid extraction (PFE), 63–65
Pretilachlor, 114–115, 116t, 121t
Proestrogens, 160
Prokaryote pesticide metabolism
atrazine metabolism, 175–176, 176f
chloroaniline cleavage by Pseudomonas, 178f, 179
dehalogenation mechanisms, 175–179, 175t
fission of aromatic ring structures in pesticides, 178–179, 178f
glutathione (GSH) conjugation, 174–175
mineralization of pesticides, 175–178
pentachlorophenol metabolism, 177–178, 177f

prokaryote genes for creating herbicide-resistant crops, 179–180, 180t
Prokaryotes, 169–171, 170t
Public health insecticides, 10
Pyrazosulfuronethyl, 114–115
Pyrethroids
　metofluthrin, 9–10
　permethrin, 235
　phenothrin, 235, 236–237
　replacement of older insecticides, 20, 235
　resistance-associated Na⁺ channel mutations, 239–241, 240f, 241f
　See also Sodium (Na⁺) channels
Pyributicarb, 114–115
Pyridalyl, 7
Pyrimidinyl carbozy herbicides, 255, 256f, 257
Pyriminobacmethyl, 114–115
Pyriproxyfen
　biological activity, 329
　concept of pyriproxyfen tape formulation, 330, 330f
　efficacy of yellow or white tape formulations, 330, 331f
　preventative vs. preventative use, 335–336, 335f, 336f
　size effect in tape formulation, 331
　structure, 7
　sustainability of yellow color of tape formulation, 332, 332f
　tape formulation application technology, 334–336
　tape formulation development, 332–334, 333f, 334f
　unhatching activity against whitefly eggs, 329
　use in integrated pest management (IPM) programs, 329, 337

Q

QoI fungicides (inhibitors of mitochondrial respiration at Qo site), 280–281, 281t, 282t, 285, 285f
Quinoclamine, 116t

R

Recovery (IPCS criteria), 124, 127, 132
Research Committee on Fungicide Resistance, 287
Residue analysis
　acrylamide in foods, 20–21
　aflatoxins, 21
　background, 14–17
　biorational insecticides, 21
　environmental safety and, 23–24
　evolution of analytical techniques, 17–19
　food safety and, 19–21
　future trends, 24–25
　methyl bromide residue distribution patterns, 22, 23t
　molinate (Ordram), 23–24
　pesticide residues in foods, 19–20
　single-residue methods (SRM), 17
　Solid Phase Extraction (SPE), 17
　supercritical fluid extraction (SFE), 18
　worker safety and, 21–22
　See also Multiresidue methods (MRMs)
Rice ALS gene numbering system, 258, 258f, 258t
Rice blast *(Magna grisea)*, 281–282
Rice culture and cropping methods, 245, 282
Rice herbicides

concentrations in Lake Kasumigaura, 117–118, 119f
concentrations in Sakura River, 117–118
herbicide susceptibility of freshwater algae and diatoms, 118–122, 120f, 121f
toxicity testing methods, 115–117
use in rice paddies, 113
See also specific herbicides
Risk assessment
AIC (Akaike Information Criteria), 141
BMD (Benchmark Dose), 141
BMDL (lower confidence limit of BMD), 141
chronic toxicity, 141–142, 152
deterministic (point estimate) model, 139, 146, 149
developmental neurotoxicity, 144–146, 145t, 151, 152–153
developmental toxicity, 143–146, 145t, 147t, 151
dietary exposure, 146, 148t, 149, 153–155
dietary exposure estimation, 139–140
Hill model, 141–142
LOEL (lowest observed effect level), 141
maternal toxicity, 144, 145t, 151
MOS (margin of safety), 146, 148t, 149–151, 150t
NOEL (no observed effect level), 139, 141–142
probabilistic (Monte Carlo) model, 140, 146
reproductive toxicity, 143
TASS and DEEM computer models, 139–140, 146, 148t, 149, 154
tolerance assessment, 140, 149–151, 157
uncertainty factor, 141
Roll Back Malaria project, 10

S

Safeners
chemical and physical properties, 227–228, 228t
chloquintocet-mexyl, 228, 228t
enhanced rate of herbicide detoxification, 201, 218
fluxofenim, 223, 228, 228t
herbicide selectivity and, 218–219
induced expression of GST genes, 219, 225–227
induced expression of P-450 genes, 219
prevention of toxic effects of chloracetamide (alachlor), 221–223, 222f
regulation of detoxification pathway, 220
Sakura River, 114, 117–118, 119f
See also Rice herbicides
SAP (Science Advisory Panel), 126, 134
SC. See Suspension concentrate (SC) formulations
Science Advisory Panel (SAP), 126, 134
Scripus juncoides, 257t
Shikimate, 276–278, 277t
Silent Spring, 16
Simetryn, 114–115, 116t, 121t
Sisymbrium orientale, 257t
SL. See Aqueous solution (SL) formulations
Sodium (Na^+) channels
kdr (knockdown-resistance) mutation, 235–236
neuronal target of DDT and pyrethroids, 235
open reading frame (ORF) in body louse cDNA, 239
reduced Na^+ channel sensitivity in pyrethroid-resistant head lice, 236

resistance-associated mutations in lice, 239–241, 240f, 241f
resistance-associated structural changes in insects, 235–236, 235f
super-kdr (knockdown-resistance) mutation, 235–236
Sprayers and spraying
 communication and training, 326
 drift control, 324, 324f
 in-field cleaning, 325, 325f
 loading of pesticides into sprayer, 322, 322f
 optimizing nozzle use, 323f
 orchard spraying, dermal/respiratory exposure for applicators, 21–22, 22t
 record keeping, 326, 327f
 regulation and standards, 321, 323, 326
 safe use of equipment, 321
SSF. *See* Structured surfactant (SSF) formulations
Strength of Association (IPCS criteria), 124, 127, 132
Structured surfactant (SSF) formulations, 302–306, 304f
SU. *See* Sulfonylurea (SU)
Sulfonylurea (SU)
 biological activity and application rates, 257
 degradation in rice paddies, 51
 resistance in *Lindernia micrantha*, 245–246, 245f, 249–251, 253
 structure, 256f
 See also Acetolactate synthase *(ALS);* Azimsulfuron
Sumitomo Chemical Co., 5–10
Surfactants
 adsorption and, 292, 293f, 294–295
 depletion effect, 295
 effect on zeta potential, 292, 293f, 295
 effects on dispersion stability, 292, 293f, 294–296
 general surfactant aqueous equilibrium phase behavior, 303–304, 303f

polyethylene diethylenetriamine dialkylamide (POE-DETADAA), 290, 291, 293f
structured surfactant (SSF) formulations, 302–306, 304f
Suspension concentrate (SC) formulations, 291, 298, 302
Suspoemulsion (SE) formulations, 298
Sustainable agriculture, 3
Sustainable development, 2–3
Syngenta, 9

T

Tank mix additives (TMA), 298, 307
Telone, 22
Temporality (IPCS criteria), 124, 127, 132
Thiobencarb, 114–115, 116t
Thiolactic acid conjugated metabolites of carbamates, 205–214, 209f, 212f
Triallate
 accuracy of predicting volatilization rate, 108–109, 108t
 application and measurement methods, 102–103
 numerical model of fate, transport, and volatilization, 103–109
 properties, 103t
 volatilization and atmospheric stability, Φ_m, 105–106, 106f
 volatilization and time after application, 105, 106f
 volatilization simulation, with isothermal conditions, 107–109, 107f, 108f
 volatilization simulation, with micrometerological data, 107–109, 107f, 108f
 volatilization simulation, with solar-driven temperature changes, 107–109, 107f, 108f

Triazolopyrimidine sulfonamides (TP), 255, 256f, 257
Trifluralin, 74, 75f
Triticum tauschii, 221–223, 222f

U

United Nations Conference on Environment and Development (UNCED), 3
U.S. Department of Agriculture (USDA), 139
U.S. Environmental Protection Agency (EPA), 16, 126, 134, 139

V

Volatilization
  accuracy of predicting volatilization rate, 108–109, 108t
  application and measurement methods, 102–103
  atmospheric stability (Φ_m) and, 105–106, 106f
  background, 101–102
  numerical model of fate, transport, and volatilization, 103–109
  simulation with isothermal conditions, 107–109, 107f, 108f
  simulation with micrometerological data, 107–109, 107f, 108f
  simulation with solar-driven temperature changes, 107–109, 107f, 108f
  time after application and, 105, 106f

W

Water dispersible granule (WG) formulations, 298, 306

Water in oil emulsion (EO, OD) formulations, 300
Weeds, herbicide resistance in. *See* Herbicide resistance in weeds
West Nile Virus, 3
Whitefly
  attraction to yellow color, 329
  attraction to yellow tape formulation, 330, 331f
  concept of pyriproxyfen tape formulation, 330, 330f
  efficacy of yellow or white pyriproxyfen tape formulations, 330, 331f
  pyriproxyfen tape formulation application technology, 334–336
  pyriproxyfen tape formulation development, 332–334, 333f, 334f
  pyriproxyfen unhatching activity against whitefly eggs, 329
  size effect in pyriproxyfen tape formulation, 331
Worker safety, improvements in, 21–22
World Commission on Environment and Development (WCED), 2–3
World Health Organization (WHO), 10
World population, 3
World Summit on Sustainable Development (WSSD), 3

X

Xanthium strumarium, 257t
Xenopus laevis, 127f

Z

Zea mays (maize, corn), 262t
Zeta potential, 292, 293f, 295